THE STEPHEN BECHTEL FUND

IMPRINT IN ECOLOGY AND THE ENVIRONMENT

The Stephen Bechtel Fund has

established this imprint to promote

understanding and conservation of

our natural environment.

The publisher gratefully acknowledges the generous contribution to this book provided by the Stephen Bechtel Fund.

Marine Historical Ecology in Conservation

Marine Historical Ecology in Conservation

Applying the Past to Manage for the Future

Edited by   JOHN N. KITTINGER, LOREN MCCLENACHAN,
KERYN B. GEDAN, and LOUISE K. BLIGHT

Foreword by   Daniel Pauly

UNIVERSITY OF CALIFORNIA PRESS

University of California Press, one of the most distinguished university presses in the United States, enriches lives around the world by advancing scholarship in the humanities, social sciences, and natural sciences. Its activities are supported by the UC Press Foundation and by philanthropic contributions from individuals and institutions. For more information, visit www.ucpress.edu.

University of California Press

Library of Congress Cataloging-in-Publication Data

Marine historical ecology in conservation : applying the past to manage for the future / edited by John N. (Jack) Kittinger, Loren McClenachan, Keryn B. Gedan, Louise K. Blight.
    p. cm.
Includes bibliographical references and index.
ISBN 978-0-520-27694-9 (cloth : alk. paper)
ISBN 978-0-520-95960-6 (e-book)
    1. Marine ecology. 2. Human ecology. I. Kittinger, John Nils, editor, contributor. II. McClenachan, Loren, 1977—editor, contributor. III. Gedan, Keryn B., 1980—editor, contributor. IV. Blight, Louise K., 1962—editor, contributor.
    QH541.5.S3M2828 2015   2014011259
    577.7—dc23

Manufactured in the United States of America

23   22   21   20   19   18   17   16   15
10   9   8   7   6   5   4   3   2   1

The paper used in this publication meets the minimum requirements of ANSI/NISO Z39.48–1992 (R 2002) (*Permanence of Paper*).

Cover image: Once depleted, populations of the humpback whale (*Megaptera novaeangliae*) are showing signs of recovery due to effective conservation measures. Photograph by John Weller, www.johnbweller.com.

I dedicate this book to the three generations of incredible women in my family, whom I love and cherish without qualification. To my late maternal grandmother, Glenda Sue Cooper Pehrson, whose perseverance, independence, and fierce intelligence have always inspired me to achieve; to my mother, Sandra Pehrson Kittinger, whose warmth, love, and gentle guidance have been a constant source of renewal; and to my wife, Daniela Spoto Kittinger, for her compassion, friendship, and enduring love.

JACK KITTINGER

Jeremy Jackson first hooked me on the detective work of historical ecology and continues to inspire. Sonora Neal was born days after the 2011 meeting that motivated this book; her love for the ocean is already clear. I hope the world she grows up into includes an ocean full of big fish and people who value them.

LOREN MCCLENACHAN

I dedicate this book to Mark Bertness, my graduate advisor and friend, who, with fish totes, mud boots, and a shovel, inaugurated me into salt marsh ecology and the rich history—scientific, stratigraphical, and human—that accumulates there. Mark's love of the coast, anthropology, and history first inspired me to explore the legacies of generations and civilizations in coastal ecosystems.

KERYN GEDAN

For my friend, mentor, and colleague David Ainley, whose Antarctic work and ever perceptive observations started me on the journey of thinking about intact marine ecosystems and the nature of long-term change.

LOUISE BLIGHT

CONTENTS

CHAPTER CONTRIBUTORS

DALAL AL-ABDULRAZZAK
Fisheries Centre
University of British Columbia
Vancouver, BC V6T 1Z4, Canada

KAREN ALEXANDER
Department of Environmental
    Conservation
University of Massachusetts
Amherst, MA 01003, USA

SHANKAR ASWANI
Department of Anthropology
Rhodes University
Grahamstown 6140, South Africa

NATALIE C. BAN
School of Environmental Studies
University of Victoria
Victoria BC V8W 2Y2, Canada

LOUISE K. BLIGHT
WWF-Canada
409 Granville Street
Vancouver, BC V6C 1T2, Canada
and
Procellaria Research & Consulting
944 Dunsmuir Road
Victoria, BC V9A 5C3, Canada

TODD J. BRAJE
Department of Anthropology
San Diego State University
San Diego, CA 92182-6040, USA

DENISE L. BREITBURG
Smithsonian Environmental Research
    Center
Edgewater, MD 21037, USA

ROBERT D. BRUMBAUGH
Global Marine Team
The Nature Conservancy
127 Industrial Road, Suite D
Big Pine Key, FL 33043, USA

JOSHUA E. CINNER
Australian Research Council Centre
    of Excellence for Coral Reef
    Studies
James Cook University
Townsville, QLD 4811, Australia

LARRY B. CROWDER
Center for Ocean Solutions
Stanford University
99 Pacific Street, Suite 555E
Monterey, CA 93940, USA

(continued)

LARRY B. CROWDER (CONTINUED)
Hopkins Marine Station
Stanford University
120 Oceanview Blvd.
Pacific Grove, CA 93950, USA

ROBERT L. DELONG
Marine Mammal Laboratory
Alaska Fisheries Science Center
National Oceanic and Atmospheric
    Administration
Seattle, WA 98115, USA

JON M. ERLANDSON
Museum of Natural and Cultural History
University of Oregon
Eugene, OR 97403-1224, USA

SIAN EVANS
Ffilmcompany, Inc.
30 Mayo Street
Belfast, ME 04915-6052, USA

FRANCESCO FERRETTI
Hopkins Marine Station
Stanford University
120 Oceanview Blvd.
Pacific Grove, CA 93950, USA

ALAN M. FRIEDLANDER
Pristine Seas
National Geographic Society
Washington, DC 20036, USA
and
Fisheries Ecology Research Lab
Department of Biology
University of Hawai'i at Mānoa
Honolulu, HI 96822, USA

KERYN B. GEDAN
Sustainable Development and Conservation
    Biology Program
Department of Biology
University of Maryland

College Park, MD 20742, USA
and
Smithsonian Environmental Research
    Center
Edgewater, MD 21037, USA

ROBIN M. GROSSINGER
Historical Ecology Program
San Francisco Estuary Institute
Richmond, CA 94804, USA

SIMON HART
Institute for Integrative Biology
ETH Zurich
Zurich, Switzerland

JOHN N. KITTINGER
Conservation International
Gordon and Betty Moore Center for Science
    and Oceans
Honolulu, HI 96825, USA
and
Center for Ocean Solutions
Stanford Woods Institute for the
    Environment
Stanford University
99 Pacific Street, Suite 555E
Monterey, CA 93940, USA

HARUKO KOIKE
Fisheries Ecology Research Lab
Department of Biology
University of Hawai'i at Mānoa
Honolulu, HI 96822, USA

HEIKE K. LOTZE
Department of Biology
Dalhousie University
Halifax, NS B3H 4R2, Canada

MATT J. LYBOLT
Tetra Tech, Inc.
759 S. Federal Hwy, Suite 314
Stuart, FL 34994, USA

J. B. MACKINNON
Independent author and science journalist
Vancouver, BC, Canada

CATHERINE MARZIN
Office of the National Marine Sanctuaries
1305 East-West Highway, 11th Floor
Silver Spring, MD 20910, USA

LOREN MCCLENACHAN
Environmental Studies Program
Colby College
Waterville, ME 04901, USA

FIORENZA MICHELI
Hopkins Marine Station
Stanford University
120 Oceanview Blvd.
Pacific Grove, CA 93950, USA

JOSH NOWLIS
Bridge Environment
9721 20th Avenue NE
Seattle, WA 98115, USA

JOHN M. PANDOLFI
Centre for Marine Science
Australian Research Council Centre of
    Excellence for Coral Reef Studies
School of Biological Sciences
University of Queensland
Brisbane, QLD 4072, Australia

DANIEL PAULY
Fisheries Centre
University of British Columbia
Vancouver, BC V6T 1Z4, Canada

ROBERT L. PRESSEY
Australian Research Council Centre of
    Excellence for Coral Reef Studies
James Cook University
Townsville, QLD 4811, Australia

TORBEN C. RICK
Program in Human Ecology and
    Archaeobiology
Department of Anthropology
National Museum of Natural History
Smithsonian Institution
Washington, DC 20013-7012, USA

MARK D. SPALDING
Global Marine Team
The Nature Conservancy
Department of Zoology
University of Cambridge
Downing Street
Cambridge, CB2 3EJ, UK

RUTH H. THURSTAN
The Centre for Marine Science
Australian Research Council Centre
    of Excellence for Coral Reef
    Studies
School of Biological Sciences
University of Queensland
Brisbane, QLD 4072, Australia

ALAN T. WHITE
Asia Pacific Program
The Nature Conservancy
923 Nu'uanu Avenue
Honolulu, HI 96817, USA

DIRK ZELLER
Fisheries Centre
University of British Columbia
Vancouver, BC V6T 1Z4, Canada

PHILINE S. E. ZU ERMGASSEN
Department of Zoology
University of Cambridge
Cambridge CB2 3EJ, United Kingdom

VIEWPOINT CONTRIBUTORS

KATIE ARKEMA
Senior Scientist
The Natural Capital Project
Stanford Woods Institute for the
    Environment
Stanford University
Seattle, WA, USA

DANIEL J. BASTA
Director of the Office of National Marine
    Sanctuaries
National Ocean Service
National Oceanic and Atmospheric
    Administration
Washington, DC, USA

JOHANN BELL
Honorary Professorial Fellow
Australian National Centre for Ocean
    Resources and Security
University of Wollongong, NSW,
    Australia

LOUISE K. BLIGHT
Senior Scientist
Procellaria Research & Consulting
Victoria, BC, Canada

BILLY CAUSEY
Superintendent
Florida Keys National Marine Sanctuary
National Oceanic and Atmospheric
    Administration
Key West, FL, USA

WILLIAM CHEUNG
Associate Professor
Fisheries Centre
University of British Columbia
Vancouver, BC, Canada

CHARLES (BUD) EHLER
President
Ocean Visions Consulting, Paris, France

STEVEN D. EMSLIE
Professor
Department of Biology and Marine
    Biology
University of North Carolina, Wilmington
Wilmington, NC, USA

JAMES A. ESTES
Professor of Ecology and Evolutionary
    Biology
University of California, Santa Cruz
Santa Cruz, CA, USA

FRANCESCO FERRETTI
Postdoctoral Scholar
Hopkins Marine Station
Stanford University
Pacific Grove, CA, USA

ROD FUJITA
Director of Research and Development
Environmental Defense Fund
San Francisco, CA, USA

PAOLO GUIDETTI
Professor of Ecology
University of Nice
Nice, France

BEN HALPERN
Professor
Bren School of the Environment
University of California, Santa Barbara
Santa Barbara, CA, USA
and
Chair in Marine Conservation
Imperial College London

JOHN HENDERSCHEDT
Executive Director
Fisheries Leadership and Sustainability
    Forum
and
Vice Chair
North Pacific Fishery Management Council
Seattle, WA, USA

JOHN ODIN JENSEN
Marine Protected Areas Federal Advisory
    Committee member
and
Associate Professor of Maritime Studies
    and Ocean Policy
Sea Education Association

Research Associate Professor of History
    and Coastal Maritime Heritage
University of Rhode Island
Kingston, RI, US

JILL JOHNSON
Exhibit Developer, Sant Ocean Hall
National Museum of Natural History
Smithsonian Institution
Washington, DC, USA

RANDALL KOSAKI
Deputy Superintendent
Papahānaumokuākea Marine National
    Monument
Honolulu, HI, USA

LOREN MCCLENACHAN
Assistant Professor
Environmental Studies Program
Colby College
Waterville, ME, USA

MATTHEW MCKENZIE
Associate Professor
Department of History
University of Connecticut
Storrs, CT, USA

FIORENZA MICHELI
Professor
Hopkins Marine Station
Stanford University
Pacific Grove, CA, USA

STEPHEN PALUMBI
Harold A. Miller Professor in Marine
    Sciences
Stanford University
Pacific Grove, CA, USA

TONY PALUMBI
Writer
San Mateo, California

DANIEL PAULY
Professor
Fisheries Centre
University of British Columbia
Vancouver, BC, Canada

TONY J. PITCHER
Professor, and Director
Policy and Ecosystem Restoration in
    Fisheries Research Unit
Fisheries Centre
University of British Columbia
Vancouver, BC, Canada

STEVE ROADY
Managing Attorney for Oceans
Earthjustice, Washington, DC
and
Adjunct Faculty Member
Nicholas School of the Environment
Duke University

RAFE SAGARIN
Program Manager
Institute of the Environment and
    Biosphere 2
University of Arizona
Tucson, AZ, USA

PETER F. SALE
Assistant Director
United Nations University
Institute for Water, Environment and Health

and
University Professor Emeritus
University of Windsor
Windsor, ON, Canada

JIM TOOMEY
Syndicated Cartoonist
Creator of *Sherman's Lagoon*
Annapolis, MD, USA

ANDREA TREECE
Staff Attorney
Earthjustice, San Francisco, California

JOELI VEITAYAKI
Associate Professor and Head
School of Marine Studies
University of the South Pacific
Suva, Fiji

GEERAT J. VERMEIJ
Distinguished Professor
Department of Geology
University of California, Davis
Davis, CA, USA

DEAN WENDT
Dean of Research
California Polytechnic State University
San Luis Obispo, CA, USA

ʻAULANI WILHELM
Superintendent
Papahānaumokuākea Marine National
    Monument
Honolulu, HI, USA

FOREWORD

Marine Historical Ecology in Conservation, the title of this book, may be hard on potential readers, in that each of its two nouns and two adjectives can be seen as potential challenges:

- "Ecology," because some find it difficult to distinguish the scientific discipline of ecology from the passion of environmentalism;
- "Historical," because until recently, many academic ecologists suffering from physics envy were attempting to ban history and contingency from ecology;
- "Marine," because we are air-breathing, terrestrial animals with a strong bias against the watery world that covers most of the surface of our ill-named planet; and finally,
- "Conservation," because the word implies, for still too many, a departure from what scientists are supposed to do (describe our world, as opposed to changing it, or in this case, developing the tools to prevent it from being dismantled).

Why do we need marine historical ecology and conservation? The fact is that since Darwin's *On the Origin of Species,* we have become quite good at inferring what existed—in terms of animals and plants—if only because we have (a) fossils and (b) a powerful theory which allows, nay demands, that we interpolate between the forms we know existed, because we have fossils, and the forms for which we have no direct evidence but which we can link to present forms, including us humans.

Thus, in a sense, we know most of what *was there* since the Cambrian, and this knowledge becomes more precise and accurate the closer we come to the present. However, we don't know *how much* of what was there actually was there, and this may be seen as the defining feature of historical ecology and its potential use in marine conservation.

One way to view this is that while evolution's "central casting" provides us with a reliable stable of actors (e.g., a wide range of dinosaurs in the Triassic or a flurry of mammals in the

Pleistocene), it is for historical ecology to give them roles to play. (Note that these examples imply that historical ecology should mean the ecology of past systems and not only past ecology as recoverable through written documents, as one could assume when relying on a narrow interpretation of the word "history".)

Thus, an ecosystem with, say, sea turtles in it will function in a radically different way if these turtles are very abundant (as they appear to have been, e.g., in the pre-Columbian Caribbean) than it will where sea turtles are marginal, as is now the case in the Caribbean.

The Earth's ecosystems have all been modified by human activities, and this applies also to essentially all marine ecosystems, which whaling and hunting of other marine mammals, and later fishing, have reduced to shadows of their former selves in terms of the larger organisms they now support and the benefits they can provide us.

Some of these ecosystem modifications were unavoidable, as humans living on coastlines are largely incompatible with large populations of, say, sturgeons, sea turtles, or pinnipeds, and our appetite for fish implies that some fish populations will have to be reduced by fishing. But to a large extent, the depredations that we have imposed on the oceans have been entirely gratuitous: we need not have eradicated the great auk (*Pinguinus impennis*) or the Caribbean monk seal (*Monachus tropicalis*) to satisfy our seafood requirements, and thus it is perfectly reasonable to ask ourselves how we could prevent such catastrophes in the future (each species loss is a catastrophe) and whether we can rebuild now depleted populations of marine organisms so as to reduce the risk of this occurring again, and to have more to enjoy.

This is what marine historical ecology in conservation is for: to inform us about what these populations have been in the past, and under which conditions these populations could flourish so that we can start helping them do so. This is what the neat book you have in your hands is about.

Daniel Pauly
Vancouver
August 2013

ACKNOWLEDGMENTS

A volume of this scope and ambition is not possible without the guidance, encouragement, and contributions of a great many individuals. The editors—who had all recently completed their PhDs when embarking on this project—benefited in particular from several experienced mentors who directed us to key resources, were enthusiastic about our ideas for this project, and pushed us to expand our thinking about the nature and impact of this volume.

First, we thank the participants of our 2011 symposium at the International Marine Conservation Congress in Victoria, British Columbia, Canada, including Dalal Al-Abdulrazzak, Rich Aronson, Tyler Eddy, Jon Erlandson, Francesco Ferretti, Jeremy B.C. Jackson, John O. Jenson, Divya Karnad, Randy Kosaki, Heike Lotze, and Dana Miller. This volume was developed out of the thoughtful presentations and inspired discussion throughout that symposium, with contributions from many of the symposium participants. Their enthusiasm for our symposium concept gave us the confidence that our book project was possible and would have an impact. Jon Erlandson, in particular, was a continual source of guidance, helping our group initiate and plan this volume. We also thank Jesse Ausubel from the Alfred P. Sloan Foundation, who advised Louise and Jack on next steps after we fortuitously first met at the History of Marine Animal Populations meeting in Dublin, Ireland, in November 2010. Louise gratefully acknowledges the Koerner Foundation for their support of her doctoral research, and she particularly thanks Steve Koerner for engaging conversations on the value of long-term baselines, and ideas about unlikely sources of historical data.

The editors are grateful for financial support from the Census of Marine Life, and in particular Kristen Yarincik (now with the Consortium for Ocean Leadership), for supporting our first editors' meeting at the University of British Columbia Fisheries Centre in November 2011. We thank Daniel Pauly, Rashid Sumaila, and Dalal Al-Abdulrazzak for hosting that initial workshop, and for their guidance and expert input during the initial stage as we mapped out the book, potential contributors, and an ambitious timeline for completion. We are similarly grateful to Stanford University's Center for Ocean Solutions, and Meg Caldwell and Larry Crowder, for supporting our second editors' workshop at Stanford University in

August 2012, which allowed us to refine a critical focus of this volume: its explicit focus on applied solutions. We also thank John Weller and Cassie Brooks for the stunning imagery that graces the cover of this book and for their important marine conservation work.

We thank Blake Edgar, Merrik Bush-Pirkle, and the dedicated team at the University of California Press. Blake understood our vision for this volume at the outset, and his team worked tirelessly to see it through. We also thank the three reviewers, whose recommendations increased the scope, clarity, and consistency of our work.

Finally, we gratefully acknowledge the many authors of chapters and viewpoint boxes in this volume. It goes without saying that this book would not have been possible without their active involvement. This group of scholars, conservation practitioners, managers, and innovative thinkers were true partners in the book's development. We thank them for their excellent contributions.

ONE

Managing Human Legacies in a Changing Sea

An Introduction

JOHN N. KITTINGER, LOUISE K. BLIGHT,
KERYN B. GEDAN, and LOREN MCCLENACHAN

In 1938, Howard Granville Sharpe was working on his small ranch, 13 miles south of Carmel on the Big Sur coast in California, when he spied something strange in the kelp beds offshore. A longtime native of the area, Mr. Sharpe was no stranger to the Big Sur coast, yet he and his ranch hands were perplexed to find a group of sleek animals lazing around the kelp beds offshore of Bixby Creek. Two days later, he drove north to Stanford University's Hopkins Marine Station, where he was politely rebuffed after reporting to the marine scientists there that he had discovered a species of sea otter *(Enhydra lutris)*. Entreaties to the local press and scientists at the California Fish and Game Commission were met with similar amusement and skepticism.

A few days later, Fish and Game officials agreed to travel south to Sharpe's Rainbow Headlands ranch, where they were amazed to find the first family of sea otters observed in nearly a century in California. Professor Harold Heath from the Hopkins Marine Station later remarked, "Had you reported dinosaurs or ichthyosaurs running down your canyon, swimming about, we couldn't have been more utterly dumbfounded" (Sharpe 1989).

With this observation, a new chapter of natural history was written on the California coast. Sea otters had occasionally been observed since the late 1800s but were widely believed to be regionally extinct after 200 years of hunting for the lucrative fur industry. Within a few decades of their rediscovery, however, otter populations spread northward up the coast, repopulating their previous range.

The recovery of otters was not met with universal enthusiasm. In an early case of shifted baselines, sea otters were viewed as a new arrival by coastal California residents whose perspectives of the coast were formed over a shorter period of time than the otters' history of decline and recovery (see Box 2.1 by Jim Estes, in chapter 2). Urchin and abalone fishermen viewed the animals as competitors, which led to conflicts between otter-friendly coastal residents and those who viewed the species as a threat to their livelihood (Cicin-Sain et al. 1982). However, their return also heralded the regrowth of kelp forests, as dense aggregations of kelp-eating urchins fed the otters' voracious appetites. Protection and active management fostered growth of both otter populations and kelp forests, ecosystems that today support a diversity of species and provide social benefits in the form of fisheries and tourism. Indeed, the iconic kelp forests now common along northern California shores can be almost entirely attributed to the recovery of otters (Estes and Palmisano 1974).

This story of the return of the sea otter mirrors other emerging stories of recovery in marine environments around the globe (Figure 1.1). In the Pacific Ocean, egg and feather hunters reduced the short-tailed albatross *(Phoebastria albatrus)* to near extinction by the early twentieth century, with an estimated 5 million birds taken from one colony alone. As with sea otters in California, it was thought that the species had been eradicated until a small breeding colony of about 10 birds was discovered on the Japanese island of Torishima in 1951. Because albatrosses spend the first several years of their life at sea, these few individuals had escaped the final depredations of the feather hunters and formed the core of a population that continues to grow to this day. Other examples of recovery include the striped bass *(Morone saxatilis)* along the east coast of the United States, which demonstrates that the effects of overfishing can be reversed. In some cases, human actions have aided recovery; for example, coastal marshes, which are fundamental to estuarine ecosystems and were badly abused in centuries past, are now the focus of intensive restoration efforts, revealing the value that society has begun to place on the important functions and benefits these systems convey. Across the Pacific, the renaissance of traditional management systems based on historical practices has increased the biomass of target reef fish populations and provided social benefits to the communities that rely on these fisheries resources.

If there are universal lessons to be learned in these recovery stories, they are that the seeds of recovery and resilience can be found in surprising places and that *we have choices about the future of the oceans*. The lessons embedded in these historical recoveries also empower our generation of conservation scientists and ocean enthusiasts with the means (and perhaps the responsibility) to create an alternative future—one with healthy ocean ecosystems and resilient coastal communities. As Peter Sale writes in Box 1.1, nature is indifferent to the path we choose to take, but people care deeply about the state of nature; the abundance of marine species and the services provided by intact ecosystems greatly affect our quality of life and, indeed, our long-term survival.

These examples also teach us that history matters. There have been great losses in the global oceans, but as societies change the way they interact with marine ecosystems, so too do we change the environmental outcomes of these interactions. Species that were former targets of hunters and fishers have gained protection. Habitats that were once dredged and filled have become recognized for their role in coastal defense and fisheries production. Historical information sources that were once ignored have gained new life as data sources to

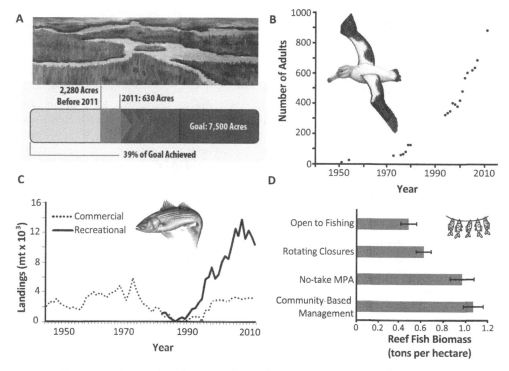

FIGURE 1.1 Four emerging stories of recovery in marine environments around the globe.
(A) Progress toward salt pond restoration in San Francisco's South Bay (from South Bay Salt Pond Restoration Project Annual Report 2012). (B) Counts of breeding short-tailed albatross at Torishima Island, Japan, 1951–2011, following rediscovery (figure based on unpublished data from the Yamashina Institute and H. Hasegawa, Toho University, Japan; U.S. Fish and Wildlife Service 2008, Agreement on the Conservation of Albatrosses and Petrels 2009). (C) Increased striped bass landings demonstrate population recovery along the U.S. Atlantic coast (from Atlantic States Marine Fisheries Commission 2014). (D) Fish biomass under community-based management is not statistically different ($P > 0.05$) from no-take marine protected areas (MPAs) and is more effective than open access areas and zones managed with rotating annual closures (partial protection) (data from Friedlander et al. 2013; based on 1,344 surveys at 143 locations: open to fishing, $n = 94$; no-take MPA, $n = 9$; community-based, $n = 18$; partial protection, $n = 22$).

understand baselines and make well-informed conservation decisions. Active management of species and habitats has certainly not always guaranteed recovery, but increasing attention to the historical dynamics of decline and recovery continues to reveal how we can use the past to better manage for the future.

In this volume, we define marine historical ecology broadly as the study of past human–environmental interactions in coastal and marine ecosystems, and the ecological and social outcomes associated with these interactions. Marine historical ecology developed out of the growing realization that humans have altered marine ecosystems over very long time scales, and that historical data often are needed to understand the true magnitude of human-induced

BOX 1.1   Viewpoint: Coral Reefs, Conservation, and Historical Ecology
Peter F. Sale

Ecologists who study coral reefs should be predisposed to the importance of history because they study a built ecosystem, one entirely assembled by its resident species as they build their skeletons from basic chemical building blocks. That reef ecologists mingle with reef geologists should aid this predisposition because geologists are far more aware of deep time than are ecologists. We should have been predisposed, but in the early days we mostly did not think of history at all. Coral reef ecology is very young, only beginning in the 1950s. At that time, the ecological paradigm was that enough time had usually passed to ensure that ecosystems were at or near an equilibrium state and, therefore, that history did not matter. I know I began my PhD research in the mid-1960s confident in my knowledge of the evolutionary history of coral reefs, but I did not view reef ecology from a historical perspective. Reefs were the way they were, and my job was to figure out how these amazing ecosystems functioned now. My colleagues mostly thought the same way. Within a decade, strongly influenced by that giant, Joseph H. Connell, who approached the challenges of both reef and rainforest by monitoring individual organisms' struggles to survive, grow, and reproduce over many years, I was using the monitoring of individual assemblages of reef fishes over as much as a 10-year span, as a major tool in my efforts to understand coral reefs.

Ten years is a very short time in the life of a coral reef, and long-term monitoring studies are difficult to sustain, but one learns that each year can be different (even in the tropics), and

there are other ways to extend knowledge into the past. However, the recent history of coral reefs can be depressing. Long before we started to learn about their ecology, coral reefs were being degraded by human activities, first in some places, then in more, and now, through our releases of CO_2, throughout the tropics. If we do not alter our behavior significantly, the reefs I knew in the 1960s will have disappeared completely by midcentury. Most students today study human impacts on reefs, something my generation did not think about, yet I worry that they still do not always appreciate the extent and speed of historical change.

Conservation science is a challenging field, with immense, unspoken value judgments. Nature does not care if a coral reef becomes an algae-covered bench; it is people who care, and conservation science struggles to make this caring suitably objective, focusing on loss of ecosystem goods and services or loss of ecological function. We also seem to have decided that sustainable use must be compatible with ecosystem sustainability and with the economic, cultural, and societal success of people who do the using, because we want it to be so. In fact, historical ecology seems to be telling us that humans, in any numbers at all, making any substantial use of coral reefs, routinely overharvest fisheries and cause substantial destruction to the reefs themselves. And yet, history also sometimes reveals success stories where reefs are managed sustainably over long periods of time. These are the stories which give hope for a future, changed but still livable, and with a few coral reefs still present.

Peter F. Sale is Assistant Director at United Nations University, Institute for Water, Environment and Health. He is also Professor Emeritus at University of Windsor.

changes. People working in marine historical ecology (including the authors who contributed to this book) come from a variety of fields, including marine biology, fisheries science, archaeology, geography, history, and more. These researchers also use information from diverse sources. Shell middens, oral histories, climate records, log books, restaurant menus, and handwritten letters in dusty museum basements all have had stories to tell about human–ocean relationships. Some marine historical studies stretch back a few decades, while others span millennia or longer (Box 1.2). All research in this area has a common goal of establishing a deeper understanding of how human societies have affected marine ecosystems through time.

While ecologists and biologists were instrumental in first describing many of the long-term anthropogenic changes to marine ecosystems, marine historical ecology has become increasingly more interdisciplinary in scope, and it will require an even greater collaborative effort to apply these findings to conservation and management. The interdisciplinary nature of this field has attracted numerous researchers and fostered cross-disciplinary collaborations, leading to more integrative approaches. For example, in the Gulf of Maine, fisheries scientists worked together with historians to estimate cod abundances in the 1850s (Rosenberg et al. 2005). In Hawaii, geographers, ecologists, and archaeologists collaborated to reconstruct the history of coral reef ecosystems and identify key social drivers associated with these changes (Kittinger et al. 2011). And a panel at the 2011 International Marine Conservation Congress brought together marine biologists, fisheries scientists, archaeologists, geographers, and others to explore ways in which history can help shape the management of marine ecosystems, launching this collaboration and edited volume. These multidisciplinary collaborations are increasingly common because they embody the potential for innovative ways of understanding long-term change, but also because interdisciplinary analyses can reframe these problems in new ways and offer new solutions to restore degraded ocean ecosystems and rebuild depleted resources.

Fueled by recognition of innovative scholarship and increased engagement by researchers and institutions, the past few decades have seen tremendous growth in this field. Marine historical ecology research now spans a growing variety of disciplines and has been published in the highest impact scientific journals. In the past decade, scholars have also developed major initiatives in historical ecology that have significantly advanced the field, including the History of Marine Animal Populations (HMAP) project in the Census of Marine Life, the Integrated History and Future of People on Earth (IHOPE) project, and a series of working groups organized by Jeremy Jackson and others at the National Center for Ecological Analysis and Synthesis in California. Large-scale regional initiatives have also been developed, such as the San Francisco Estuary Institute's historical ecology project and the Mannahatta project in New York. This growth in the field demonstrates broad appeal, due in part to recognition that despite the limitations of historical data, discounting the long-term perspectives they provide can lead to inappropriate conservation actions and unintended negative consequences for ocean environments and coastal communities.

Increased interest in marine historical ecology in the research community corresponds with increased attention from the general public. Findings from marine historical ecology

Geerat J. Vermeij

In this age when change is so rapid that most of us have detected it in our lifetimes, it is tempting to think that no amount of historical understanding can help illuminate Earth's current unique transformation. Confronted with the reality that our own species dominates the biosphere as no species before us has ever done, most conservationists and policymakers have sought to comprehend and manage our relations with the rest of the living world by considering only the present and the immediate future. If history enters the picture at all (Willis and Birks 2006), it is limited to the postglacial period, when humans were already affecting our planet's climate and biota. With this short-term perspective, the primary application of history is to recreate ecosystems that existed before the advent of human hegemony.

Just as the history of civilization can inform present-day human affairs, so the history of life as chronicled by fossils offers us a long time scale as we grapple with the crisis of our planet-wide ecological monopoly (Vermeij and Leigh 2011). The Earth has withstood catastrophes and periods of rapid change before, and our deepening understanding of the circumstances of life in the past and of mechanisms of evolution not only expands insights into how the modern biosphere works, but often alters conceptions that were founded only on the world as it is today.

Ecologists were not alone in being slow to recognize this potential. For centuries, paleontology was a descriptive science, whose practitioners were content to name species, infer evolutionary lines of descent, and, beginning in the 1950s, describe the composition of ancient communities. It was not until the 1970s that these accounts were complemented by a more analytical approach, in which processes and interactions affecting living systems were traced back to the distant past. It soon became clear that the types and intensities of competition, consumption, production, extinction, nutrient cycling, mutualistic association, species movements, and the regulation of atmospheric and oceanic chemistry have changed dramatically over time, in accordance with previously underappreciated evolutionary innovations set against a backdrop of ceaseless mountain building, erosion, and tectonic rearrangements of land masses.

projects continue to gain a strong following at national and global scales. Nonfiction works in marine historical ecology have become popular books—for example, Jared Diamond's *Collapse,* Mark Kurlansky's *Cod,* Callum Roberts's *The Unnatural History of the Sea,* and James MacKinnon's *The Once and Future World*)—and environmental reporting and journalism has turned its attention to historical topics (Weiss et al. 2006). Marine education and outreach programs have also started including historical content, such as the Sant Ocean Hall at the Smithsonian Institution in Washington, D.C., which features an exhibit on long-term changes to marine fish populations, and the U.S. National Marine Sanctuaries Program, which has brought historical ecology into its programmatic goals. Collectively, these examples point to a broad public interest in the ocean's past and what it can tell us about current challenges in environmental sustainability.

How can this knowledge help all of us who seek to protect Earth's living resources? The fossil record offers the only long-term insights about how ecosystems recover and reassemble after great crises, what are the enduring effects of warming and ocean acidification, what happens when species from different parts of the world come together, how ecosystems create and accommodate tipping points in composition and organization, how natural systems have resolved tragedies of the commons that stem from the inner workings of the system itself, and how the biosphere's chemical environment is affected by innovations and disruptions. It documents economic trends in life's history, which strikingly parallel developments in our own short history (Vermeij and Leigh 2011). Knowledge of the past reminds us that the courses of evolution and the history of ecosystems exhibit predictable properties, all of which indicate the universality of change. Whatever equilibrium is achievable in the short run, it is upended by evolution from within and by disruptions that emanate from outside the realm of organisms. New species and ecosystems complement or supplant older ones, novel ecological relationships and criteria for natural selection become established, and systems collapse and recover, all thanks to adaptation or its absence.

And adaptation is, of course, the key to coping with change. The ability of living things to respond to and cause change in ways that benefit them is central to the persistence and evolution of life on Earth. It is this capacity to adapt that human dominance has all but eliminated from most species. History teaches that providing species everywhere with the resources, space, and time to adapt is the single most important condition for maintaining a viable, productive, and responsive biosphere.

The American historian Gordon Wood (2008:14) cautioned that "history tends to inculcate skepticism about our ability to manipulate and control purposefully our destinies." In the present context of our stranglehold over the world's ecosystems, it may be better to enable millions of species to adapt to change than to manage the world's affairs all on our own.

Geerat J. Vermeij is Distinguished Professor in the Department of Geology, University of California, Davis.

FOUR CRITICAL CHALLENGES IN MARINE CONSERVATION

Marine historical ecology is increasingly oriented toward real-world applications, and researchers and practitioners are exploring tangible policy, management, and conservation strategies based on knowledge of the past. At the same time, marine conservation programs and practitioners worldwide struggle to meet the immense challenges of safeguarding biological diversity and maintaining the ecosystem services upon which society depends.

In this book, we use four parts to focus on four key challenges that confront marine conservation: (1) recovering endangered species, (2) conserving fisheries, (3) restoring ecosystems, and (4) engaging the public. These four distinct areas represent specific challenges and opportunities, where marine historical ecology is distinctly poised to help address the implementation gap—or the distance between conservation science and policy actions and desired social and environmental outcomes. By providing real-world examples of applied approaches, as well as options for potential use and application, each of these sections advances concepts and tools that can be implemented in management and policy. Taken together, the sections offer a blueprint for using marine historical ecology to confront the challenges of ocean conservation in a rapidly changing world.

Recovering Endangered Species

Endangered species protection and recovery has always been a central part of modern efforts to conserve and manage nature, in terms of public perception, science, and on-the-ground action. Similarly, estimating historical baselines for endangered species has long been a focus of marine historical ecology. These efforts have demonstrated that human exploitation has reduced the population abundance of many large marine animals over long time scales and has compromised the role of top predators and keystone species in ocean ecosystems. Some species, such as the great auk *(Pinguinus impennis)* and Steller's sea cow *(Hydromalis gigas)*, are now gone forever, while others, such as the northern elephant seal *(Mirounga angustirostris)*, have dramatically recovered from near extinction. The fate of some species, such as certain whales, still hangs in the balance. In this section, authors examine ways in which historical ecological research can contribute to modern efforts to recover marine species, many of which have endured centuries of exploitation. These authors go beyond documenting decline and show how historical reconstructions can help set realistic recovery targets, highlighting actions that have aided species in need of protection, or even helped turn endangered species back from the brink of extinction.

Conserving Fisheries

Fisheries worldwide face critical challenges in sustainability, and marine historical ecology has played an important role in defining the extent of changes in fish populations globally. Daniel Pauly's now famous concept of "shifting baselines" was first conceived in the context of fisheries, and since that time considerable historical evidence has helped define the current status of, and trends in, fisheries. Fisheries sustainability, however, means moving beyond quantifying impacts and scales of loss and toward developing a portfolio of potential solutions. In this section, authors advance novel ways to apply historical data to the challenge of managing fisheries and describe a series of cases where these nonconventional datasets and approaches are resulting in real-world successes. For example, coastal and island communities are integrating historically based management practices into place-based resource stewardship efforts, preserving fish populations and ensuring ecological benefits from marine ecosystems. Additionally, stock assessment practices, which are difficult in data-poor fisheries contexts across the globe, are being modified to include historical data, providing more accurate baselines of fish populations and historically based recovery targets. These examples and others in this section point to a broad range of applied roles for marine historical ecology in fisheries conservation and management.

Restoring Ecosystems

Restoring ecosystems to a healthy and resilient state is a fundamental goal of marine conservation, and marine historical ecology has played an important role in helping scholars and practitioners understand the nature of healthy ecosystems as they existed in the past. Authors in this section show us how historical information on the distribution and condition of habitats, as well as the historical production of social benefits from these systems (known as ecosystem services), can guide modern restoration efforts. For example, historical reconstructions can illuminate past ecosystem states and current population trends, highlighting the key drivers or processes (such as predation) that may be acted on to achieve positive change. Historical studies can also provide environmental baselines against which to measure the effectiveness of conservation actions. Finally, this section also examines how marine historical ecology can reveal the dynamic nature of marine ecosystems and ecosystem responses to past eras of environmental change (especially in studies over evolutionary and geologic timescales; Box 1.2). Such efforts are increasingly relevant to restoration efforts striving to protect ecosystem integrity and resilience in the face of a globally changing environment.

Engaging the Public

The real-world application of results from marine historical ecology would be impossible without public engagement, because decisions about endangered-species recovery, fisheries conservation, and ecosystem restoration ultimately play out in a public sphere. Stories about the historical abundance of marine animals, the past bounty of fisheries, and the healthy functioning of intact ecosystems inspire wonder about the potential of the natural world to sustain and support humans and other life and, in doing so, influence policy debates. However, challenges also exist in this realm—for example, the uptake of conservation messages by a media-saturated public can be limited, and stakeholders can respond in a range of ways (sometimes unpredictably) to historical information. As authors in this section discuss, historical ecology can provide compelling narratives about the past and can also inspire alternative visions for the future. A diversity of stories may lead to conflict over desired outcomes of conservation and management, but they can also empower communities to effect change or advance potential solutions. Engaging the public around the history of people and life in the sea enriches these important discussions and is essential to developing and implementing conservation actions.

GOALS FOR THIS VOLUME

We have two overarching goals for this volume. First, we hope to provide impetus for a vibrant, transdisciplinary discussion on using insights from historical ecology to improve the management and conservation of marine ecosystems and species. In essence, we hope to show—through tangible examples—how the research community can develop better, more viable science-based solutions, and highlight practical ways to enable their uptake in the policy and conservation realm. Second, it is our intention to showcase practical examples of how historical data can be used in the conservation of marine ecosystems. Throughout the book, authors provide real-world and hypothetical examples of management strategies, policy levers, and conservation actions and perspectives, drawing on a diverse set of case studies from around the globe. Additionally, we have supplemented each chapter with "Viewpoint" boxes that contain reflections from policymakers, managers, and leading scientists about how the concepts presented can be engaged in real-world applications.

We hope this book will be of interest to a broad range of stakeholders working in the multidisciplinary fields of marine science and conservation, including academic researchers, educators, students, policy specialists, environmental managers, marine protection organizations, and others. We developed this book with this diverse ocean-minded community in mind, knowing that strong and diverse knowledge-to-action partnerships are necessary to work collectively toward a promising future for the ocean.

GOING BEYOND THE SCIENCE: LINKING KNOWLEDGE TO ACTION

Understanding how to move beyond the science and to real-world results is increasingly important in a world where the integrity of marine environments is challenged by a growing number and intensity of human drivers. The future of these systems is as much about the plants, animals, and habitats that compose these ecosystems as it is about us. As historian David McCullough has eloquently written, "History is a guide to navigation in perilous times. History is who we are and why we are the way we are" (McCullough 1984).

Our generation of marine scientists does not have the benefit of seeing firsthand the intact ecosystems of the past—we work in highly altered environments. The changes to these systems were first described in detail by the generation of scientists who came before us, many of whom witnessed these changes over a lifetime of work. These groundbreaking researchers—some of whom contributed to this volume (including Boxes 1.1 and 1.2)—inspired us and prompted our initial interest in and commitment to the field of marine historical ecology. In science, it is often said that one stands on the shoulders of giants, and we gratefully acknowledge the important and transformative work of these scholars.

Along with altered ecosystems, we inherited the knowledge and skills to understand the long history of change in marine environments. The majority of the initial work in marine historical ecology focused on quantifying, reconstructing, and characterizing long-term change. The task now is to take marine historical ecology beyond the initial step of reconstructing and understanding change and apply it toward the significant conservation questions of the future. For example, how can long-term baselines best be used to plan for and recover depleted and endangered marine species? Can fisheries be productive enough to support growing populations and also be environmentally sustainable? What do we stand to gain by recovering coastal ecosystems? How can the vision of past ecosystems be used to inspire the public toward conservation action? And how does a historical understanding of past changes shape a collective vision for a sustainable future and the options for getting there?

These questions place us at a critical juncture of applying an amassed knowledge base to problems facing the real world. The central challenge is clear: it is one of going beyond the descriptive science of marine historical ecology and toward identifying tangible solutions. With that in mind, we strived to create not a volume *of* baselines, but rather a volume on *how to use them*. We sincerely hope the findings presented herein embolden readers with new ideas and tools to restore healthy ocean environments and build resilient coastal communities. After all, it is in this century that we must learn from the past to secure the future of our blue planet.

REFERENCES

Agreement on the Conservation of Albatrosses and Petrels (2009) ACAP species assessment: short-tailed albatross *Phoebastria albatrus*. Available at www.acap.aq.

Atlantic States Marine Fisheries Commission (2014) Atlantic striped bass. www.asmfc.org /species/atlantic-striped-bass.

Cicin-Sain, B., Grifman, P. M., and Richards, J. B. (1982) Social science perspectives on managing conflicts between marine mammals and fisheries. Marine Policy Program, University of California at Santa Barbara and University of California Cooperative Extension, Santa Barbara and San Luis Obispo, CA.

Estes, J. A., and Palmisano, J. F. (1974) Sea otters: their role in structuring nearshore communities. *Science* 185, 1058–1060.

Friedlander, A. M., Shackeroff, J. M., and Kittinger, J. N. (2013) Customary marine resource knowledge and use in contemporary Hawaii. *Pacific Science* 67, 441–460.

Kittinger, J. N., Pandolfi, J. M., Blodgett, J. H., et al. (2011) Historical reconstruction reveals recovery in Hawaiian coral reefs. *PLoS ONE* 6, e25460.

MacKinnon, J. B. (2013) *The Once and Future World: Nature As It Was, As It Is, As It Could Be*. Random House Canada, Toronto, ON.

McCullough, D. (1984) Commencement address at Wesleyan University, June 3.

Rosenberg, A. A., Bolster, W. J., Alexander, K. E., et al. (2005) The history of ocean resources: modeling cod biomass using historical records. *Frontiers in Ecology and the Environment* 3, 78–84.

Sharpe, H. G. (1989) The discovery of the "extinct" sea otters. In *A Wild Coast & Lonely—Big Sur Pioneers* (R. S. Wall, Ed.). Wide World Publishing/Tetra, San Carlos, CA.

South Bay Salt Pond Restoration Project (2012) Annual Report 2011. www.southbayrestoration. org/documents/technical/2011sbspannualreport.FINAL.pdf.

U.S. Fish and Wildlife Service (2008) Short-tailed albatross recovery plan. Anchorage, AK.

Vermeij, G. J., and Leigh, E. G., Jr. (2011) Natural and human economies compared. *Ecosphere* 2, Article 39.

Weiss, K. R., McFarling, U. L., and Loomis, R. (2006) Altered oceans [article series]. *Los Angeles Times*, July 30–August 3.

Willis, K. J., and Birks, H. J. B. (2006) The need for a long-term perspective in biodiversity conservation. *Science* 315, 1261–1265.

Wood, G. S. (2008) *The Purpose of the Past: Reflections on the Uses of History*. Penguin, New York, NY.

PART ONE

RECOVERING ENDANGERED SPECIES

Lead Section Editor: LOUISE K. BLIGHT

Conservation practice is most often carried out at the level of individual species and their recovery. While this species-centric approach is often deemed wrongheaded—because species require ecosystems to survive and because focusing on single species or populations can be an inefficient and expensive way to conduct conservation activities—there are often practical reasons for a single-species focus. For example, trend data are more easily obtained for a species or population than are data on past ecosystem states; the use of single-species models requires relatively few assumptions; and public support is more easily garnered for species-focused conservation efforts, particularly if the creature in question is charismatic. Individual species may also play key roles in marine ecosystem structure and function, for example as habitat (e.g., architectural species like oyster reefs and corals) or as apex predators, grazers, or prey. As this section highlights, there are now many examples worldwide of such important species being depleted or endangered in marine systems.

Historically, many large marine species were hunted as a source of food or other resources or persecuted as real or apparent competitors with humans for their prey. Most of the great whales were reduced to a fraction of their former abundance by centuries or decades of whaling (with industrial whaling emerging in Europe in the 1600s); and in the 1900s, basking sharks (*Cetorhinus maximus*) in Canada's Pacific waters were killed on sight as a matter of federal policy. Rammed with special blade-equipped boats because they were deemed a nuisance to fishermen, this species is now all but extirpated in the region (Wallace and Gisborne 2006). In the 1800s, tens of thousands of seabird eggs were collected annually at California's Farallon Islands to feed a rapidly growing San Francisco, a practice that continued for nearly 50 years (Doughty 1971). This caused declines that arguably affected population trajectories well into the twentieth century. If we wish to recover these and other populations—for their own benefit or to help restore ecosystem function—much could be learned from the circumstances that surrounded their decline or revival.

For example, in many instances, society has responded to such wholesale slaughter by implementing legislative and policy solutions, allowing for targeted species to begin their journey to recovery. In this section Heike Lotze explores how such regulatory tools have succeeded (or failed) where implemented in response to overharvesting and other historical drivers of marine population declines. She uses case studies and long-term datasets to examine common patterns in the historical recoveries or nonrecoveries of species and explores the main drivers of positive population response. These case studies thus provide information on the type of data and management approaches required to improve existing conservation and management strategies for long-lived marine species.

In a complementary chapter, Jon Erlandson and colleagues tackle the thorny question that eventually plagues most practitioners attempting to compile long-term datasets in order to determine a historical baseline—what is "natural"? Using data from archaeological sites along the Pacific coast of North America, these authors reconstruct the biogeography of the region's pinniped fauna and argue that humans have been affecting populations of seals and sea lions for thousands of years: a comparison of archaeological records suggests that the behavior, distribution, and abundance of some species have altered significantly over millennia, likely in response to hunting pressure by early North Americans.

Finally, in the last chapter of this section, Francesco Ferretti and coauthors explore innovative ways to locate disparate and sometimes obscure sources of historical information and how best to overcome the analytical challenges of extracting usable data from such qualitative or fragmented records. These authors contend that Bayesian and meta-analytical approaches are highly conducive to rendering difficult datasets analyzable and thus suitable for incorporation into studies using modern, more quantitative data. Highlighting their earlier work on reconstructing the population trends of large sharks in the Mediterranean— where historical data analyzed in this manner indicated that shark populations had declined by 96–99% over 50–200 years—they show how such data were instrumental in getting natural resource managers and policy makers at the European Commission to develop and adopt the first Plan of Action for the Conservation and Management of Sharks for the region.

Describing the long-term trends of marine animal populations is challenging, and determining causes even more so. However, looking back into the past through the multiple perspectives provided by historical ecology can highlight the actions that have resulted in extinctions, provide more accurate baselines, and improve understanding of the approaches that might help recover species and populations.

REFERENCES

Doughty, R. W. (1971) San Francisco's nineteenth-century breadbasket: the Farallons. *Geographical Review* 61, 554–572.
Wallace, S., and Gisborne, B. (2006) *Basking Sharks: The Slaughter of BC's Gentle Giants.* New Star Books, Vancouver, BC.

TWO

What Recovery of Exploited Marine Animals Tells Us about Management and Conservation

HEIKE K. LOTZE

Over the past centuries and decades, high exploitation pressure has led to strong declines in a wide range of marine mammal, bird, reptile, and fish populations. Today, many species are at low abundance levels, endangered, or extinct on a regional or global scale. Yet throughout history, people have responded to declining resource abundance by implementing management and conservation measures. Sometimes these measures were successful and resulted in recovery, and other times they failed. Such successes and failures can serve as guides for conservation and management efforts aimed at preventing further biodiversity loss and restoring functioning ecosystems. This chapter highlights the role of marine historical ecology in assessing the magnitude and success of recovery by analyzing long-term population trends. I present an overview of trajectories of recovery and nonrecovery across a range of species and taxonomic groups, and the major drivers enabling recovery. I also assess how common recovery is among depleted populations and what the timeline and magnitude of recoveries have been. Finally, I evaluate how knowledge from marine historical ecology can be applied to improve current management strategies and future conservation planning to foster further recoveries in marine ecosystems.

INTRODUCTION

A growing number of studies in marine historical ecology have reported strong declines in animal populations over recent decades and centuries, largely due to overexploitation, habitat loss, and pollution (Jackson et al. 2001, 2011; Pandolfi et al. 2003; Airoldi and Beck 2007; Rick and Erlandson 2008; Starkey et al. 2008; Lotze and Worm 2009). Many species are now at low abundance levels, and several are listed as endangered or extirpated or have become globally extinct (Dulvy et al. 2003, 2009; Vié et al. 2009; Harnik et al. 2012). Such declines and losses have altered the structure and function of many marine ecosystems and, consequently, affected the ecological goods and services that help provide for human well-being (Lotze et al. 2006, 2011a; Worm et al. 2006). Yet not all populations are at their historical low point; several studies have shown recent increases in the abundance of previously depleted marine populations, often as a result of enhanced management and conservation measures in the twentieth century (Rosenberg et al. 2006, Worm et al. 2009, Lotze et al. 2011b, Ward-Paige et al. 2012). Other populations, however, have remained at low abundance levels or continue to decline despite considerable management and conservation efforts.

Recovery can be measured and interpreted in many different ways. Despite increasing interest, there is no standard definition of recovery (Redford et al. 2011). According to the Oxford Dictionary, recovery means "a return to a normal state of health, mind, or strength" (www.oxforddictionaries.com). Unfortunately, such a "normal" or "natural" state is largely undefined for marine populations that have been exploited for decades or centuries (Jackson et al. 2001, Lotze and Worm 2009). Often, recovery is simply seen as an increase in abundance. Yet when will a population be declared to have successfully recovered: when it has reached a perceived level of high abundance, its biological carrying capacity, or a historical baseline? Here, historical ecology can play an important role in assessing the magnitude and success of recovery compared to historical abundance levels, thereby helping to set meaningful management and conservation targets (Lotze et al. 2011b).

Below, I first present selected case studies of recoveries and nonrecoveries across different species groups to highlight the variety of population trends and recovery scenarios. This includes the management and conservation measures involved in recovery as well as the environmental and anthropogenic factors influencing it. Where available, I also provide more synthetic estimates about the frequency of recoveries among depleted populations and their magnitude and timeline. I then evaluate how this knowledge can be applied to improve management and conservation planning to foster further recoveries among marine populations and ecosystems.

RECOVERY AND NONRECOVERY ACROSS DIFFERENT SPECIES GROUPS
Marine Mammals

Throughout history, marine mammals have been hunted for their meat, oil, blubber, baleen, fur, or ivory, which reduced many populations to very low abundance and drove some to regional or global extinction as early as 500 to 1,000 years ago (Dulvy et al. 2003, McClenachan and Cooper 2008, Rick and Erlandson 2008, Lotze and Worm 2009). By the late nineteenth century, several whale species and fur- or ivory-bearing seals, sea otters, and walruses were so decimated that governments began to take measures to prevent further depletion and extinction. With more declines in the twentieth century, such management and conservation measures increased and, in many cases, have resulted in at least partial recoveries (Lotze et al. 2006, 2011b).

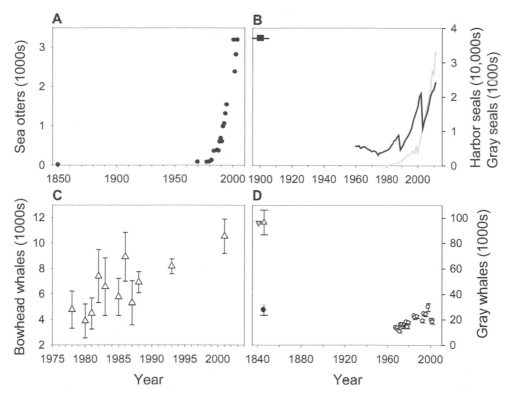

FIGURE 2.1 Examples of marine mammal recoveries. (A) Recovery of sea otters in British Columbia (data from Nichol et al. 2005, Gregr et al. 2008, Magera et al. 2013). (B) Recovery of harbor seals (black line) and gray seals (gray line) in the Wadden Sea (southern North Sea), with a historical population estimate (black square) for harbor seals (data from Reijnders 1992, Lotze 2005, Trilateral Seal Expert Group 2011). (C) Recovery of western Arctic bowhead whales (means and SD; data from Gerber et al. 2007). (D) Recovery of gray whales in the northeastern Pacific (open circles; means and SE; data from Rugh et al. 2005, Magera et al. 2013), with different historical population estimates from catch data (filled circle; Rugh et al. 2005), genetic studies (white triangle; Alter et al. 2007), and habitat availability (gray triangle; Pyenson and Lindberg 2011).

One early conservation measure was the Fur Seal Treaty of 1911, which prohibited commercial hunting of fur seals and sea otters (*Enhydra lutris*) in the North Pacific and enabled several populations to recover from the brink of extinction (Lotze et al. 2011b, Magera et al. 2013). In California, sea otters recovered from 50–100 survivors in 1914 to 2,000 in the 1990s, and they continue to be protected by the U.S. Marine Mammal Protection Act and Endangered Species Act (U.S. Fish and Wildlife Service 2003; but see Box 2.1). By contrast, sea otters were extirpated in British Columbia in 1929 but then reintroduced from Alaska between 1969 and 1972. The population increased from 89 translocated individuals to 3,000 in 2004, repopulating 25–33% of the species' former range (Figure 2.1A; Gregr et al. 2008). There are no good estimates of historical population size in either region; however, Gregr et al. (2008) used habitat availability to estimate a coast-wide carrying capacity

BOX 2.1   Viewpoint from a Practitioner: Sea Otters and Kelp Forests

James A. Estes

When viewed from the perspective of a world in which so many large predators are depleted and continuing to decline, the conservation and management of the sea otter (*Enhydra lutris*) might appear to be a resounding success. These animals, which abounded in coastal waters of the North Pacific Ocean and southern Bering Sea until the mid-1700s, had been hunted to near extinction during the Pacific maritime fur trade. Following protection afforded in 1911 under the North Pacific Fur Seal Convention, otter populations have recovered, or are recovering, across much of their historical range in Canada, the United States, and Russia.

In North America, at least, these recoveries are not universally embraced as a good thing, because the spreading otter populations conflict with commercial and recreational shellfisheries. Such conflicts have arisen repeatedly, in numerous areas and over a variety of shellfish species. The depletion of shellfish by sea otters is a very real problem. But some of the resulting difficulty stems from the failure of fisheries and the management agencies to consider historical baselines in their formulation of policy and in determining what the proper management of sea otters ought to be.

The loss of sea otters from coastal ecosystems caused their various macroinvertebrate prey species—sea urchins, snails, crabs, clams, and the like—to increase, often spectacularly. This condition defined the state of coastal ecosystems as they were first seen by modern human societies, and shaped the perception of the natural ecological baseline. The real ecological baseline, in which otters abounded and shellfish were comparatively rare, was thus seen as an undesirable and unnatural state as otters recovered. The vast shellfish resources conveyed a clear and simple value to these altered systems. The benefits of sea otters, mostly defined in economic terms by their more complex positive influence on kelp and the associated ecosystem services (e.g., finfish enhancement and carbon sequestration), have been more difficult for most people to grasp.

These various issues are at the center of two ongoing conflicts in the United States—one a lawsuit filed by shellfishers against the U.S. Fish and Wildlife Service (FWS) for recently deciding to allow sea otters to reoccupy southern California waters, and the other resulting from rumblings of FWS support for the control of sea otters in southeast Alaska in the interest of local shellfisheries. The irony of the FWS's overall position is that it embraces both sides of the conflict, in apparent deference to differing political pressures in California and Alaska.

This unfortunate situation is a clear testimony to why history matters in natural resource management. More attention to the historical baseline, and all that it says about ecological process and associated ecosystem services, would aid FWS in formulating more rational (i.e., less politically based) and more consistent policies on the management of sea otters, whatever those policies might eventually be.

James A. Estes is Professor in the Department of Ecology and Evolutionary Biology, University of California, Santa Cruz.

of 52,459 sea otters in British Columbia, placing the current population at <6% of its potential abundance.

Another example of pinniped recoveries comes from the Wadden Sea (in the southern North Sea), where harbor seals (*Phoca vitulina*) and gray seals (*Halichoerus grypus*) have increased since the 1970s and 1980s, respectively (Figure 2.1B; Lotze 2005). Although gray seals were historically the most common seals in the Wadden Sea, they were essentially absent after ~1500 AD (Wolff 2000). However, twentieth-century conservation efforts in Great Britain resulted in increased populations of gray seals in the North Sea, and the species naturally recolonized and established two small, but permanent and growing, colonies in the German and Dutch Wadden Sea in the 1960s and 1980s, respectively. It is unknown what their historical population size may have been. In comparison, reconstruction of harbor seal abundance on the basis of hunting records suggests that there were ~37,000 individuals in 1900 (Figure 2.1B; Reijnders 1992). After hunting reduced the population to 3,000 in 1974, exploitation bans and increased habitat protection enabled an increase to 24,000 animals in 2011, almost 65% of the 1900 estimate (Lotze 2005, Trilateral Seal Expert Group 2011). Similar cases of increase following anthropogenic decline can be found in many pinniped populations around the world, such as New Zealand fur seals (*Arctocephalus forsteri*; Smith 2005) and northern elephant seals (*Mirounga angustirostris*; Carretta et al. 2005; also see chapter 3, this volume), while others have remained at low population abundance after declines (e.g., Baltic ringed seal, *P. hispida botnica*; Kokko et al. 1999) or became extinct (e.g., Caribbean monk seal, *Monachus tropicalis*; McClenachan and Cooper 2008).

The first large-scale effort to protect cetaceans came from the League of Nations in the 1930s, which banned commercial whaling of right, bowhead, and gray whales. This allowed some populations to markedly increase in abundance, including the western Arctic bowhead whale (*Balaena mysticetus*; Figure 2.1C; Gerber et al. 2007) and the eastern North Pacific gray whale (*Eschrichtius robustus*; Rugh et al. 2005), whereas others increased only slightly (e.g., southern right whale, *Eubalaena australis*; Baker and Clapham 2004) or remained at low population levels (e.g., Davis Strait bowhead whale; Shelden and Rugh 1995). In 1986, the International Whaling Commission expanded the commercial whaling moratorium to all great whales, leading to some increases in other species, such as sperm whales (*Physeter macrocephalus*; Whitehead 2002) and blue whales (*Balaenoptera musculus*; Branch et al. 2004). As with pinnipeds and otters, there are rarely accurate historical or pre-exploitation estimates of abundance for cetaceans, which limits our ability to judge recovery success (but see Box 2.2). For the North Pacific gray whale (Figure 2.1D), reconstructions based on catch records resulted in a prewhaling population estimate of 19,480–35,430 individuals, which suggests that the current population of 18,000–29,000 has almost completely recovered (Rugh et al. 2005). By contrast, genetic analyses estimated a Holocene pre-exploitation abundance of North Pacific gray whales of 76,000–118,000 (Alter et al. 2007), and habitat availability analyses suggest 96,000 individuals (Pyenson and Lindberg 2011).

Overall, the International Union for the Conservation of Nature (IUCN) classified 12% of all 127 marine mammal species as increasing, whereas remaining taxa were decreasing, stable, or data deficient (Vié et al. 2009). In comparison, a recent study on marine mammal recoveries revealed that 42% of 92 distinct populations worldwide experienced significant increases over three generations (Magera et al. 2013). This estimate excluded many cryptic and offshore populations because of insufficient data, which probably explains the higher proportion of recoveries. Interestingly, recovery was more common among pinnipeds (50%)

BOX 2.2 Viewpoint from Practitioners: Thar She Goes

Stephen Palumbi and Tony Palumbi

Whaling has been a human occupation for thousands of years (Alter et al. 2012), yet it was only in the twentieth century that the Antarctic's fertile whaling grounds were finally tapped. Despite their late start, hunters fell on the whales of the Antarctic like ravenous hyenas. From 1907 to 1985, humans killed more than a million blue whales (*Balaenoptera musculus*), fin whales (*B. physalus*), sei whales (*B. borealis*), and humpback whales (*Megaptera novaeangliae*) in the Southern Ocean (Hilborn et al. 2003, Clapham and Baker 2009). These whales ate countless truckloads of krill every year, and so many were taken that a surge in krill abundance may well have emerged in their absence (Fraser et al. 1992). In a self-serving extension of this "krill surplus hypothesis," the Japanese Foreign Ministry contended that the killing of great whales led to such an overabundance of small baleen whales like the minke (*B. bonaerensis*, spared from major hunting until the 1980s) that they had become like "cockroaches of the sea" (C. S. Baker, personal communication). On the basis of this assumption, Japan has asserted that "culling of minke whales may greatly help the recovery of the larger baleen whales" (Ministry of Foreign Affairs of Japan 2013). If true—if great whale populations really are limited by the amount of krill eaten by overabundant minke whales—this would provide a scientific justification for the continued Japanese hunting of Antarctic minkes.

Until recently, the claim that minke whales were more abundant than usual could not be directly tested. No hard population numbers exist for this species before the twentieth century. In 2010, a new kind of DNA analysis refuted the "cockroach" claim and provided a fresh glimpse into the Southern Ocean's ecological history. The technique is based on the tendency of bigger populations to contain higher levels of genetic variation than smaller ones: by measuring present variation, and by knowing the rate at which genetic variation is generated by mutation, we can get an estimate of past population sizes.

When this approach was applied to Antarctic minke meat—ironically purchased from the Japanese whaling industry—the results put the lie to Japanese whaling claims. There were no more minke whales in 2010 than before whaling's golden age: the average population size in the deep past was estimated at ~700,000 animals, similar to or higher than numbers estimated today (Ruegg et al. 2010). Without evidence of recent overpopulation, there's no need to kill minke whales (Lenfest Ocean Program 2010). The krill surplus hypothesis remains reasonable as an ecological explanation for krill abundance but does not explain the abundance of minke whales or, indeed, the evidence for their recent decline over the past few decades. It's possible that minke numbers are determined by winter mortality or by the loss of sea ice over the past 50 years (Fraser et al. 1992). Either way, there is no scientific basis for culling minke whales, and it took the approach of applying a long-term perspective to the data to show this.

Stephen Palumbi is Harold A. Miller Professor in Marine Sciences at Stanford University. Tony Palumbi is a novelist and technology writer in San Mateo, California. His work has appeared in *The Atlantic, Natural History, National Geographic,* and many other publications. Most recently, Stephen and Tony are coauthors of *The Extreme Life of the Sea* (Princeton University Press, 2014); this Viewpoint is adapted from an excerpt of that book.

than among cetaceans (31%) and, within cetaceans, more common among coastal (58%) than among offshore (13%) populations (Magera et al. 2013). Across 47 populations with adequate historical abundance estimates, recovery reached, on average, 61% of historical abundance, albeit with large variability (zero to 100%). Large historical population declines were related to low increases, with some notable exceptions (Magera et al. 2013). As long-lived species, many marine mammals have taken several decades before starting to show recovery from low population numbers (Baker and Clapham 2004, Branch et al. 2004, Magera et al. 2013).

Birds

Marine and coastal birds have also experienced a long history of exploitation for their meat, eggs, feathers, and oil, which decimated many populations and drove several species to extinction, particularly in the nineteenth and twentieth century, but in some instances much earlier (Lotze and Milewski 2004, Lotze 2005, Lotze et al. 2006). Several conservation measures were implemented in the early twentieth century to prevent further declines and possible extinctions. For example, the U.S. Lacey Act in 1900 likely prevented the near extinction of the great blue heron (*Ardea herodias*) and several other beautifully plumed birds by prohibiting the trade of their highly valued feathers (Lotze 2010). In 1916, this was followed by the Migratory Birds Convention (known as the Migratory Bird Treaty in the United States) between the USA and Great Britain (on behalf of Canada), which protected a wide range of migratory birds from hunting, egg collection, and nest destruction (Lotze 2010). Similar conventions were implemented in many countries in the twentieth century to ban or restrict bird exploitation and egg collection and to protect nesting and staging habitats (Oro and Ruxton 2001, Lotze and Milewski 2004, Lotze 2005). This enabled many decimated populations to increase, albeit rarely to historical levels (Lotze et al. 2006).

In the German Wadden Sea, for example, seven waterfowl species, including common eider (*Somateria mollissima*) and common shelduck (*Tadorna tadorna;* Figure 2.2A); five seabirds, including European herring gull (*Larus argentatus*) and black-headed gull (*L. ridibundus;* Figure 2.2B); and six shorebirds, including Eurasian oystercatcher (*Haematopus ostralegus*) and pied avocet (*Recurvirostra avosetta;* Figure 2.2C) have shown increasing breeding populations since the 1950s (Behm-Berkelmann and Heckenroth 1991). By contrast, declining populations have been observed in three species of waterfowl, three seabirds (e.g., little tern, *Sterna albifrons;* Figure 2.2D) and two shorebirds (e.g., Kentish plover, *Charadrius alexandrinus;* Figure 2.2D) despite protection efforts. A global comparison of waterbirds showed that among 1,200 populations with known abundance trends (out of 2,305 identified populations), only 17% were increasing while 40% were decreasing (Delany and Scott 2006).

For most birds, there are no accurate estimates of historical population size; thus, estimating the magnitude and success of recovery is difficult. In some cases, rough estimates of former abundance exist, such as for Galveston Bay, Texas. There, Pelican Island had been noted as an important nesting site for brown pelicans (*Pelecanus occidentalis*) in 1820, and there were ~5,000 pelicans breeding on the Texas coast in 1918 (Lester and Gonzalez 2002). Relentless killing by fishermen and pesticide pollution in the 1960s reduced their numbers

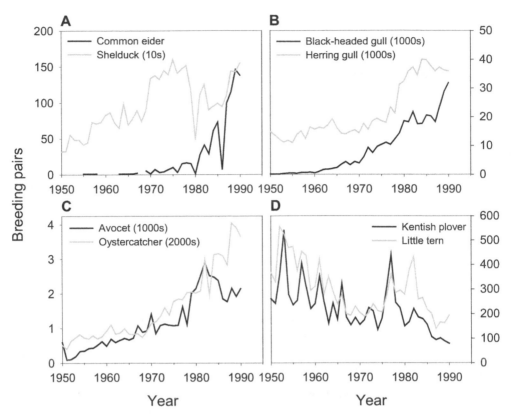

FIGURE 2.2 Examples of bird recoveries and nonrecoveries in the Wadden Sea of Lower Saxony, Germany, since 1946: recovery of selected (A) waterfowl, (B) seabirds, and (C) shorebirds, and nonrecovery of (D) two endangered marine birds (data from Behm-Berkelmann and Heckenroth 1991; figure redrawn with permission from Lotze 2005).

to <10 breeding pairs in 1967–1974. Stricter pesticide controls and legal protection enabled brown pelicans to reestablish their nesting colonies in Galveston Bay as of 1993, and the population grew to 800 breeding pairs in 2000 (Lester and Gonzalez 2002, Lotze 2010). Pesticide control, particularly the ban of DDT in the 1970s, was an important factor in the recovery of many bird species worldwide, especially for birds of prey. Ospreys (*Pandion haliaetus*) were reduced to 11 breeding pairs in Massachusetts in 1963, largely owing to DDT, but recovered to ~350 breeding pairs in 2000 (Massachusetts Division of Fisheries and Wildlife 2009, Lotze 2010).

Although many coastal and marine bird species were locally or regionally extirpated, some naturally recolonized their former habitat, although often only after decades or centuries. For example, the great white egret (*Egretta alba*) was extirpated in the Netherlands in the fourteenth century and returned to breed in 1978 after a 600-year absence (van Eerden 1997, Lotze 2005). In the Outer Bay of Fundy, the northern gannet (*Morus bassanus*) recolonized its former breeding colony after 133 years of absence and the common murre (*Uria aalge*) did so after 45 years (Lotze and Milewski 2004). In several cases, eradication of terrestrial, human-introduced predators, such as rats, foxes, or raccoons, was necessary to restore seabird colonies to former numbers, and some species have needed assisted reintroduction or have formed new breeding colonies at suitable sites (Oro and Ruxton 2001, Lotze and Milewski 2004, Jones 2010).

Reptiles

Sea turtles and alligators have been exploited for their meat, eggs, shells, and skins for many centuries, and the disturbance and transformation of breeding beaches and habitats has contributed to severe population declines globally (McClenachan et al. 2006, Chaloupka et al. 2008, Lotze 2010). For example, green turtles (*Chelonia mydas*) and hawksbill turtles (*Eretmochelys imbricata*) in the Caribbean have been reduced to <1% of their abundance prior to European arrival (McClenachan et al. 2006). Only after the protection of nesting beaches in the 1970s did some local populations increase. Worldwide, six major nesting populations of green turtles in Japan (Figure 2.3A), Australia, Hawaii, Florida, and Costa Rica have increased by 3.8–13.9% per year since the 1970s as a result of protection from exploitation and habitat disturbance (Chaloupka et al. 2008). Other sea turtle species have also shown recent recoveries at specific nesting locations, such as olive ridley (*Lepidochelys olivacea*) at Oaxaca, Pacific Mexico, and Kemp's ridley (*L. kempii*) in Rancho Nuevo, Atlantic Mexico (Márquez et al. 1998); giant leatherback (Figure 2.3B; *Dermochelys coriacea*) in St. Croix, U.S. Virgin Islands (Dutton et al. 2005); and loggerhead (*Caretta caretta*) in Brazil (Marcovaldi and Chaloupka 2007). By contrast, several populations of loggerhead and leatherback turtles spending extensive time in offshore waters have experienced strong declines due to bycatch in high-seas fisheries (Figure 2.3C, D; Lewison et al. 2004). Crowder et al. (1994) found that loggerhead turtles in the southern United States continued to decline despite protection of nesting beaches, and a stage-based population model suggested that reduction of juvenile bycatch with turtle excluder devices would be the most efficient conservation measure to enable recovery.

Because of their slow growth rate and late onset of reproduction, sea turtles might take many decades to recover from suppressed population numbers, and long-term data series of at least 20–25 years are needed to provide reliable trends in abundance (Bjorndal et al. 2005, Chaloupka et al. 2008). Yet McClenachan et al. (2006) found that data series covering 40 years, spanning more than one turtle generation, likely show long-term declines compared with recent short-term trends. This indicates the importance of historical ecology in providing population trends that are long enough to accurately assess recent recovery successes.

There are also cases of recovery and nonrecovery among crocodilians. For example, the American alligator (*Alligator mississippiensis*) has shown remarkable recovery in the southeastern United States (Brandt 1991, Lotze 2010). Alligators were hunted extensively for their highly prized belly skin and were severely depleted by the early twentieth century. Although hunting was prohibited in the 1960s, illegal poaching continued into the 1970s, and only legal protection through the U.S. Endangered Species Act and hunting and trade bans

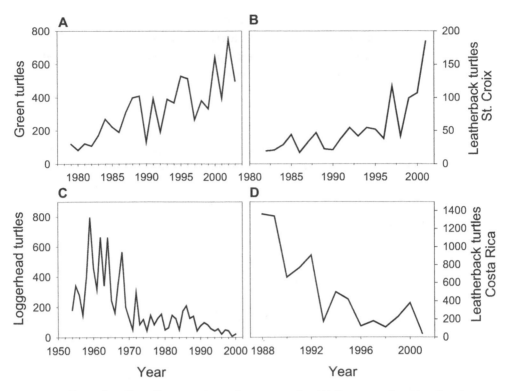

FIGURE 2.3 Examples of reptile recoveries and nonrecoveries. (A) Recovery of nesting female green turtles since 1979 at the Ogasawara rookery on Chichi-jima, Japan (data from Chaloupka et al. 2008). (B) Recovery of female leatherback turtles nesting at the Sandy Point rookery in St. Croix, U.S. Virgin Islands, 1982–2001 (data from Dutton et al. 2005). (C) Trends of nesting loggerhead turtles at Kamouda Beach, Japan (data from Lewison et al. 2004). (D) Leatherback turtles in Playa Grande, Costa Rica (data from Lewison et al. 2004).

through the Convention on International Trade in Endangered Species (CITES, Appendix II) enabled the population to recover (Brandt 1991, Lotze 2010). Today, the entire population is estimated at 1 million (Britton 2009). In comparison, the Chinese or Yangtze alligator (*A. sinensis*) continues to decline despite its listing on CITES (Appendix I). Only restoration of natural habitat and introduction of captive-bred individuals may spare the population from extinction (Thorbjarnarson et al. 2002).

Fishes

Although the status of global fisheries has been widely debated, there is no doubt that many stocks have been severely overexploited and are in need of recovery (Rosenberg et al. 2006, Worm et al. 2009). Thus, many management agencies have implemented stricter harvest controls to enable rebuilding of viable stocks. Such measures include effort and quota controls, bans of certain gear, and temporal or spatial closures. For example, strongly reduced harvest rates enabled rebuilding of northwest Atlantic swordfish (*Xiphias gladius;* Rosenberg

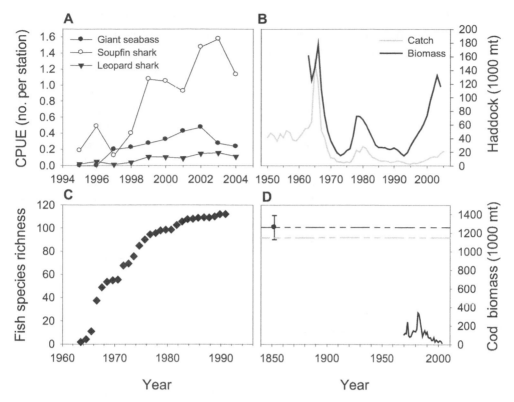

FIGURE 2.4 Examples of fish recoveries and nonrecoveries. (A) Catch per unit effort (CPUE) of giant sea bass, soupfin shark, and leopard shark from a monitoring program after the ban of gill nets in 1994 in the Southern California Bight (data from Pondella and Allen 2008). (B) Recovery of haddock biomass and catch on Georges Bank (data from Worm et al. 2007). (C) Cumulative increase in fish species richness after restoration of water quality in the Thames Estuary, UK (data from Cloern 2001). (D) Recent trends in cod biomass on the Scotian Shelf compared to historical baseline estimates from nineteenth-century log books (solid circle and dashed black line) and carrying capacity (dashed gray line) (data from Rosenberg et al. 2005)

et al. 2006) and of fish biomass in the California Current (Worm et al. 2009). The ban of gill nets in the Southern California Bight in 1994 resulted in recovery of white sea bass (*Atractoscion nobilis*), a severely depleted target species, as well as other predatory fish such as soupfin shark (*Galeorhinus galeus*) and leopard shark (*Triakis semifasciata;* Figure 2.4A; Pondella and Allen 2008). The banning of beach seine nets, together with the implementation of closed areas, resulted in marked increases in fish abundance in Kenya (McClanahan et al. 2008), and a large area closure on Georges Bank (USA) in 1992 helped rebuild Atlantic haddock (*Melanogrammus aeglefinus*) biomass back to its late-1960s level (Figure 2.4B; Worm et al. 2007).

Recoveries of fish abundance in marine reserves can have wide-ranging consequences, from increased species richness, secondary productivity, and ecosystem stability to the reestablishment of food-web structure and predatory interactions, as well as enhanced economic revenues from direct fishing and recreational diving (Micheli et al. 2004; Worm et al. 2006; Guidetti and Sala 2007; McClanahan et al. 2007, 2008; Libralato et al. 2010). Temporal fishing closures, both unintentional and intentional, have also resulted in some fish recoveries, as was the case around World War II, a period of greatly reduced fishing pressure (Caddy and Agnew 2004). Similarly, after the profitability of foreign fishing ceased, northwest Atlantic porbeagle sharks (*Lamna nasus*) strongly increased in abundance until renewed Canadian fisheries depleted the population again in the 1990s (Dulvy et al. 2008).

For many estuarine and diadromous fishes, not only overfishing but other anthropogenic threats such as habitat loss, river damming, and pollution need to be addressed to enable recovery (Kappel 2005, Lotze et al. 2006). For example, controls on water pollution and installation of functioning fish ladders enabled greater returns of gaspereau (*Alosa* spp.) and Atlantic salmon (*Salmo salar*) in the St. Croix River, Canada, during the 1980s before some dams were closed again in 1995 (Lotze and Milewski 2004). Similarly, strong returns of several diadromous fish species were observed after the recent removal of dams on the Kennebec River in Maine (Cane 2009). In the Thames River, United Kingdom, the implementation of pollution controls in the 1960s enabled an increase in water quality and oxygen availability, followed by the return of 110 fish species over the following three decades (Figure 2.4C; Cloern 2001).

In many cases, a single management or conservation measure may not be sufficient to ensure recovery and multifaceted strategies are needed. In the southeastern United States, a combination of stricter harvest controls, establishment of protected areas and habitat restoration zones, increased enforcement, fisheries-independent monitoring and assessment, and education and outreach has been implemented to help the recovery of elasmobranchs (i.e., cartilaginous fishes such as sharks and rays; Ward-Paige et al. 2012). Several species have responded positively to these measures, including the smalltooth sawfish (*Pristis pectinata;* Carlson et al. 2007), scalloped hammerhead (*Sphyrna lewini;* Hayes et al. 2009), and tiger shark (*Galeocerdo cuvier;* Baum and Blanchard 2010). Yet across 40 increasing elasmobranch populations worldwide, only 25% of increases were attributed to management-induced decreases in fishing mortality, while the majority were attributed to release from predation (Ward-Paige et al. 2012).

Despite these positive examples, there are many cases of nonrecovery of fish stocks. Cod (*Gadus morhua*) in the northwest Atlantic has seen no or only slight biomass increases since a fishing moratorium was implemented in the early 1990s (Frank et al. 2011) and is far from its historical abundance in the 1850s or its carrying capacity (Figure 2.4D; Myers et al. 2001, Rosenberg et al. 2005). The goliath grouper (*Epinephelus itajara*) provides another example of how the timeline for assessment may change the conclusion of recovery success. In southern Florida, it was thought to have recovered to 31–36% of its former abundance on the basis of data from the 1970s to 1980s, but an extension of historical data to the 1920s showed long-term declines that call into question whether recent increases represent significant recovery (McClenachan 2009).

A more general overview on fish recoveries showed that only 12% of 232 strongly depleted fish stocks from around the world had fully recovered 15 years after collapse, whereas 40% did not show any recovery (Hutchings and Reynolds 2004). As with marine mammals, larger population declines were related to weaker recoveries: fish stocks with 90% declines recovered less than those with 70% declines. Even for fish stocks requiring recovery under the U.S. Magnuson-Stevens Act, only half of 74 stocks showed an increase in abundance from 1996 to 2004 and only three stocks reached their recovery target (Rosenberg et al. 2006). For long-lived, slowly reproducing fish species, recovery might take decades. The porbeagle shark has an estimated recovery time of 70–100 years to previous abundance (Dulvy et al. 2008). Caddy and Agnew (2004) showed that observed recoveries ranged from 3 to 30 years, with demersal fish taking longer than pelagic species. Similarly, Hutchings and Reynolds (2004) found that gadoids showed no or only slow recovery 15 years after depletion, whereas clupeids recovered within 5–10 years.

Invertebrates

Many marine invertebrates have also been exploited throughout history, including mussels, snails, lobsters, crabs, sponges, corals, and sea urchins. Several high-value species have experienced strong long-term population declines (Lotze 2005, 2010; Lotze et al. 2011a), yet population assessment, management, and conservation of many invertebrate species lag behind that of finfish (Anderson et al. 2011). However, a few examples of recovery of exploited and depleted marine invertebrates exist. One example comes from Georges Bank, where a large-scale fishing closure in the 1990s, intended to enable groundfish recovery, resulted in strong recovery of scallops (*Placopecten magellanicus;* Figure 2.5A; Rosenberg et al. 2006). In New Zealand, reduction of fishing effort and shortening of fishing seasons resulted in a substantial increase in rock lobsters (*Jasus edwardsii*), as well as increases in fisheries catch rates and profitability (Figure 2.5B; Breen and Kendrick 1997). And in Chile, abundance and mean size of locos (*Concholepas concholepas*) strongly increased after implementation of comanagement and property rights, which solved many overexploitation issues (Figure 2.5C; Castilla and Fernandez 1998).

Recovery has been more difficult for complex, reef forming species such as oysters, but restoration efforts in Chesapeake Bay and Pamlico Sound in the United States, the Strangford Lough in Northern Ireland, and the Limfjord in Denmark show some potential for the rebuilding of native oyster reefs (Beck et al. 2011; also see chapter 8, this volume). Yet in Strangford Lough, recovery has been undermined by unregulated harvesting and requires stricter enforcement and legislation (Smyth et al. 2009).

APPLYING HISTORICAL KNOWLEDGE TO RECOVERY PLANNING

There has been much trial and error involved in finding the best management and conservation actions for enabling populations to recover. The above collection of case studies and syntheses provides important lessons that can be learned from marine historical ecology to better inform current and future recovery planning and prevent further population declines

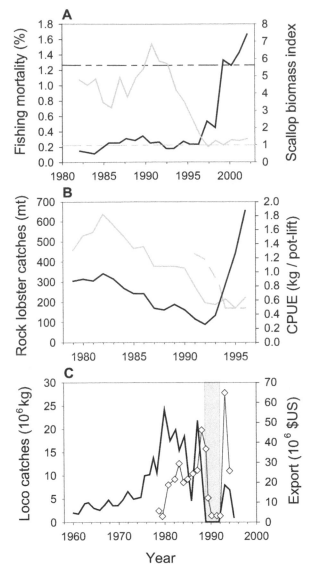

FIGURE 2.5 Examples of invertebrate recoveries and nonrecoveries. (A) Recovery of scallop biomass (solid black line) on Georges Bank after the reduction of fishing mortality (solid gray line) in relation to the management targets of biomass (B_{msy}, dashed black line) and fishing mortality (F_{msy}, dashed gray line) (data from Rosenberg et al. 2006). (B) Increase of rock lobster fisheries catch rates (CPUE, black line) after declining overall catches (solid gray line) and total allowable catches (TAC, dashed gray line) through reductions in fishing effort, fishing season, and quotas (data from Breen and Kendrick 1997). (C) Recovery of loco catches (thick black line) and exports (diamonds) after the implementation of a multiyear fishing closure (gray bar) as well as comanagement regimes and property rights after the reopening of the fishery (data from Castilla and Fernandez 1998).

and extinctions. In the following, I outline four different ways in which marine historical ecology can be used to inform recovery planning.

Better Assessment of Long-term Population Trends

Historical ecology can reconstruct long-term population changes over historical timescales and provide a more complete picture of past abundances and current states. Ideally, the magnitude of recovery can then be judged against a natural or historical baseline as full or partial recovery, no recovery, or further decline (Figure 2.6A). A "historical" baseline, however, can

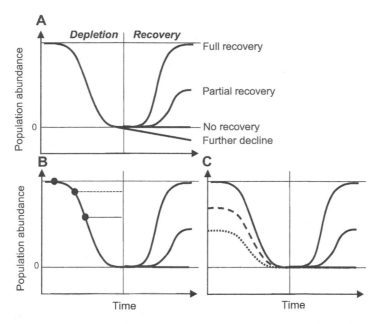

FIGURE 2.6 Conceptual diagrams of different recovery scenarios and historical baseline estimates. (A) Full recovery, partial recovery, no recovery, and further decline compared to an ideal historical baseline before depletion. However, the estimated historical baseline can vary with (B) different temporal ranges of estimates (filled circles), such as 100 years ago, pre-exploitation, or prehistoric; or (C) different reconstruction methods (dashed and dotted lines), such as those based on catches, genetic analysis, or habitat availability. These different estimates will influence the assessment of the magnitude or success of recovery.

refer to 50, 100, or 200 years ago, or to pre-exploitation, pre-European, or prehistoric times, and therefore varies depending on the time span used (Figure 2.6B; Pauly 1995, McClenachan 2009). Also, different reconstruction methods can provide different results (Figure 2.6C), as with reconstructions based on catch data, genetic analyses, or habitat availability for North Pacific gray whales (Figure 2.1D); thus, whether they are deemed fully or only partially recovered depends on which historical estimate is used. By contrast, for Scotian Shelf cod, reconstructions of historical biomass and carrying capacity revealed similar and thus more reliable baselines (Figure 2.4D), making the judging of recovery more straightforward. "Natural" baselines of abundance can also vary with changes in environmental conditions such as climate, productivity, habitat availability, or predator–prey interactions. As long as there are discrepancies among different baseline estimates, the success of recovery can only be judged in the appropriate temporal, methodological, or environmental context. Nonetheless, long-term historical population trends provide a much better idea of the general and species-specific magnitude and timeline of depletion and recovery. Such historical estimates should therefore be used to determine species status, such as in endangered-species, population, and fish-stock assessments (McClenachan 2009, McClenachan et al. 2012).

Setting More Meaningful Management and Conservation Targets

Today, most conservation and management targets are based on monitoring and assessment data spanning the past 10 to 60 years. Yet during this time, most marine populations were already severely depleted (e.g., Jackson et al. 2001, Pandolfi et al. 2003, Lotze et al. 2006). Therefore, historical reference points, such as pre-exploitation abundance or historical carrying capacity (see above; Figure 2.6), are essential for setting ecologically meaningful conservation and management targets. For example, recovery of the goliath grouper in south Florida has been substantial in relation to data from the 1970s to 1980s but minor compared with the 1920s (McClenachan 2009). Similarly, recovery of sea otters in British Columbia has been strong but still represents only 6% of estimated coast-wide carrying capacity (Gregr et al. 2008; but see Box 2.1). In both cases, further species protection may thus be necessary to allow more complete recovery. In the Wadden Sea, harbor seals were seen as "unnaturally" abundant in the 1990s to 2000s and therefore vulnerable to disease outbreaks; however, historical reconstruction revealed they had only reached about half of their estimated historical abundance in 1900 (Figure 2.1B; Lotze 2005). And whether the ban on commercial whaling of North Pacific gray whales could be lifted partly depends on which historical estimate is chosen to judge recovery success (see above; Figure 2.1D). These examples illustrate that historical reference points can alter the assessment of a species' abundance or conservation status and should therefore be used to set meaningful management and recovery targets.

Shaping Realistic Expectations for Recovery

The above case studies demonstrate that recovery of depleted marine populations is possible, providing a promising outlook on the future of marine biodiversity and ocean ecosystems. However, so far only a fraction of species have shown recovery, many species need decades to more than a hundred years to do so, and only a few species have recovered to historical levels of abundance. Worldwide, 12–42% of marine mammals (Vié et al. 2009, Magera et al. 2013), 17% of waterbirds (Delany and Scott 2006), and 12% of collapsed fish stocks (Hutchings and Reynolds 2004) showed some recovery. In 12 estuarine and coastal ecosystems, 14% of depleted species showed some recovery, but only 2% increased to 50% or more of former abundance, and in most cases it took decades to centuries for long-lived species to start rebuilding (Lotze et al. 2006). Across 256 depleted populations of large marine animals, 15.6% experienced some recovery, with average abundance increasing from 13% to 39% of historical levels (Lotze and Worm 2009). This historical context for the frequency, magnitude, and timeline of population recoveries can help in shaping realistic expectations for the recovery of marine species and ecosystems that are free from either an over-idealization of past riches or a doom-and-gloom view of the future. Such historical context should therefore be part of management and conservation plans.

Planning for Better Management Strategies

The historical review of marine species recoveries provides important lessons in terms of the different factors that have enabled or hampered recovery in the past. First, long-lived species can take decades to centuries to recover from depletion, and natural recolonization can take much longer than recovery from remnant but still existing local populations; thus, recovery planning needs to be long term (e.g., Lotze and Milewski 2004, Lotze 2005, Chaloupka et al. 2008, Dulvy et al. 2008). Second, the magnitude of recovery can depend on the magnitude of former population depletion in terms of numbers (Hutchings and Reynolds 2004, Magera et al. 2013) or geographic range (Abbitt and Scott 2001); therefore, severe population depletion should be avoided at all costs. Both the time span and the magnitude of recovery are linked to internal factors such as life history, genetic diversity, and population structure (Hutchings and Reynolds 2004, Hughes and Stachowicz 2011, Lotze et al. 2011b), which need to be considered in management plans. Third, environmental or biological factors can influence recovery, including climate, productivity, food and habitat availability, species interactions, and diversity (Worm et al. 2006, Lotze et al. 2011b). In order to recover, a species requires fulfillment of its basic needs and a functioning ecosystem in which to live; these should be priorities in recovery planning.

Finally, managing anthropogenic factors, including direct human impacts on the species or its ecosystem as well as indirect societal and economic factors, is important for enabling recovery. The reduction of those human impacts that caused the depletion is the concern of most management and conservation actions (Lotze et al. 2011b). In estuarine and coastal ecosystems, reduced or banned exploitation contributed to 95% of recoveries, followed by habitat protection (72%) and pollution controls (8%) (Lotze et al. 2006, Lotze 2010). In 78% of the cases, these measures were most successful when implemented in combination, which highlights the importance of cumulative factors. Implementation of such controls, however, requires public and political awareness and support for management and conservation actions (Forbes and Jermier 2002, Worm et al. 2009, Beck et al. 2011). Often, legal protection through endangered species legislation or trade bans has been essential (Lotze 2010, Lotze et al. 2011b), as has stringent enforcement of management and conservation plans (Rosenberg et al. 2006, Smyth et al. 2009, Worm et al. 2009). Economic factors can also influence recovery, such as reduced profit from targeting depleted stocks or new and cheaper products replacing old ones, such as mineral oil replacing whale oil in the early twentieth century (Schneider and Pearce 2004, Lotze 2010). These lessons from historical drivers of recovery can inform current and future management and conservation planning.

CONCLUSIONS

Overall, this chapter provides a historical overview on the variety, common patterns, and major drivers of population recoveries in the ocean. This information can guide management and conservation planning by providing a context for the expectations and potentials for overall and species-specific recoveries. The good news is that recovery of exploited marine animals is definitely possible and occurring, although it takes considerable time and effort. The bad news is that so far, only a fraction of depleted marine species have actually recovered, and often only to a small extent compared with their historical abundance. Moreover, marine mammals and birds on the IUCN Red List Index have been declining more rapidly over the past 10–20 years than their terrestrial counterparts (Hoffmann et al. 2010), indicating that marine conservation still lags behind terrestrial conservation. Understanding which factors have been most crucial for marine population recoveries so far would be of great help to management and conservation agencies, which usually need to carefully evaluate where to invest limited resources. In the past, the best strategies for successful recoveries included raising public and political awareness to gain support for recovery planning; taking legal action to protect endangered species; addressing major threats and cumulative human impacts to enable recovery; enforcing management plans to be effective; maintaining or restoring biodiversity and functioning ecosystems that support marine populations; and planning for the long term, because the recovery of long-lived species takes time. Lastly, historical information, long-term population trends, and reference points can be used to set more meaningful recovery targets and management strategies. These lessons from the history of marine animal recoveries could help prevent further population declines and losses.

ACKNOWLEDGMENTS

My sincere thanks go to the many colleagues who contributed to stimulating discussions on marine recoveries at the Beyond Obituaries Symposium and the International Marine Conservation Congress in Victoria in May 2011 and at the Oceans Past II conference in Vancouver in May 2009, with special thanks to N. Knowlton, K. Gedan, L. Blight, J. Kittinger, L. McClenachan, L. Airoldi, M. Coll, A. Magera, and C. Ward-Paige. Financial support was provided by the Natural Sciences and Engineering Research Council of Canada.

REFERENCES

Abbitt, R. J. F., and Scott, J. M. (2001) Examining differences between recovered and declining endangered species. *Conservation Biology* 15, 1274–1284.

Airoldi, L., and Beck, M. W. (2007) Loss, status, and trends for coastal marine habitats of Europe. *Oceanography and Marine Biology: An Annual Review* 45, 345–405.

Alter, S. E., Newsome, S. D., and Palumbi, S. R. (2012) Pre-whaling genetic diversity and population ecology in eastern Pacific grey whales: insights from ancient DNA and stable isotopes. *PLoS ONE* 7, e35039.

Alter, S. E., Rynes, E., and Palumbi, S. R. (2007) DNA evidence for historic population size and past ecosystem impacts of gray whales. *Proceedings of the National Academy of Sciences USA* 104, 15162–15167.

Anderson, S. C., Mills Flemming, J., Watson, R., and Lotze, H. K. (2011) Rapid global expansion of invertebrate fisheries: trends, drivers, and ecosystem effects. *PLoS ONE* 6, e14735.

Baker, C. S., and Clapham, P. J. (2004) Modelling the past and future of whales and whaling. *Trends in Ecology & Evolution* 19, 365–371.

Baum, J. K., and Blanchard, W. (2010) Inferring shark population trends from generalized linear mixed models of pelagic longline catch and effort data. *Fisheries Research* 102, 229–239.

Beck, M. W., Brumbaugh, R., Airoldi, L., et al. (2011) Oyster reefs at risk and recommendations for conservation, restoration, and management. *BioScience* 61, 107–116.

Behm-Berkelmann, K., and Heckenroth, H. (1991) Uebersicht der Brutbestandsentwicklung ausgewaehlter Vogelarten 1900–1990 an der niedersaechsischen Nordseekueste. *Naturschutz- und Landschaftspflege in Niedersachsen* 27, 1–97.

Bjorndal, K. A., Bolten, A. B., and Chaloupka, M. (2005) Evaluating trends in abundance of immature green turtles, *Chelonia mydas*, in the greater Caribbean. *Ecological Applications* 15, 304–314.

Branch, T. A., Matsuoka, K., and Miyashita, T. (2004) Evidence for increases in Antarctic blue whales based on Bayesian modelling. *Marine Mammal Science* 20, 726–754.

Brandt, L. A. (1991) Long-term changes in a population of *Alligator mississippiensis* in South Carolina. *Journal of Herpetology* 25, 419–424.

Breen, P. A., and Kendrick, T. H. (1997) A fisheries management success story: the Gisborne, New Zealand, fishery for red rock lobsters (*Jasus edwardsii*). *Marine and Freshwater Research* 48, 1103–1110.

Britton, A. (2009) *Alligator mississippiensis*. Crocodilian species list. Florida Natural History Museum, Gainesville, FL. http://crocodilian.com/cnhc/csp_amis.htm.

Brooke, M. de L., Butchart, S. H. M., Garnett, S. T., et al. (2008) Rates of movement of threatened bird species between IUCN Red List categories and toward extinction. *Conservation Biology* 22, 417–427.

Caddy, J. F., and Agnew, D. J. (2004) An overview of recent global experience with recovery plans for depleted marine resources and suggested guidelines for recovery planning. *Reviews in Fish Biology and Fisheries* 14, 43–112.

Cane, J. (2009) "Setting the river free": the removal of the Edwards dam and the restoration of the Kennebec River. *Water History* 1, 131–148.

Carlson, J. K., and Brusher, J. H. (1999) An index of abundance for coastal species of juvenile sharks from the northeast Gulf of Mexico. *Marine Fisheries Review* 61, 37–45.

Carlson, J. K., Osborne, J., and Schmidt, T. W. (2007) Monitoring the recovery of smalltooth sawfish, *Pristis pectinata*, using standardized relative indices of abundance. *Biological Conservation* 136, 195–202.

Carretta, J. V., Forney, K. A., Muto, M. M., et al. (2005) U.S. Pacific marine mammal stock assessments: 2004. NOAA Technical Memorandum NMFS-SWFSC-375.

Castilla, J. C., and Fernandez, M. (1998) Small-scale benthic fisheries in Chile: on co-management and sustainable use of benthic invertebrates. *Ecological Applications* 8, S124-S132.

Chaloupka, M., Bjorndal, K. A., Balazs, G. H., et al. (2008) Encouraging outlook for recovery of a once severely exploited marine megaherbivore. *Global Ecology and Biogeography* 17, 297–304.

Clapham, P. J., and Baker, C. S. (2009) Modern whaling. In *Encyclopedia of Marine Mammals*, vol. 2, 2nd ed. (W. F. Perrin, B. Würsig, and J. G. M. Thewissen, Eds.). Academic Press, New York, NY. pp. 1328–1332.

Cloern, J. E. (2001) Our evolving conceptual model of the coastal eutrophication problem. *Marine Ecology Progress Series* 210, 223–253.

Crowder, L. B., Crouse, D. T., Heppell, S. S., and Martin, T. H. (1994) Predicting the impact of turtle excluder devices on loggerhead sea turtle populations. *Ecological Applications* 4, 437–445.

Delany, S., and Scott, D. (2006) *Waterbird Population Estimates*. Wetlands International, Wageningen, The Netherlands.

Dulvy, N. K., Baum, J. K., Clarke, S., et al. (2008) You can swim but you can't hide: the global status and conservation of oceanic pelagic sharks and rays. *Aquatic Conservation: Marine and Freshwater Ecosystems* 18, 459–482.

Dulvy, N. K., Pinnegar, J. K., and Reynolds, J. D. (2009) Holocene extinctions in the sea. In *Holocene Extinctions* (S. T. Turvey, Ed.). Oxford University Press, Oxford, UK. pp. 129–150.

Dulvy, N. K., Sadovy, Y., and Reynolds, J. D. (2003) Extinction vulnerability in marine populations. *Fish and Fisheries* 4, 25–64.

Dutton, D. L., Dutton, P. H., Chaloupka, M., and Boulon, R. H. (2005) Increase of a Caribbean leatherback turtle *Dermochelys coriacea* nesting population linked to long-term nest protection. *Biological Conservation* 126, 186–194.

Forbes, L., and Jermier, J. (2002) The institutionalization of bird protection: Mabel Osgood Wright and the early Audubon movement. *Organization & Environment* 15, 458–465.

Frank, K. T., Petrie, B., Fisher, J. A. D., and Leggett, W. C. (2011) Transient dynamics of an altered large marine ecosystem. *Nature* 477, 86–89.

Fraser, W. R., Trivelpiece, W. Z., Ainley, D. G., and Trivelpiece, S. G. (1992) Increases in Antarctic penguin populations: reduced competition with whales or a loss of sea ice due to environmental warming? *Polar Biology* 11, 525–531.

Gerber, L. R., Keller, A. C., and DeMaster, D. P. (2007) Ten thousand and increasing: is the western Arctic population of bowhead whale endangered? *Biological Conservation* 137, 577–583.

Gregr, E. J., Nichol, L. M., Watson, J. C., et al. (2008) Estimating carrying capacity for sea otters in British Columbia. *Journal of Wildlife Management* 72, 382–388.

Guidetti, P., and Sala, E. (2007) Community-wide effects of marine reserves in the Mediterranean Sea. *Marine Ecology Progress Series* 335, 43–56.

Harnik, P. G., Lotze, H. K., Anderson, S. C., et al. (2012) Extinctions and ancient and modern seas. *Trends in Ecology & Evolution* 27, 608–617.

Hayes, C. G., Jiao, Y., and Cortes, E. (2009) Stock assessment of scalloped hammerheads in the western North Atlantic Ocean and Gulf of Mexico. *North American Journal of Fisheries Management* 29, 1406–1417.

Hilborn, R., Branch, T. A., Ernst, B., et al. (2003) State of the world's fisheries. *Annual Review of Environment and Resources* 28, 359–399.

Hoffmann, M., Hilton-Taylor, C., Angulo, A., et al. (2010) The impact of conservation on the status of the world's vertebrates. *Science* 330, 1503–1509.

Hughes, A., and Stachowicz, J. (2011) Seagrass genotypic diversity increases disturbance response via complementarity and dominance. *Journal of Ecology* 99, 445–453.

Hutchings, J. A., and Reynolds, J. D. (2004) Marine fish population collapses: consequences for recovery and extinction risk. *BioScience* 54, 297–309.

Jackson, J. B. C., Alexander, K. E., and Sala, E. (Eds.) (2011) *Shifting Baselines: The Past and the Future of Ocean Fisheries.* Island Press, Washington, DC.

Jackson, J. B. C., Kirby, M., Berger, W. H., et al. (2001) Historical overfishing and the recent collapse of coastal ecosystems. *Science* 293, 629–638.

Jones, H. P. (2010) Seabird islands take mere decades to recover following rat eradication. *Ecological Applications* 20, 2075–2080.

Kappel, C. V. (2005) Losing pieces of the puzzle: threats to marine, estuarine, and diadromous species. *Frontiers in Ecology and the Environment* 3, 275–282.

Kokko, H., Helle, E., Lindstroem, J., et al. (1999) Back-casting population sizes of ringed and grey seals in the Baltic and Lake Saimaa during the 20th century. *Annales Zoologici Fennici* 36, 65–73.

Lenfest Ocean Program (2010) Press release: new study suggests minke whales are not preventing recovery of larger whales. www.lenfestocean.org/press-release/new-study-suggests-minke-whales-are-not-preventing-recovery-larger-whales-0.

Lester, J., and Gonzalez, L. (2002) *The State of the Bay—A Characterization of the Galveston Bay Ecosystem*, 2nd ed. Galveston Bay Estuary Program, Houston, TX. http://gbic.tamug.cdu/publications.htm.

Lewison, R. L., Freeman, S. A., and Crowder, L. B. (2004) Quantifying the effects of fisheries on threatened species: the impact of pelagic longlines on loggerhead and leatherback sea turtles. *Ecology Letters* 7, 221–231.

Libralato, S., Coll, M., Tempesta, M., et al. (2010) Food-web traits of protected and exploited areas of the Adriatic Sea. *Biological Conservation* 143, 2182–2194.

Lotze, H. K. (2005) Radical changes in the Wadden Sea fauna and flora over the last 2000 years. *Helgoland Marine Research* 59, 71–83.

Lotze, H. K. (2010) Historical reconstruction of human-induced changes in U.S. estuaries. *Oceanography and Marine Biology: An Annual Review* 48, 267–338.

Lotze, H. K., Coll, M., and Dunne, J. (2011a) Historical changes in marine resources, food-web structure and ecosystem functioning in the Adriatic Sea, Mediterranean. *Ecosystems* 14, 198–222.

Lotze, H. K., Coll, M., Magera, A. M., et al. (2011b) Recovery of marine animal populations and ecosystems. *Trends in Ecology & Evolution* 26, 595–605.

Lotze, H. K., Lenihan, H. S., Bourque, B. J., et al. (2006) Depletion, degradation, and recovery potential of estuaries and coastal seas. *Science* 312, 1806–1809.

Lotze, H. K., and Milewski, I. (2004) Two centuries of multiple human impacts and successive changes in a North Atlantic food web. *Ecological Applications* 14, 1428–1447.

Lotze, H. K., and Worm, B. (2009) Historical baselines for large marine animals. *Trends in Ecology & Evolution* 24, 254–262.

Magera, A. M., Flemming, J. M., Kaschner, K., et al. (2013) Recovery trends in marine mammal populations. *PLoS ONE* 8, e77908.

Marcovaldi, M. A., and Chaloupka, M. (2007) Conservation status of the loggerhead sea turtle in Brazil: an encouraging outlook. *Endangered Species Research* 3, 133–143.

Márquez, R., Jiménez, M., Carrasco, M., and Villanueva, N. (1998) Comments on the populations trends of sea turtles of the *Lepidochelys* genus, after total ban of 1990. *Oceánides* 13, 41–62.

Massachusetts Division of Fisheries and Wildlife (2009) Species conservation. MDFW, Westborough, MA. www.mass.gov/dfwele/dfw/nhesp/conservation/conservation_home.htm.

McClanahan, T. R., Graham, N. A. J., Calnan, J. M., and MacNeil, M. A. (2007) Toward pristine biomass: reef fish recovery in coral reef marine protected areas in Kenya. *Ecological Applications* 17, 1055–1067.

McClanahan, T. R., Hicks, C. C., and Darling, E. S. (2008) Malthusian overfishing and efforts to overcome it on Kenyan coral reefs. *Ecological Applications* 18, 1516–1529.

McClenachan, L. (2009) Historical declines of goliath grouper populations in south Florida, USA. *Endangered Species Research* 7, 175–181.

McClenachan, L., and Cooper, A. B. (2008) Extinction rate, historical population structure and ecological role of the Caribbean monk seal. *Proceedings of the Royal Society of London Series B* 275, 1351–1358.

McClenachan, L., Ferretti, F., and Baum J. K. (2012) From archives to conservation: why historical data are needed to set baselines for marine animals and ecosystems. *Conservation Letters* 5, 349–359.

McClenachan, L., Jackson, J. B., and Newman, M. J. (2006) Conservation implications of historic sea turtle nesting beach loss. *Frontiers in Ecology and the Environment* 4, 290–296.

Micheli, F., Halpern, B. S., Botsford, L. W., and Warner, R. R. (2004) Trajectories and correlates of community change in no-take marine reserves. *Ecological Applications* 14, 1709–1723.

Ministry of Foreign Affairs of Japan (2013) Minke whales are increasing. www.mofa.go.jp/policy /economy/fishery/whales/iwc/minke.html.

Myers, R. A., MacKenzie, B. R., Bowen, K. G., and Barrowman, N. J. (2001) What is the carrying capacity of fish in the ocean? A meta-analysis of population dynamics of North Atlantic cod. *Canadian Journal of Fisheries and Aquatic Sciences* 58, 1464–1476.

Nichol, L. M., Watson, J. C., Ellis, G. M., and Ford, J. K. B. (2005) An assessment of abundance and growth of the sea otter population (*Enhydra lutris*) in British Columbia. *Canadian Science Advisory Secretariat Research Document* 2005/094, 1–22.

Oro, D., and Ruxton, G. D. (2001) The formation and growth of seabird colonies: Audouin's gull as a case study. *Journal of Animal Ecology* 70, 527–535.

Pandolfi, J. M., Bradbury, R. H., Sala, E., et al. (2003) Global trajectories of the long-term decline of coral reef ecosystems. *Science* 301, 955–958.

Pauly, D. (1995) Anecdotes and the shifting baseline syndrome of fisheries. *Trends in Ecology & Evolution* 10, 430.

Pondella, D. J., and Allen, L. G. (2008) The decline and recovery of four predatory fishes from the Southern California Bight. *Marine Biology* 154, 307–313.

Pyenson, N. D., and Lindberg, D. R. (2011) What happened to gray whales during the Pleistocene? The ecological impact of sea-level change on benthic feeding areas in the North Pacific Ocean. *PLoS ONE* 6, e21295.

Redford, K. H., Amato, G., Baillie, J., et al. (2011) What does it mean to successfully conserve a (vertebrate) species? *BioScience* 61, 39–48.

Reijnders, P. J. H. (1992) Retrospective population analysis and related future management perspectives for the harbour seal *Phoca vitulina* in the Wadden Sea. *Netherlands Institute for Sea Research Publication Series* 20, 193–197.

Rick, T. C., and Erlandson, J. M. (2008) *Human Impacts on Ancient Marine Ecosystems: A Global Perspective*. University of California Press, Berkeley, CA.

Rosenberg, A. A., Bolster, W. J., Alexander, K. E., et al. (2005) The history of ocean resources: modeling cod biomass using historical records. *Frontiers in Ecology and the Environment* 3, 84–90.

Rosenberg, A. A., Swasey, J. H., and Bowman, M. (2006) Rebuilding US fisheries: progress and problems. *Frontiers in Ecology and the Environment* 4, 303–308.

Ruegg, K., Anderson, E., Baker, C. S., et al. (2010) Are Antarctic minke whales unusually abundant because of 20th century whaling? *Molecular Ecology* 19, 281–291.

Rugh, D., Hobbs, R. C., Lerczak, J. A., and Breiwick, J. M. (2005) Estimates of abundance of the eastern North Pacific stock of gray whales (*Eschrichtius robustus*) 1997–2002. *Journal of Cetacean Research and Management* 7, 1–12.

Schneider, V., and Pearce, D. (2004) What saved the whales? An economic analysis of 20th century whaling. *Biodiversity and Conservation* 13, 543–562.

Shelden, K. E. W., and Rugh, D. J. (1995) The bowhead whale, *Balaena mysticetus*: its history and current status. *Marine Fisheries Review* 57, 1–20.

Smith, I. (2005) Retreat and resilience: fur seals and human settlement in New Zealand. In *The Exploitation and Cultural Importance of Sea Mammals. Proceedings of the 9th Conference of the International Council of Archaeozoology* (G. Monks, Ed.). Oxbow Books, Oxford, UK. pp. 6–18.

Smyth, D., Roberts, D., and Browne, L. (2009) Impacts of unregulated harvesting on a recovering stock of native oysters (*Ostrea edulis*). *Marine Pollution Bulletin* 58, 916–922.

Starkey, D. J., Holm, P., and Barnard, M. (2008) *Oceans Past: Management Insights from the History of Marine Animal Populations*. Earthscan, London, UK.

Swartz, S. L., Taylor, B. L., and Rugh, D. J. (2006) Gray whale *Eschrichtius robustus* population and stock identity. *Mammal Review* 1, 66–84.

Thorbjarnarson, J., Wang, X., Ming, S., et al. (2002) Wild populations of the Chinese alligator approach extinction. *Biological Conservation* 103, 93–102.

Trilateral Seal Expert Group (2011) Aerial surveys of harbour seals in the Wadden Sea in 2011. Common Wadden Sea Secretariat, Wilhelmshaven, Germany. www.waddensea-secretariat .org/news/news/Seals/Annual-reports/seals2011.html.

U.S. Fish and Wildlife Service (2003) Final revised recovery plan for the southern sea otter (*Enhydra lutris nereis*). U.S. Fish and Wildlife Service, Portland, OR.

van Eerden, M. R. (1997) Patchwork: patch use, habitat exploitation and carrying capacity for water birds in Dutch freshwater wetlands. PhD dissertation, Rijksuniversiteit Groningen, The Netherlands.

Vié, J.-C., Hilton-Taylor, C., and Stuart, S. N. (2009) *Wildlife in a Changing World—An Analysis of the 2008 IUCN Red List of Threatened Species*. IUCN, Gland, Switzerland.

Ward-Paige, C. A., Keith, D. M., Worm, B., and Lotze, H. K. (2012) Recovery potential and conservation options for elasmobranchs. *Journal of Fish Biology* 80, 1844–1869.

Whitehead, H. (2002) Estimates of the current global population size and historical trajectory for sperm whales. *Marine Ecology Progress Series* 242, 295–304.

Whitehead, H. (2010) Conserving and managing animals that learn socially and share cultures. *Learning & Behavior* 38, 329–336.

Wolff, W. J. (2000) The south-east North Sea: losses of vertebrate fauna during the past 2000 years. *Biological Conservation* 95, 209–217.

Worm, B., Barbier, E. B., Beaumont, N., et al. (2006) Impacts of biodiversity loss on ocean ecosystem services. *Science* 314, 787–790.

Worm, B., Barbier, E. B., Beaumont, N., et al. (2007) Biodiversity loss in the ocean: how bad is it? *Science* 316, 1282–1285.

Worm, B., Hilborn, R., Baum, J. K., et al. (2009) Rebuilding global fisheries. *Science* 325, 578–585.

THREE

Natural or Anthropogenic?

Novel Community Reassembly after Historical Overharvest
of Pacific Coast Pinnipeds

JON M. ERLANDSON, TODD J. BRAJE, ROBERT L. DELONG,
and TORBEN C. RICK

In this chapter, we examine the process of novel community reassembly following historical overharvest, through an examination of the historical ecology of pinnipeds along North America's Pacific coast. We compare changes in the biogeography of ancient versus modern pinniped populations, and the implications for the use of archaeological records in conservation biology and environmental management. Driven to the brink of extinction by commercial hunting in historical times, several Pacific coast pinniped species have recovered dramatically under federal and state protection. Pacific coast archaeological records show that humans hunted pinnipeds for at least the past 12,000 years and that the ancient distribution and abundance of northern elephant seals and Guadalupe fur seals differed significantly from today. Knowledge of such long-term anthropogenic changes, along with a dearth of data about the "natural" state of Pacific coast pinniped populations, raises interesting questions about the nature and sustainability of conservation efforts.

INTRODUCTION

With the emergence of the first truly global industrialized economies, many pinnipeds, cetaceans, and other marine mammals were relentlessly hunted to the brink of extinction (or beyond) worldwide. . . . This decimation fundamentally altered the structure and distribution of marine mammal populations and their recovery over the past century has sometimes occurred under fundamentally different ecological conditions from the more 'natural' conditions in which they evolved. Modern communities of pinnipeds and [other] sea mammals are a product, therefore, of severe historic over-hunting and their subsequent

protection and recovery in a demographic vacuum, often in food webs and
ecosystems altered by overfishing, pollution, climate change, and other
anthropogenic disruptions. (BRAJE et al. 2011:274)

Along California's central coast, thousands of people stop each year to view hundreds of
northern elephant seals (*Mirounga angustirostris*) hauled out on mainland beaches. Awed by
the size and strangeness of these sea creatures, most human viewers leave thinking they
have witnessed a marvel of the natural world. A marvel these elephant seals may be, but
such mainland haul-outs are almost certainly a historical anomaly, an unnatural phenomenon that has likely not been seen in tens of thousands of years. Since at least the Late Pleistocene, large terrestrial predators such as sabertooth cats, grizzly bears, and humans would
have ensured the destruction of such mainland colonies. Only after the demise of such predators, and a population recovery facilitated by federal and state laws, were such colonies able
to establish themselves, largely unmolested, on easily accessible mainland beaches. What do
such recent anomalies—and the archaeological and paleontological records that precede
them—tell us about what is "natural" for pinniped populations along North America's west
coast? And what are the implications of this for the restoration and management of seal and
sea lion populations in the eastern North Pacific?

The recovery of Pacific coast pinniped populations over the past century is a major success story for conservation and restoration biology, but it has not been without tension. In
some cases, restoration efforts have been so successful that pinniped encounters with boat-
ers, fishers, and sightseers have become a serious concern (see Fritz et al. 1995, DeMaster et
al. 2001). Growing sea lion populations have preyed upon threatened or endangered salmon
stocks, elephant seals crossing coastal highways have caused occasional traffic accidents,
and pinnipeds are damaging or destroying a growing number of archaeological sites (Rick
et al. 2009b, Braje et al. 2011). Such tensions increase political pressure to "manage" pinniped populations to preserve public safety or other endangered resources.

Here, we review and synthesize what is currently known about the historical ecology of
pinnipeds along the Pacific coast of North America and the history of their interaction with
humans over the millennia. In the process, we examine their resilience, recovery, and reorganization under nonanalog conditions, and the utility of archaeology and paleontology in
evaluating the natural versus anthropogenic status of recovered populations. We begin by
summarizing the biology, biogeography, and current status of pinnipeds along the west
coast of North America. We then compare these recent records to a deeper history of Pacific
coast pinniped populations revealed by paleontological, genomic, and archaeological records.

PACIFIC COAST PINNIPEDS: BIOGEOGRAPHY,
BEHAVIOR, AND NATURAL HISTORY

Some of the richest marine habitats in the world are located off North America's Pacific
coast, including kelp forests, rocky coasts and reefs, sandy beaches, estuaries, and deep-
water pelagic, midwater, and benthic communities. Marine upwelling and extensive kelp

FIGURE 3.1 North Pacific pinnipeds discussed in the text: (A) northern elephant seal, (B) harbor seal, (C) California sea lion, (D) Steller sea lion, (E) northern fur seal, and (F) Guadalupe fur seal (photos from National Marine Mammal Laboratory, U.S. National Marine Fisheries Service; composite by T. Braje).

forests drive high marine productivity and biodiversity, supporting a variety of relatively large marine predators. From south-central Alaska to Baja California, these include six pinniped species (Figure 3.1): two phocid ("true") seals, the massive northern elephant seal and the relatively diminutive harbor seal (*Phoca vitulina*); and four otariids ("eared" seals, i.e., fur seals and sea lions), the Steller sea lion (*Eumetopias jubatus*), California sea lion (*Zalophus californianus*), northern fur seal (*Callorhinus ursinus*), and Guadalupe fur seal (*Arctocephalus townsendi*). Other pinnipeds, such as the ribbon seal (*Histriophoca fasciata*), occasionally venture into these waters (Scammon 1874:140) but are generally confined to areas north of the Aleutians. McLaren and Smith (1985) identified three primary drivers of pinniped distribution and density: food availability, terrestrial breeding space, and predation. These factors—along with interspecific competition, evolutionary forces, and long-term climatic and geographic shifts—resulted in the development of unique natural histories and feeding, breeding, and fight–flight behaviors that may help explain the archaeological record of pinniped exploitation along the Pacific coast (Table 3.1).

The smallest of the North Pacific pinnipeds, harbor seals, are found in coastal waters from northern Japan to Baja California, including bays, estuaries, and river mouths. Females generally give birth to single pups on land, but because they breed in the water, they do not establish terrestrial rookeries or breeding colonies the way other Pacific coast pinnipeds do. Harbor seal pups can also swim shortly after birth and nurse for just 4–6 weeks, often in the water.

Relatively skittish on land, harbor seals flee quickly into the water when disturbed. These reproductive and flight behaviors may help protect them from terrestrial predators, but the fact that they live and forage year round in coastal waters may have made them vulnerable to maritime hunters.

Northern elephant seals forage widely, diving deep in pelagic waters from Alaska to the North Pacific transition zone. They haul out for extended periods, mostly on islands off Baja and Alta California, to breed, pup, and molt. Onshore, elephant seals show little or no fear of humans, and mothers leave helpless pups ashore after weaning at ~30 days of age, making them especially vulnerable to large terrestrial predators. In the nineteenth century, elephant seals were hunted relentlessly for their oil and were thought to be extinct until an 1892 Smithsonian expedition found eight individuals on Mexico's remote Guadalupe Island, killing seven for the museum's collections (Ellis 2003:193). Under federal protection, elephant seals have recovered dramatically, from a few survivors to a population of ~200,000. Their breeding range has also expanded from Mexico to California's Channel Islands and into the Pacific Northwest as far north as southern British Columbia.

A predominantly southern species, Guadalupe fur seals forage from Baja California to British Columbia. These small seals are far more common within their southern range, where they breed or pup only on Mexico's Guadalupe Island and San Benito Islands. Guadalupe fur seals were also thought to be extinct until a small population was identified on Guadalupe Island in the early twentieth century. Giving birth in June and July, mothers nurse their pups for 8–9 months but leave them ashore alone for up to 13 days while feeding at sea. Because they are generally nonmigratory, Guadalupe fur seals were relatively accessible and vulnerable to human predation (Braje and DeLong 2009). With a current population of ~10,000 individuals, they are gradually expanding their range northward and are more frequently seen on the Channel Islands.

Northern fur seals are broadly distributed across the North Pacific, with adults spending ≥250 days a year feeding in pelagic waters far from shore. They return each year to breeding rookeries in the Kuril, Commander, Pribilof, and Bogoslof islands, as well as California's Farallon Islands and San Miguel Island. Today, up to 600,000 northern fur seals breed on the Pribilofs annually (Allen and Angliss 2012), with males establishing breeding territories

TABLE 3.1 Modern Biogeography and Ecology of North America's Pacific Coast Pinnipeds

Common and Scientific Names	Historical/Modern Feeding Range	Breeding Habit and Size (males/females)	Mother and Newborn Behaviors	Pacific Coast Conservation Status
Guadalupe fur seal (*Arctocephalus townsendi*)	Baja California to British Columbia	Onshore; 180/50 kg	Pups nurse for 8–9 months; mothers leave for up to 13 days to feed	Threatened/Depleted: extirpated from much of range; population (~10,000) slowly growing
Northern fur seal (*Callorhinus ursinus*)	Japan to Baja California	Onshore; 270/60 kg	Pups helpless for several weeks, weaned after ~4 months	Depleted: population decline from ~2,100,000 to 665,000
Steller sea lion (*Eumetopias jubatus*)	Sea of Japan to Alta California	Onshore; 1,120/350 kg	Pups swim at birth; mothers leave on foraging trips	Western stock is Endangered: population (~45,000) has declined by ~85%
California sea lion (*Zalophus californianus*)	British Columbia to Mexico	Onshore; 450/110 kg	Pups left for extended periods while mothers feed at sea	Protected/Managed: population (~238,000) growing
Northern elephant seal (*Mirounga angustirostris*)	British Columbia to Baja California	Onshore; 2,000/600 kg	Pups helpless on land for ~4 months; weaned at 1 month	Protected: major population (~200,000) and range recovery
Harbor seal (*Phoca vitulina*)	Northern Japan to Southern California	In water; 120/120 kg	Pups able to swim; nurse for 4–6 weeks	Protected: population (250,000) expanding, except in Alaska

Sources: Data are compiled from Belcher and Lee (2002); Braje et al. (2011); Carretta et al. (2006); Estes (1980); Loughlin et al. (1987); Schoenherr et al. (1999); Stewart and Huber (1993); and the National Marine Fisheries website (www.nmfs.noaa.gov/pr/species/mammals/pinnipeds).
Notes: Size is given as average weights for adults. Conservation status is under current U.S. Federal Law.

in late May and June and females arriving in June or July to give birth and breed, after which females alternate between feeding at sea and nursing on shore until pups are weaned in October or November. When females forage, pups play and sleep in groups, primarily onshore. After weaning, adult females and juveniles from the Pribilofs migrate through eastern Aleutian Islands passes, radiating across the North Pacific (Ream et al. 2005). Pups, females, and juveniles arrive off the northern Washington coast in January and off the Channel Islands by February. They remain in such wintering areas until March and April, when adult females begin migrating northward to return to breeding rookeries. Most pups and juveniles continue to forage in the North Pacific until they are 3 years old. By age 3 or 4, juveniles return to rookery islands, where males occupy hauling areas and females enter reproductive territories to breed. Fur seals normally feed well offshore, near the edge of the continental shelf or over deeper pelagic waters, but they forage close to shore when preferred prey (e.g., Pacific herring, *Clupea pallasii*) aggregate in coastal waters.

Steller sea lions are also widely distributed, with rookeries in Kamchatka, the Aleutian Islands, Gulf of Alaska, southeast Alaska, British Columbia, Oregon, and northern California. Large and sexually dimorphic, Steller sea lions aggregate in breeding rookeries on offshore islands and rocks, where dominant adult males form reproductive territories for 1–2 months. Adult females give birth to one pup per year, between mid-May to mid-July. After 1–2 weeks of nursing, mothers spend roughly equal amounts of time hunting at sea and nursing pups on land. Males and adult females typically leave rookeries when pups are 2–3 months of age, travel several hundred kilometers, and occupy hauling sites near concentrations of marine prey through the autumn and winter. Females continue to nurse their pups for much of the first year.

California sea lions have a more southerly range, inhabiting coastal waters, estuaries, and some large rivers from central Mexico to southeast Alaska. Their primary breeding range is more limited, from California's Farallon Islands and Channel Islands to Mexico. On land they are highly social, sometimes forming groups of several hundred animals. In breeding rookeries, mothers usually give birth in June or July. Adult females reside on rookery islands for up to 11 months of the year, alternating between 3–5 days at sea feeding and 2 days ashore to nurse their pups. After the breeding season, male California sea lions travel to haul-out spots from central California to British Columbia. Since the 1950s, the California sea lion population has expanded from an estimated 10,000 individuals to some 238,000 today.

ORIGINS AND ASSEMBLY OF PINNIPED DIVERSITY

Paleontological and molecular genetic data suggest that pinnipeds diverged from primitive arctoid carnivores (weasels, bears, etc.) about 40–50 mya (Figure 3.2; Berta 2002, Arnason et al. 2006). There are three extant pinniped lineages: phocids, otariids, and odobenids (walruses, *Odobenus rosmarus*). The first known phocids are from Upper Oligocene (23.0–28.4 mya) deposits of California and South Carolina. By ~22 mya, this lineage diverged into two main groups: phocines that radiated into high northern latitudes; and monachines that

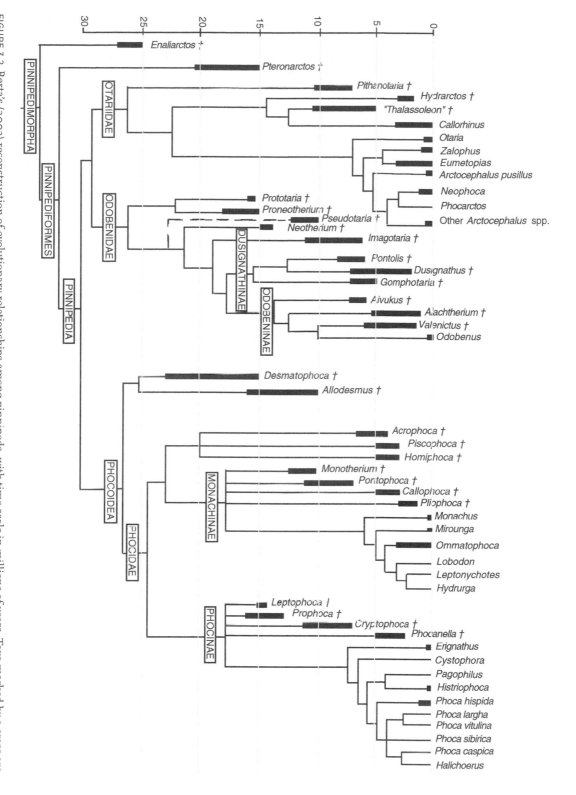

FIGURE 3.2 Berta's (2002) reconstruction of evolutionary relationships among pinnipeds, with time scale in millions of years. Taxa marked by a cross are known only from the fossil record. (Figure reproduced with permission from the author.)

radiated more widely through the tropics, midlatitudes of the northern and southern hemispheres, and high latitudes of the southern hemisphere. Modern North Pacific phocines diversified in the Arctic with late Cenozoic cooling and the associated development of a polar ice cap. Harbor seals apparently reinvaded temperate zones during the Pleistocene, with their presence in the North Pacific and North Atlantic resulting from trans-Arctic interchange during interglacial periods (Vermeij 1991).

Today, monachines are represented by the tropical monk seals, temperate and sub-Antarctic elephant seals, and Antarctic ice seals. Lineages leading to the modern elephant seals developed by the middle Miocene. Elephant seals radiated into the temperate North Pacific (northern elephant seal) and later dispersed across the equator to temperate and polar waters of the southern hemisphere (southern elephant seal, *M. leonina*). The otariids shared a common ancestor with the walruses ~20 mya. Evolutionary relationships among otariid species that occur strictly in the North Pacific (*Callorhinus, Eumetopias,* and *Zalophus*) and the Guadalupe fur seal are poorly resolved.

As Steven Emslie notes in Box 3.1, the nature of Late Pleistocene pinniped populations along the Pacific coast is also poorly understood, partly because global sea levels have risen ~120 m since the Last Glacial Maximum (LGM, ~20,000 years ago), leaving the shorelines on which pinnipeds would have hauled out deeply submerged on the continental shelves. Paleontological evidence suggests that fluctuations in sea surface temperature influenced ancient distributions of marine mammals. Cold-water-adapted species such as the Steller sea cow (*Hydrodamalis gigas*) and walrus have been identified in California deposits dating to the Pleistocene, when colder ocean temperatures extended their ranges southward (Harington 1984, Clementz 2002). It seems likely that the ranges of other pinniped species also shifted southward during the LGM. Although many Pacific coast pinnipeds today feed primarily in deeper waters, changes in the extent of nearshore kelp forests from the Late Pleistocene to the Holocene (Kinlan et al. 2005) also may have influenced the distribution and abundance of harbor seals.

ARCHAEOLOGY AND HUMAN IMPACTS ON PINNIPEDS

For most of the twentieth century, archaeologists believed that intensive coastal settlement developed late in human history, within the past 10,000 or so years (Yesner 1987, Erlandson 2001). Such theories influenced models of the peopling of the Americas for decades, which saw terrestrial hunters marching from Asia to Beringia, passing through an "ice-free" corridor, and emerging onto the plains of North America ~13,000 years ago. These Paleoindians supposedly hunted the large Pleistocene land mammals to extinction in interior regions before spreading into coastal zones where they gradually adapted to life by the sea (Erlandson and Braje 2011). Fully maritime cultures, including those engaged in intensive pinniped hunting, were thought to have developed only in the past few millennia.

A growing body of archaeological and genetic evidence now suggests that the First Americans arrived 15,000 to 16,000 years ago, including maritime peoples who followed coastlines

BOX 3.1   Viewpoint from a Practitioner: Using Modern and Ancient Tissues to Track Natural and Anthropogenic Impacts on Marine Animals through Time

Steven D. Emslie

Teasing apart natural from anthropogenic impacts on species and communities has been a difficult challenge for ecologists, paleoecologists, and conservation biologists. In past millennia, changes in species distributions, extirpations, and extinctions could be attributed largely to natural factors in the environment, including climate change. Now, however, the situation is reversed, and anthropogenic factors far outweigh natural climate cycles in their impact on species worldwide. This reversal is especially evident in the historical record over the past 300 years, when industrialization, marine exploitation, habitat loss, and increased pollution in the air, in the water, and on land have all affected ecosystems and species at an unprecedented scale. The impact that whaling and sealing had on marine environments alone was enormous and nearly caused the extinction of numerous species. These impacts began in the 1700s and accelerated well into the nineteenth and twentieth centuries.

My own research on penguins with my colleague William Patterson has documented, through stable isotope analysis of modern and ancient tissues, that the Adélie penguin (*Pygoscelis adeliae*) underwent a major dietary shift from fish to krill during this historical period (Emslie and Patterson 2007). Was this dietary shift due to a "krill surplus" caused by the removal of baleen whales and krill-eating seals that was so extensive in the Southern Ocean at that time, or was it due to another cause, perhaps natural? We are still seeking answers to this question, and it is exciting to see the work here by Jon Erlandson and colleagues in which they provide valuable insight into how pinni-

peds in the Pacific Northwest responded to both natural and anthropogenic impacts throughout the Holocene.

It is rare to have such a complete record; only polar regions may also provide the long-term preservation of fossil tissues that allow such thorough analyses, which combine archaeological and paleontological studies with recovery of ancient DNA. Erlandson and colleagues provide a compelling argument that humans were affecting pinniped populations long before the whaling and sealing era, perhaps throughout the Holocene, as demonstrated by the work on California's Channel Islands described here. Although these earlier impacts have not received much attention from paleoecologists, they can provide insightful perspectives for predicting future impacts, natural and anthropogenic, to aid in conservation decisions on these species. The possibility that Pacific Coast pinnipeds altered their behavior and ecology in response to human hunting is intriguing. As with all fossil records, there are some gaps in the pinniped data, and many questions remain unanswered, but this study is an excellent start. There is no doubt, though, that human impacts will continue to accelerate and affect pinniped populations in the future. For example, sea-level rise alone threatens critical haul-out beaches for northern elephant seals in California; Funayama and colleagues recently predicted that most of these beaches will be inundated by 2050 (Funayama et al. 2013). Thus, knowledge of long-past impacts from both anthropogenic and natural factors is becoming a valuable tool for conservation biologists to help sustain species into the future.

Steven D. Emslie is Professor in the Department of Biology and Marine Biology, University of North Carolina, Wilmington.

FIGURE 3.3 Channel Island barbed projectile points (adapted from Glassow et al. 2013; composite by M. Glassow).

from Northeast Asia to Beringia's now submerged southern margin, and down the Pacific coast of North and South America. This coastal migration theory, and an ecological correlate known as the "kelp highway hypothesis" (Erlandson et al. 2007), suggests that some Paleoindians were maritime from the start, with implications for the history of island and coastal ecosystems (Rick and Erlandson 2008). Although the earliest archaeological sites along the Pacific coast have probably been lost to rising seas, scattered Terminal Pleistocene sites show that the First Americans hunted pinnipeds in British Columbia, California's Channel Islands, and Baja California for ≥12,000 years (Erlandson et al. 2008, 2011). The First Americans may also have brought domesticated dogs with them, which would have further disrupted ecosystems, especially on islands that lacked large terrestrial predators (Rick et al. 2008).

The northern Channel Islands, settled by Paleocoastal people ≥13,000 years ago, have produced a nearly continuous record of human pinniped hunting since the Terminal Pleistocene. Pinniped bones identifiable to species from the earliest sites are rare, but analysis of ancient DNA from bone fragments may help identify the species represented at these early sites. For now, the recovery of numerous finely made Paleocoastal projectile points (Figure 3.3) suggests that marine mammal hunting was an important pursuit. Through the Holocene, there seems to have been a general increase in pinniped and other marine mammal hunting (Rick et al. 2005, Erlandson et al. 2009). Several San Miguel Island sites dated to about 1,500–1,200 years ago contain especially large amounts of marine mammal bone, suggesting that pinniped hunting intensified with the development of larger and more sophisticated boats and the introduction of the bow and arrow (Rick 2007, Braje 2010).

After Cabrillo's voyage in AD 1542–1543 and European settlement in the seventeenth and eighteenth centuries, conflict and the introduction of Old World diseases devastated Native

American populations along the Pacific coast. In response, some wildlife populations in Alta and Baja California were released from human predation and made dramatic recoveries (e.g., Ogden 1941, Preston 2002). As global commercial markets for marine mammal products (blubber, skins, etc.) developed, however, a veritable slaughter of Pacific coast pinnipeds took place (Busch 1985, Ellis 2003:139–151). Fortunately, beginning in about 1970 a series of federal, state, and provincial laws provided protection crucial to the recovery of Pacific coast pinnipeds and other marine mammals.

Scholars analyzing pinniped bones from Pacific coast archaeological sites have debated the meaning of patterns observed in zooarchaeological assemblages (see Hildebrandt and Jones 1992; Jones and Hildebrandt 1995; Lyman 1995, 2003, 2011; Colten and Arnold 1998; Erlandson et al. 1998; Porcasi et al. 2000; Etnier 2002, 2007; Walker et al. 2002; Jones et al. 2004; Gifford-Gonzalez et al. 2005; Moss et al. 2006; Newsome et al. 2007; Braje and DeLong 2009; Rick et al. 2009a, 2011; Braje and Rick 2011; Braje et al. 2011; Gifford-Gonzalez 2011; Moss and Losey 2011). Despite growing amounts of data, no consensus has been reached on the timing of intensive pinniped exploitation, the distribution of rookery locations (including the presence or absence of mainland rookeries), or the long-term effects that 10,000 years of Native American hunting had on pinniped abundance, behavior, and distributions.

For the Pacific coast as a whole, Lyman (2011) summarized data on 66,185 identifiable pinniped bones found in archaeological sites from Kodiak Island to southern California (Table 3.2). These are a valuable source of information for reconstructing the historical ecology of North Pacific pinnipeds, but the data are unevenly distributed through space and time, with just 3,762 specimens (5.7%) from Alaska and British Columbia, 8,488 (12.8%) from Alta California, and none from Baja California. Of 53,935 specimens (81.5%) from Washington and Oregon, 47,720 (88.5%) are from one Olympic Peninsula site (see Etnier 2002)—comprising 72% of the larger Pacific coast sample. The data are also skewed chronologically, with the vast majority coming from sites dating to the last millennium or two. To fully understand the deep history of pinniped–human interactions, we need data from more sites, more evenly distributed through space and time.

Despite these problems, significant patterns are apparent in the data (Table 3.2). Overall, the geographic range of many pinniped species in Late Holocene archaeological records matches reasonably well with what is known of their historical or recent range. Harbor seal, Steller sea lion, and northern fur seal bones are found in archaeological sites distributed throughout their broad historical range, for instance, whereas California sea lion and Guadalupe fur seal bones are found almost exclusively within their more restricted southerly ranges. The general correlation of prehistoric and historical pinniped ranges along the Pacific coast suggests some deep continuity, resilience, and fidelity in the biogeography of these species, at least for the past several millennia. Some long-term continuity and resilience is also evident on more local scales. For the northern Channel Islands, Braje et al. (2011) compiled data from archaeological sites containing substantial amounts of marine mammal bone. They found that the highest densities of pinniped remains cluster around

TABLE 3.2 Archaeological Distributions and Abundance of Northeastern Pacific Pinnipeds, Based on Analyzed Skeletal Remains from Archaeological Sites

Pinniped Species	Archaeological Range (Northern Latitude)	Alaska–British Columbia (%)	Washington–Oregon (%)	California (%)
Northern elephant seal	British Columbia to southern California (50–33°)	<0.1	<0.1	0.6
Steller sea lion	Alaska to southern California (58–33°)	8.2	3.4	12.8
California sea lion	British Columbia to southern California (50–33°)	0.9	0.2	12.6
Northern fur seal	Alaska to southern California (58–33°)	29.3	92.0	19.0
Guadalupe fur seal	Washington to Baja California (48–33°)	–	0.1	40.9
Harbor seal	Alaska to southern California (58–33°)	61.2	4.4	14.1

Notes: Percentages were recalculated from Lyman (2011:31), based on NISP (number of identified specimen) values. Total sample sizes = 3,762 for Alaska–British Columbia; 53,935 for Washington–Oregon; and 8,488 for California. Numbers for Washington are dominated by 47,296 northern fur seal bones from the Ozette site. Numbers for California are dominated by 3,444 Guadalupe fur seal bones from the Channel Islands and the southern California coast.

western San Miguel Island, near the modern Point Bennett rookery, utilized by 150,000 pinnipeds annually (DeLong and Melin 2002). There is no evidence that Point Bennett contained a substantial rookery prehistorically—with numerous large Island Chumash village sites found in the immediate area (see Braje and DeLong 2008, Walker et al. 2002)—but the density of marine mammal bones suggests a long-term continuity in the abundance of pinnipeds in the general area, probably on smaller islands offshore.

If the biogeography of Pacific coast pinnipeds shows some signs of deep historical continuity, recent studies have also identified some significant differences in modern versus ancient pinniped distributions and abundance. In Alta and Baja California, for instance, prehistoric pinniped hunting focused on Guadalupe fur seals (Rick et al. 2009a), even in areas where they are rare today. Clearly, Guadalupe fur seals were more abundant and broadly distributed in the past, but their abundance in California middens may also reflect the fact that they often remain in breeding and birthing areas for as much as 11 months of the year. Northern fur seals, Steller sea lions, and northern elephant seals, by contrast, all make long-range foraging trips for as much as 8 months per year and would have been less accessible to ancient hunters. In many areas, therefore, Guadalupe fur seals, harbor seals, sea otters (*Enhydra lutris*), and some cetaceans were probably the only species available nearly year round. Harbor seals were probably more difficult to hunt than Guadalupe fur seals because they breed in the water, pups can swim from birth, and mothers need not haul out to feed their young.

Analyses of pinniped remains from Channel Island and Pacific coast archaeological sites are also striking for what is not found. Today, northern elephant seals are widely distributed, with breeding populations on Guadalupe and other islands from central Baja California to the Channel Islands, on Ano Nuevo and Farallon islands in central California, and on California mainland beaches around Point Piedras Blancas, Año Nuevo Point, and Point Reyes. Low numbers of elephant seals also haul out, pup, and breed at remote sites near Coos Bay in Oregon, on Destruction and Protection islands off Washington, and at Race Rocks off British Columbia's Vancouver Island. Relatively tolerant of humans, elephant seals pup and breed on sand or gravel beaches and beach terraces. Females give birth to pups, nurse them for ~30 days, and depart the islands, thereby weaning the pups. But the pups remain on the beaches for up to 3 months before they leave for the first marine foraging phase of their lives, which extends for up to a year. Leaving helpless young unprotected for extended periods is a behavioral trait that would have made them easy prey for Native hunters. Yet elephant seal remains are rare in Channel Islands and mainland archaeological sites of any age (Rick et al. 2011). Some of the first Paleocoastal peoples on the Channel Islands may have targeted elephant seals, displacing them from the primary beaches they traditionally hauled out on and limiting their breeding colonies to remote refugia, such as Guadalupe Island, that humans reached only in historical times. Evidence for early elephant seal hunting has not yet been found, but it may have been lost to rising seas and the submergence of Late Pleistocene coastlines where early pinniped hunting and processing camps were likely to have been located.

California sea lions are also common on the Channel Islands today, with breeding colonies on San Miguel, Santa Rosa, San Nicolas, Santa Barbara, and San Clemente islands, where they haul out on beaches and upland landforms, leaving their young helpless for extended periods (Table 3.1). Although California sea lion bones are relatively common in archaeological sites postdating ~AD 500 (Braje et al. 2011), analysis suggests that nearly all are from adult males. The fact that remains of California sea lion females and pups are rare or absent in Island Chumash archaeological sites suggests that they were not breeding in large numbers on or around these islands during the past 1,500 years. Whether they did in the deeper past is currently unknown.

Finally, northern fur seal bones are widely found in Pacific coast archaeological sites, from the Aleutians to California, dating from early Holocene to historical times (Etnier 2002, Gifford-Gonzalez 2011, Lyman 2011). Some archaeologists interpret the presence of pup and female bones in middens as evidence that northern fur seals bred in all these areas, but many other scientists see these fur seal bones as evidence that Native Americans hunted migratory fur seals at sea (for a discussion, see Gifford-Gonzalez 2011:231–232). Recently, researchers have used osteometric, stable isotope, and genetic analyses to investigate the possibility that Holocene fur seals had different distributions, reproductive strategies, and feeding patterns than they have today (e.g., Burton et al. 2001, Crockford et al. 2002, Etnier 2002, Newsome et al. 2007). Gifford-Gonzalez (2011) summarized competing hypotheses over possible prehistoric northern fur seal breeding along the central California coast, suggesting several avenues for future collaborative work between biologists and archaeologists to document variation in ancient and modern northern fur seal ecology. One important line of investigation may lie in the analysis of fur seal and other pinniped bones recovered from the Farallon Islands off the mouth of San Francisco Bay, where a historical northern fur seal rookery—a colony only recently reestablished—supported harvests of 30,000 skins annually before being destroyed by modern commercial sealing (Starks 1922, Pyle et al. 2001).

Although fur seals normally feed well offshore, near the edge of the continental shelf or over deeper pelagic waters, they forage close to shore when herring or other preferred prey aggregate there. Northeast Pacific herring have many regional populations that feed offshore but spawn in shallow coastal waters, where northern fur seals feed on concentrations of adult fish for up to 2 months each winter and spring (Kenyon and Wilke 1953). Such behavior brought them into contact with coastal hunting peoples for millennia, until the last subsistence harvest of fur seals by Tlingit Indians in 1953 in Alaska's Sitka Sound (Kenyon 1955). Today, many stocks of herring, smelt (Osmeridae), sardines (*Sardinops sagax*), and other schooling fish are severely depleted, so the behavior of northern fur seals may have changed significantly from ancient times. Northern fur seals now breeding on San Miguel Island have the same reproductive timing and behavior as Pribilof Islands fur seals, but their migratory behavior is quite different—despite the fact that they are descended from more northern populations (Gifford-Gonzalez 2011). Satellite tagging studies show that fur seal females and pups leave San Miguel in autumn, migrating northward along the edge of the continental shelf as far north as Washington and British Columbia before the females return to San Miguel in June (Lea et al. 2009, Melin et al. 2012). Analyses of northern fur seal bones from archaeological sites are helping reconstruct the ecology, behavioral flexibility, and resilience of Pacific coast fur seal populations in the past, data that may contribute to a deeper understanding of their ecology and resilience today.

DISCUSSION: RESILIENT AND RECOVERING, BUT ARE THEY NATURAL?

Combining paleontological, archaeological, historical, and recent ecological records illustrates the general resilience of Pacific coast pinnipeds through deep time. Paleontological records demonstrate that pinnipeds have evolved and survived in a highly dynamic North Pacific for millions of years, including multiple glacial–interglacial cycles of the Pleistocene. Archaeological records show that Native peoples have hunted pinnipeds for 12,000 years or more, including relatively intensive harvest over the past 2,000 to 1,500 years, when large Native populations had sophisticated seafaring and maritime hunting technologies. We have highlighted some discontinuities between past and present pinniped populations, but the general ranges of most Pacific coast species show significant continuity between prehistoric and historical times. In several areas along the Pacific coast, archaeological records suggest that Native Americans hunted the same species of pinnipeds for centuries or even millennia (see Lyman 2003, Etnier 2007, Rick et al. 2009a, Moss and Losey 2011). Such continuity in harvest patterns suggests continual presence and availability of pinnipeds.

This raises several possibilities: (1) that access to some Pacific coast islands was limited by remoteness, high seas, or ancient boat technologies; (2) that hunting events recorded in middens were episodic, with recovery time (of several years) between events sufficient to sustain viable populations; (3) that occupation or exploitation of some island rookeries was intermittent, allowing pinniped populations time to recover from whatever harvesting may have occurred; and/or (4) that some ancient pinniped populations may have been sustainably harvested by Native peoples for long-term yields (see Etnier 2002, 2007). Currently, the chronological resolution obtained from dating archaeological middens is not sufficient to evaluate whether harvests by early peoples were continuous or episodic over decadal scales.

What is clear is that unregulated commercial hunting during the nineteenth and early twentieth centuries drove several North Pacific pinniped species (along with the sea otter and several large cetaceans) to the brink of extinction. The development of global markets and more efficient shipping and hunting technologies contributed to this steep and rapid decline. This underscores the obvious fact that large, *K*-selected species (slow growing, long lived) generally cannot sustain large-scale commercial hunting without severe population crashes or even extinction. Where large reductions in geographic range or population size have occurred, the deep histories of human exploitation should be taken into account if managers wish to restore a species or population to something resembling "natural" (i.e., unexploited) baselines.

Despite millennia of subsistence hunting and the devastating consequences of unregulated commercial hunting in historical times, most Pacific coast pinniped species have recovered dramatically under government protection, especially along the Alta and Baja California coasts. Many Pacific coast pinniped populations are now thriving, and expanding their populations and ranges, despite having low genetic diversity in comparison to prehistoric populations (see Hoelzel et al. 2002, Weber et al. 2004). The recovery of Pacific coast pinnipeds is heartening, but there are hints that the current abundance and distribution of some of these populations are fundamentally different than in the past 10,000 years or more (see Newsome et al. 2007; Rick et al. 2009a, 2011; Braje et al. 2011; Gifford-Gonzalez 2011). Over the past century, and even in the past decade, there have been some major changes in the breeding range and broader distribution of Pacific coast pinnipeds. Such changes suggest that the biogeography of some of these species may continue to be highly dynamic. Such dynamism also raises interesting questions about what is "natural" about North Pacific pinniped populations. Novel behaviors, such as establishing normally vulnerable mainland rookeries, are without apparent precedent. Dramatic recent reductions in northern Steller sea lion populations may or may not be driven by natural (nonanthropogenic) causes, but there is growing evidence for an anthropogenic cause (Estes et al. 1998, Paine et al. 2003, Springer et al. 2003, Atkinson et al. 2008). Even on California's Channel Islands, which today are a refuge for six pinniped species (and may also have been so in the Late Pleistocene, before the arrival of Paleocoastal peoples), long-term ecological monitoring has demonstrated changes in population levels, breeding locations, disease and mortality, and interspecies competition, all of which suggest that the recovery process is still unfolding.

BOX 3.2   Viewpoint from Practitioners: Inferring Past Temperatures from Middens and
Similar Sources of Species Assemblages

Daniel Pauly and William Cheung

Fish cannot regulate their body temperature (we include water-breathing invertebrates here but exclude tuna and a few other very active, large pelagic fishes whose existence does not affect the thrust of our argument). This implies that, to a large extent, fish must follow the average temperature of the habitat in which they have evolved, which explains the pattern of migration of many fishes [e.g., sardines *(Sardinops sagax)* along the western coast of North America].

We have recently demonstrated that temperature constraints have caused, in the past 40 years, a poleward migration of the fish exploited by commercial fisheries (Cheung et al. 2013). This migration had been documented through dozens, perhaps hundreds, of reported occurrences of low-latitude fish in the catch of high-latitude fisheries, but these anecdotes were not perceived as representative of anything but themselves. The procedure we used was simple in principle: we attributed to each

species occurring in fishery catches a single temperature representing the mode of the temperatures observed over their historical distribution range (mostly based on maps derived from FishBase; see www.fishbase.org). This temperature was then multiplied, for each species and location of interest, by the catch of that species; the species-specific products were added for all species in the catch of a given year; and the sum of these products was divided by the total catch. The result was, for each location and year, a precise estimate of the mean temperature of the catch (MTC), which we showed to have increased in most parts of the world; indeed, these tracked observed temperatures rather faithfully (Figure 3.4).

These findings, which generalized a local study by Collie et al. (2008), can be applied, in the context of historical marine ecology, to estimate the prevailing sea surface temperature at the time when subfossil or fossil assemblages of fish were generated (e.g., in middens or other

---

Given that U.S. federal laws and guidelines require some evaluation of the deep history of threatened or endangered species, as well as broader principles of ecosystem management, it is unfortunate that archaeological and paleontological data currently are of only limited help. Along the Pacific coast, paleontological records of Late Pleistocene pinniped populations before human arrival are limited by sea-level rise and changes in coastal geography over the past 18,000 years. Although archaeological records are an increasingly important source of information on the geography and genomics of Pacific coast pinniped populations, detailed data are still limited primarily to Late Holocene sites widely scattered along the Pacific coast—a period when the biogeography and behavior of pinnipeds and other marine mammals had already been shaped by ≥10,000 years of interaction with humans. At this time, without deeper and more complete historical records, we cannot fully understand the "natural states" of pinniped populations along the Pacific coast.

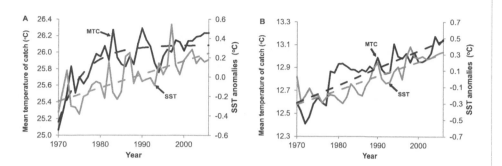

FIGURE 3.4 Average rate of change of mean temperature of catch (MTC) and sea surface temperature (SST) of (A) tropical large marine ecosystems (LMEs) and (B) nontropical LMEs. The average MTC of tropical LMEs was fitted with a logistic function, whereas the other data series were fitted with linear functions (data from Cheung et al. 2013).

deposits; Barrett et al. 2004, McKechnie 2007). All that is needed to compute their MTC is a characteristic temperature for each species in the assemblage and the proportion of each species in the assemblage, preferably as estimated by live-weight equivalent reconstructions. (This is better than counting numbers of bone fragments, as these can be seriously biased toward species with numerous small bones that do not contribute much to the biomass of the catch). We are looking forward to the first publication of an ancient-fish-based temperature.

Daniel Pauly is Principal Investigator in the Sea Around Us Project, Fisheries Centre, University of British Columbia, Vancouver, Canada. William Cheung is Associate Professor in the Changing Ocean Unit and Nereus Program, Fisheries Centre, University of British Columbia.

SUMMARY AND CONCLUSIONS

North Pacific pinnipeds have a deep history, with paleontological records and molecular genetics suggesting that the ancestors of modern pinnipeds began to adapt to life in the sea 40–50 mya. Both otariid and phocid seals have evolved and contributed to the six species of Pacific coast pinnipeds discussed here. Each of these species has followed unique evolutionary trajectories and developed distinctive morphologies, behaviors, ecologies, and adaptations. Unfortunately, Late Pleistocene fossil records for pinnipeds are limited along the Pacific coast of North America, and the nature of pinniped biogeography just prior to human arrival is poorly known.

Archaeological records show that humans began hunting pinnipeds in the northeast Pacific at least 12,000 years ago, but genetic data suggest that the Americas were settled by

John Odin Jensen

As this chapter clearly notes, growing archaeological data demonstrate that intensive human uses of coasts and oceans for migration, travel, and resource gathering have occurred for tens of thousands of years. The shift from thinking of humankind as recent maritime adaptors to regarding them as longer-term ecosystem occupants diminishes the value of incorporating "what is natural" in managing our coastal and ocean areas today. Regulatory frameworks based on returning areas to pristine, natural, or wild conditions—all ways of saying "before humans intruded"—reflect some notion of a Garden of Eden before the introduction of illegal, unregulated, or unreported apple poaching. In that sense, "natural" versus "anthropogenic" represents yet another western dualism that appears to clarify but limits deeper understanding. More important than informing philosophical debates is the capacity of historical ecology to provide robust data and cultural information regarding specific human influences on ecosystems and to make visible the social and cultural processes that drive marine resource uses and regulation. From this vantage point, the application of historical ecology is not limited to the past.

Even with the authors' clearly stated limitations, the North Pacific pinniped story offers lessons for managers that go beyond methods and data. One lesson is that in facing contemporary biocultural challenges that may seem insurmountable, small, seemingly simple changes in human systems can exercise immense influence on individual species, ecosystems, and human communities. In short, while ecological dynamics may be endlessly complex, the human systems and behaviors that affect specific ecosystems are not. Second, the relatively rapid recovery and observed changes in breeding strategies of some pinnipeds suggest that at least some species may take positive advantage of human-induced environmental changes. This opens the door to a new idea of ecological agency in developing place-based management strategies. While we may try to approximate ancient ecological conditions and biodiversity in one protected area, we also establish enabling conditions and manage spaces to recognize or even encourage the development of an internally self-reorganized version of "nature," more resilient to change and more sustainable and productive for human well-being.

Historical ecology of marine environments has had an overwhelming focus on the negative outcomes. In terms of the big lessons, the archaeological, historical, and ecological data make it blindingly obvious that abstract global markets, exponential population growth, and industrialization associated with modernity have caused nearly all of our present ecological perils. Historical ecology, its related disciplines, and increasingly non-Western knowledge systems are moving beyond those lessons to reveal how individual cultures have successfully come to terms with ecological limits and structured human behaviors. We are learning not only what we have done wrong as a species, but also the dimensions of what we need to do now. With this in mind, we cannot afford to step outside of nature at the very time when we depend on it and threaten it the most. Historical ecology and temporal human dimensions provide the connectivity we need to make good decisions about marine environments.

John Odin Jensen is a member of the Marine Protected Areas Federal Advisory Committee and Research Associate Professor of History and Coastal Maritime Heritage at University of Rhode Island.

humans as much as 16,000 years ago, meaning that the oldest archaeological records of pinniped hunting may be preceded by several millennia of human exploitation we currently know nothing about. This gap, as well as the dearth of data from Terminal Pleistocene and Early Holocene archaeological sites, limits current understanding of human impacts on Pacific coast pinniped populations. Because the vast majority of the available zooarchaeological data come from sites dating to the past millennium or two, many patterns identified by archaeologists may only describe pinniped populations already shaped by millennia of human hunting and other activities.

Paleontological and archaeological data can help decipher the deep histories of Pacific coast pinnipeds (and other species; see Box 3.2), providing valuable information about the progressive development of the anthropogenic landscapes and seascapes pervasive in today's "natural" world (Erlandson and Rick 2010). As John Jensen notes in Box 3.3, building effective management and restoration plans for Pacific coast pinnipeds requires data on historical and ancient harvests to establish baselines that account for overharvesting, ecosystem alteration, or other anthropogenic disturbances (see Pauly 1995, Dayton et al. 1998, Pauly et al. 1998, Etnier 2007). Although the ranges and abundances of Pacific coast pinniped species have been altered by historical and prehistoric human activities, evidence of long-term spatial continuity and ecological resilience provides hope for the restoration of viable, stable, and healthy pinniped communities. With additional interdisciplinary research, some of which is currently underway, there is no doubt that archaeologists and archaeological data will play a key role in better understanding the historical ecology and conservation biology of Pacific coast pinnipeds.

ACKNOWLEDGMENTS

We are deeply indebted to Jim Estes, who contributed significantly to our summary of the paleontological and genomic history of North Pacific pinnipeds and carefully read and commented on multiple drafts of this paper. Our research was supported by the National Science Foundation (grants 9731434 and 0746314 to Erlandson; 0917677 to Erlandson and Rick; 0613982 to Braje and Erlandson; and 0201668 to Rick and Erlandson), Channel Islands National Park, Western National Parks Association, Marine Conservation Biology Institute, and our home institutions. Annalisa Berta kindly provided a digital copy (and permission to use it) of her 2002 figure of pinniped evolution, and our Figure 3.3 was adapted from a composite image created by Michael Glassow. We thank Louise Blight, Jack Kittinger, and the other editors for inviting us to participate in this volume, for their patience, and for comments that helped improve the chapter. Finally, we are intellectually indebted to Paul Dayton, Jeremy Jackson, and Daniel Pauly.

REFERENCES

Allen, B. M., and Angliss, R. P. (2012) Alaska Marine Mammal Stock Assessments, 2011. U.S. Department of Commerce, NOAA Technical Memorandum NMFS-AFSC-234.

Arnason, U., Gullberg, A., Janke, A., et al. (2006) Pinniped phylogeny and a new hypothesis for their origin and dispersal. *Molecular Phylogenetics and Evolution* 41, 345–354.

Atkinson, S., DeMaster, D. P., and Calkins, D. G. (2008) Anthropogenic causes of the western Steller sea lion *Eumetopias jubatus* population decline and their threat to recovery. *Mammal Review* 38, 1–18.

Barrett, J. H., Locker, A. M., and Roberts, C. M. (2004) The origins of intensive marine fishing in medieval Europe: the English evidence. *Proceedings of the Royal Society of London Series B* 271, 2417–2421.

Belcher, R. L., and Lee, T. E., Jr. (2002) Mammalian species: *Arctocephalus townsendi*. *Mammalian Species* 700, 1–5.

Berta, A. (2002). Pinniped evolution. In *Encyclopedia of Marine Mammals* (W. F. Perrin, B. Würsig, and J. G. M. Thewissen, Eds.). Academic Press, San Diego, CA. pp. 921–929.

Braje, T. J. (2010) *Modern Oceans, Ancient Sites: Archaeology and Marine Conservation on San Miguel Island, California*. University of Utah Press, Salt Lake City, UT.

Braje, T. J., and DeLong, R. L. (2009) Ancient sea mammal exploitation on the south coast of San Miguel Island. In *Proceedings of the Seventh California Islands Symposium* (C. C. Damiani and D. K. Garcelon, Eds.). Institute for Wildlife Studies, Arcata, CA. pp. 43–52.

Braje, T. J., and Rick, T. C. (Eds.) (2011) *Human Impacts on Seals, Sea Lions, and Sea Otters: Integrating Archaeology and Ecology in the Northeast Pacific*. University of California Press, Berkeley, CA.

Braje, T. J., Rick, T. C., DeLong, R. L., and Erlandson, J. M. (2011) Resilience and reorganization: archaeology and the historical ecology of California Channel Island sea mammals. In *Human Impacts on Seals, Sea Lions, and Sea Otters: Integrating Archaeology and Ecology in the Northeast Pacific* (T. J. Braje and T. C. Rick, Eds.). University of California Press, Berkeley, CA. pp. 273–296.

Burton, R. K., Snodgrass, J. J., Gifford-Gonzalez, D., et al. (2001) Holocene changes in the ecology of northern fur seals: insights from stable isotopes and archaeofauna. *Oecologia* 128, 107–115.

Busch, B. C. (1985) *The War against the Seals: A History of the North American Seal Fishery*. McGill-Queen's University Press, Kingston, ON.

Carretta, J. V., Forney, K. A., Muto, M. M., et al. (2006) U.S. Pacific Marine Mammal Stock Assessments: 2005. U.S. Department of Commerce, NOAA Technical Memorandum NMFS.

Cheung, W., Watson, R., and Pauly, D. (2013) Signature of ocean warming in global fisheries catch. *Nature* 497, 365–368.

Clementz, M. T. (2002) The evolution of herbivorous marine mammals: ecological and physiological transitions during the evolution of the orders Sirenia and Desmostylia. PhD dissertation, University of California, Santa Barbara, CA.

Collie, J. S., Wood, A. D., and Jeffries, H. P. (2008) Long-term shifts in the species composition of a coastal fish community. *Canadian Journal of Fisheries and Aquatic Sciences* 65, 1352–1365.

Colten, R. H., and Arnold, J. E. (1998) Prehistoric marine hunting on California's northern Channel Islands. *American Antiquity* 63, 679–701.

Crockford, S. J., Frederick, S. G., and Wigens, R. J. (2002) The Cape Flattery fur seal: an extinct species of *Callorhinus* in the Eastern North Pacific. *Canadian Journal of Archaeology* 26, 152–174.

Dayton, P. K., Tegner, M. J., Edwards, P. B., and Riser, K. L. (1998) Sliding baselines, ghosts, and reduced expectations in kelp forest communities. *Ecological Applications* 8, 309–322.

DeLong, R. L., and Melin, S. R. (2002) Thirty years of pinniped research at San Miguel Island. In *The Fifth California Islands Symposium* (D. Browne, K. Mitchell, and H. Chaney, Eds.). Santa Barbara Museum of Natural History, Santa Barbara, CA. pp. 401–406.

DeMaster, D. P., Fowler, C. W., Perry, S. L., and Richlen, M. F. (2001) Predation and competition: the impact of fisheries on marine-mammal population over the next one hundred years. *Journal of Mammalogy* 82, 641–651.

Ellis, R. (2003) *The Empty Ocean*. Island Press/Shearwater Books, Washington, DC.

Emslie, S. D., and Patterson, W. P. (2007) Abrupt recent shift in $\delta^{13}C$ and $\delta^{15}N$ values in Adélie penguin eggshell in Antarctica. *Proceedings of the National Academy of Sciences USA* 104, 11666–11669.

Erlandson, J. M. (2001) The archaeology of aquatic adaptations: paradigms for a new millennium. *Journal of Archaeological Research* 9, 287–350.

Erlandson, J. M., and Braje, T. J. (2011) From Asia to the Americas by boat? Paleogeography, paleoecology, and stemmed points of the northwest Pacific. *Quaternary International* 239, 28–37.

Erlandson, J. M., Graham, M. H., and Bourque, B. J. (2007) The kelp highway hypothesis: marine ecology, the coastal migration theory, and the peopling of the Americas. *Journal of Island and Coastal Archaeology* 2, 161–174.

Erlandson, J. M., Moss, M. L., and Des Lauriers, M. (2008) Life on the edge: early maritime cultures of the Pacific coast of North America. *Quaternary Science Reviews* 27, 2232–2245.

Erlandson, J. M., and Rick, T. C. (2010) Archaeology meets marine ecology: the antiquity of maritime cultures and human impacts on marine fisheries and ecosystems. *Annual Reviews of Marine Science* 2, 165–185.

Erlandson, J. M., Rick, T. C., and Braje, T. J. (2009) Fishing up the food web? 12,000 years of maritime subsistence and adaptive adjustments on California's Channel Islands. *Pacific Science* 63, 711–724.

Erlandson, J. M., Rick, T. C., Braje, T. J., et al. (2011) Paleoindian seafaring, maritime technologies, and coastal foraging on California's Channel Islands. *Science* 441, 1181–1185.

Erlandson, J. M., Tveskov, M. A., and Byram, R. S. (1998) The development of maritime adaptations on the southern northwest coast of North America. *Arctic Anthropology* 35, 6–22.

Estes, J. A. (1980) *Enhydra lutris*. *Mammalian Species* 133, 1–8.

Estes, J. A., Tinker, M. T., Williams, T. M., and Doak, D. F. (1998) Killer whale predation on sea otters linking oceanic and nearshore ecosystems. *Science* 282, 473–476.

Etnier, M. A. (2002) The effects of human hunting on northern fur seal (*Callorhinus ursinus*) migration and breeding distributions in the late holocene. PhD dissertation, University of Washington, Seattle, WA.

Etnier, M. A. (2007) Defining and identifying sustainable harvests of resources: archaeological examples of pinniped harvests in the eastern North Pacific. *Journal for Nature Conservation* 15, 196–207.

Fritz, L. W., Ferrero, R. C., and Berg, R. J. (1995) The threatened status of Steller sea lions, *Eumetopias jubatus*, under the Endangered Species Act: effects on Alaska groundfish fisheries management. *Marine Fisheries Review* 57, 14–27.

Funayama, K., Hines, E., Davis, J., and Allen, S. (2013) Effects of sea-level rise on northern elephant seal breeding habitat at Point Reyes Peninsula, California. *Aquatic Conservation: Marine and Freshwater Ecosystems* 23, 233–245.

Gifford-Gonzalez, D. (2011) Holocene Monterey Bay fur seals: distribution, dates, and ecological implications. In *Human Impacts on Seals, Sea Lions, and Sea Otters: Integrating Archaeology and Ecology in the Northeast Pacific* (T. J. Braje and T. C. Rick, Eds.). University of California Press, Berkeley, CA. pp. 221–242.

Gifford-Gonzales, D., Newsome, S. D., Koch, P. L., et al. (2005) Archaeofaunal insights on pinniped–human interactions in the northeastern Pacific. In *The Exploitation and Cultural Importance of Sea Mammals* (G. G. Monks, Ed.). Oxbow Books, Oxford, UK. pp. 19–38.

Glassow, M. A., Erlandson, J. M., and Braje, T. J. (2013) Channel Island barbed points: shape and size variation within a Paleocoastal projectile type. *Journal of California and Great Basin Anthropology* 33, 185–195.

Harington, C. R. (1984) Quaternary marine and land mammals and their paleoenvironmental implications—some examples from northern North America. In *Contributions in Quaternary Vertebrate Paleontology: A Volume in Memorial to John E. Guilday* (H. H. Genoways and M. R. Dawson, Eds.). Carnegie Museum of Natural History Special Publication No. 8. pp. 5511–5525.

Hildebrandt, W. R., and Jones, T. L. (1992) Evolution of marine mammal hunting: a view from the California and Oregon coasts. *Journal of Anthropological Archaeology* 11, 360–401.

Hoelzel, A. R., Fleischer, R. C., Campagna, C., et al. (2002). Impact of a population bottleneck on symmetry and genetic diversity in the northern elephant seal. *Journal of Evolutionary Biology* 15, 567–575.

Jackson, J. B. C., Kirby, M. X., Berger, W. H., et al. (2001) Historical overfishing and the recent collapse of coastal ecosystems. *Science* 293, 561–748.

Jones, T. J., and Hildebrandt, W. R. (1995) Reasserting a prehistoric tragedy of the commons: reply to Lyman. *Journal of Anthropological Archaeology* 14, 78–98.

Jones, T. J., Hildebrandt, W. R., Kennett, D. J., and Porcasi, J. F. (2004) Prehistoric marine mammal overkill in the northeastern Pacific: a review of new evidence. *Journal of California and Great Basin Anthropology* 24, 69–80.

Kennett, D. J. (2005) *The Islands Chumash: Behavioral Ecology of a Maritime Society.* University of California Press, Berkeley, CA.

Kenyon, K. W. (1955) Last of the Tlingit sealers. *Natural History* (June), 295–298.

Kenyon, K. W., and Wilke, F. (1953) Migration of the northern fur seal, *Callorhinus ursinus. Journal of Mammalogy* 34, 86–98.

Kinlan, B. P., Graham, M. H., and Erlandson, J. M. (2005) Late Quaternary changes in the size and shape of the California Channel Islands: implications for marine subsidies to terrestrial communities. In *Proceedings of the Sixth California Islands Symposium* (D. K. Garcelon and C. A. Schwemm, Eds.). Institute for Wildlife Studies, Arcata, CA. pp. 131–142.

Lea, M.-A., Johnson, D., Ream, R., et al. (2009) Extreme weather events influence dispersal of naïve northern fur seals. *Biology Letters* 5, 252–257.

Loughlin, T. R., Perez, M. A. and Merrik, R. L. (1987) *Eumetopias jubatus. Mammalian Species* 283, 1–7.

Lyman, R. L. (1995) On the evolution of marine mammal hunting on the west coast of North America. *Journal of Anthropological Archaeology* 14, 45–77.

Lyman, R. L. (2003) Pinniped behavior, foraging theory, and the depression of metapopulations and nondepression of a local population on the southern northwest coast. *Journal of Anthropological Research* 22, 376–388.

Lyman, R. L. (2011) A history of paleoecological research on sea otters and pinnipeds of the eastern Pacific Rim. In *Human Impacts on Seals, Sea Lions, and Sea Otters: Integrating Archaeology and Ecology in the Northeast Pacific* (T. J. Braje and T. C. Rick, Eds.). University of California Press, Berkeley, CA. pp. 19–40.

McKechnie, I. (2007) Investigating the complexities of sustainable fishing at a prehistoric village on western Vancouver Island, British Columbia, Canada. *Journal for Nature Conservation* 15, 208–222.

McLaren, I. A., and Smith, T. G. (1985) Population ecology of seals: retrospective and prospective views. *Marine Mammal Science* 1, 54–83.

Melin, S. R., Sterling, J. T., Ream, R. R., et al. (2012) A tale of two stocks: studies of northern fur seals breeding at the northern and southern extent of the range. Research Feature, Alaska Fisheries Science Center, Quarterly Report, April–June. www.afsc.noaa.gov /Quarterly/amj2012/amj12.htm.

Moss, M. L., Lang, D. Y., Newsome, S. D., et al. (2006) Historical ecology and biogeography of North Pacific pinnipeds: isotopes and ancient DNA from three archaeological assemblages. *Journal of Island and Coastal Archaeology* 1, 165–190.

Moss, M. L., and Losey, R. J. (2011) Native American use of seals, sea lions, and sea otters in estuaries of northern Oregon and southern Washington. In *Human Impacts on Seals, Sea Lions, and Sea Otters: Integrating Archaeology and Ecology in the Northeast Pacific* (T. J. Braje and T. C. Rick, Eds.). University of California Press, Berkeley, CA. pp. 167–195.

Newsome, S. D., Etnier, M. A., Gifford-Gonzalez, D., et al. (2007) The shifting baseline of northern fur seal ecology in the northeast Pacific Ocean. *Proceedings of the National Academy of Sciences USA* 104, 9709–9714.

Ogden, A. (1941) *The California Sea Otter Trade 1784–1848.* University of California Press, Berkeley, CA.

Paine, R. T., Bromley, D. W., Castellini, M. A., et al. (2003) *Decline of the Steller Sea Lion in Alaskan Waters: Untangling Food Webs and Fishing Nets.* National Academies, Press, Washington, DC.

Pauly, D. (1995) Anecdotes and the shifting baselines syndrome of fisheries. *Trends in Ecology & Evolution* 10, 430.

Pauly, D. (1998) Tropical fishes: patterns and propensities. *Journal of Fish Biology* 53 (Supplement A), 1–17.

Pauly, D., Christensen, V., Dalsgaard, J., et al. (1998) Fishing down marine food webs. *Science* 279, 860–863.

Porcasi, J. F., Jones, T. L., and Raab, L. M. (2000) Trans-Holocene marine mammal exploitation on San Clemente Island, California: a tragedy of the commons revisited. *Journal of Anthropological Archaeology* 19, 200–220.

Preston, W. L. (2002) Post-Columbian wildlife irruptions in California: implications for cultural and environmental understanding. In *Wilderness and Political Ecology: Aboriginal Influences and the Original State of Nature* (C. E. Kay and R. T. Simmons, Eds.). University of Utah Press, Salt Lake City, UT. pp. 111–140.

Pyle, P., Long, D. J., Schonewald, et al. (2001) Historical and recent colonization of the South Farallon Islands, California, by northern fur seals (*Callorhinus ursinus*). *Marine Mammal Science* 17, 297–402.

Ream, R. R., Sterling, J. T., and Loughlin, T. R. (2005) Oceanographic features related to northern fur seal migratory movements. *Deep Sea Research Part II: Topical Studies in Oceanography* 52, 823–843.

Rick, T. C. (2007) *The Archaeology and Historical Ecology of Late Holocene San Miguel Island.* Cotsen Institute of Archaeology, University of California, Los Angeles.

Rick, T. C., DeLong, R., Erlandson, J. M., et al. (2009a) A trans-Holocene archaeological record of Guadalupe fur seals (*Arctocephalus townsendi*) on the California Coast. *Marine Mammal Science* 25, 487–502.

Rick, T. C., DeLong, R., Erlandson, J. M., et al. (2011) Where were the northern elephant seals? Holocene archaeology and biogeography of *Mirounga angustirostris*. *Holocene* 21, 1159–1166.

Rick, T. C., and Erlandson, J. M. (Eds.) (2008) *Human Impacts on Ancient Marine Ecosystems: A Global Perspective.* University of California Press, Berkeley, CA.

Rick, T. C., and Erlandson, J. M. (2009) Coastal exploitation. *Science* 325, 952–953.

Rick, T. C., Erlandson, J. M., Braje, T. J., and DeLong, R. L. (2009b) Seals, sea lions, and the erosion of archaeological sites on California's Channel Islands. *Journal of Island and Coastal Archaeology* 4, 125–131.

Rick, T. C., Erlandson, J. M., Vellanoweth, R. L., and Braje, T. J. (2005) From Pleistocene mariners to complex hunter-gatherers: the archaeology of the California Channel Islands. *Journal of World Prehistory* 19, 169–228.

Rick, T. C., Walker, P. L., Willis, L., et al. (2008) Dogs, humans, and island ecosystems: the distribution, antiquity, and impacts of domestic dogs (*Canis familiaris*) on California's Channel Islands. *Holocene* 18, 1077–1087.

Scammon, C. M. (1874) *The Marine Mammals of the Northwestern Coast of North America*. Putnam, New York, NY [Reprinted by Dover, New York, NY, 1968].

Schoenherr, A. A., Feldmeth, C. R., and Emerson, M. J. (1999) *Natural History of the Islands of California*. University of California Press, Berkeley, CA.

Springer, A. M., Estes, J. A., van Vliet, G. B., et al. (2003) Sequential megafaunal collapse in the North Pacific Ocean: an ongoing legacy of industrial whaling? *Proceedings of the National Academy of Sciences USA* 100, 12223–12228.

Starks, E. C. (1922) Records of the capture of fur seals on land in California. *California Fish and Game* 8, 155–160.

Stewart, B. S., and Huber, H. R. (1993) Mammalian species: *Mirounga angustirostris*. *American Society of Mammalogists* 449, 1–10.

Vermeij, G. J. (1991) Anatomy of an invasion: the trans-arctic interchange. *Paleobiology* 17, 281–307.

Walker, P. L., Kennett, D. J., Jones, T. L., and DeLong, R. (2002) Archaeological investigations at the Point Bennett pinniped rookery on San Miguel Island. In *Proceedings of the Fifth California Channel Islands Symposium* (D. Browne, K. Mitchell, and H. Chaney, Eds.). Santa Barbara Museum of Natural History, Santa Barbara, CA. pp. 628–632.

Weber, D. S., Stewart, B. S., Schienman, J., and Lehman, N. (2004) Major histocompatability complex variation at three class II loci in the northern elephant seal. *Molecular Ecology* 13, 711–718.

Yesner, D. R. (1987) Life in the "Garden of Eden": constraints of marine diets for human societies. In *Food and Evolution* (M. Harris and E. Ross, Eds.). Temple University Press, Philadelphia, PA. pp. 285–310.

FOUR

Using Disparate Datasets to Reconstruct Historical Baselines of Animal Populations

FRANCESCO FERRETTI, LARRY B. CROWDER, and FIORENZA MICHELI

When reconstructing long-term changes in marine ecosystems and populations of marine animals, historical data are needed to encompass the natural scale of population dynamics, disentangle short-term variability from longer fluctuations, and describe events that occurred decades or centuries ago. Historical data, however, are often difficult to obtain, vary greatly in format and quality, and were less consistently collected than most modern quantitative data. Concern for incorrectly integrating such different sources of information across long periods means that many historical datasets are used only in part or not at all. However, for many locations, such datasets provide the only sources of information on changes to populations or ecosystems. In this chapter, we review methods for accessing and incorporating disparate forms of historical data into quantitative historical reconstructions for marine species. We show how reconstructing historical baselines and documenting long-term changes can provide a powerful means to engage the public and motivate and inform policy reform. Our examples include Mediterranean fisheries and historical analyses of sharks and rays, a region and species group characterized by long histories of exploitation.

INTRODUCTION

Historical baselines of species abundance and ecosystem structure are often unknown because of limited human observations and a paucity of records, with those records that do exist often extending back only several decades (Bonebrake et al. 2010). Especially in the ocean, where ecological processes are generally concealed from direct observation, there is a continuous intergenerational loss of information on the "natural" structure of marine ecosystems (Pauly 1995, Pauly et al. 2005). For large marine animals (e.g., sharks, pinnipeds, seabirds, cetaceans, marine turtles), the effect of this information loss can be particularly severe because these species are long lived, slow to reproduce, and highly migratory (Musick 1999, Pulliam 2000, Collie et al. 2004). Our scientific understanding of the ecological processes that occur across these spatial and temporal scales is still fragmentary (e.g., Myers and Worm 2003, Myers et al. 2007, Block et al. 2011).

For these reasons, describing long-term population trends and ecological processes for large marine vertebrates is difficult, and approaches using multiple sources of historical information are increasingly being used (Myers and Worm 2003, Myers et al. 2007, Baum and Worm 2009). This chapter uses real-world examples to review the technical and analytical challenges of using disparate historical datasets for assessing long-term changes in marine species and ecosystems. We outline analytical methods and case studies that show the utility of integrating multiple heterogeneous datasets to estimate baselines of population abundance and ecosystem structure. Historical data provide great challenges and opportunities for reconstructing long-term baselines of animal populations. We show that virtually any form of data can be incorporated into quantitative assessments, given certain caveats, and that such integration can provide important guidance for marine management, conservation, and understanding of ecological processes. We draw primarily on examples related to long-lived species such as elasmobranchs (sharks and rays), showing that vulnerable marine animals can be amenable to approaches that reveal historical population changes. The overarching goal of this chapter is to explore the challenges and possibilities of using multiple datasets and data-gathering techniques to generate such trends and infer ecosystem change, and to illustrate the utility of these long-term historical analyses for motivating and informing policy changes.

CHALLENGES AND OPPORTUNITIES OF A HISTORICAL ECOLOGICAL APPROACH

Comparing Apples and Oranges

Historical data vary greatly in scale, nature, and quality. Technological advances and development in study design and analytical capability have allowed for the gathering of progressively more detailed and complex marine data that differ even from those collected a few decades ago. A widespread approach to dealing with information of varying quality is to select the data with the highest quality and consistency of collection and disregard the rest (McClenachan et al. 2012). Thus, many researchers discard potentially useful datasets to avoid comparing "apples and oranges." For example, when evaluating the status of exploited commercial fishery resources, the Scientific, Technical and Economic Committee for Fisheries of the European Commission often uses only high-resolution survey data (Anonymous 2007). This is the case even when historical information from similar monitoring programs exists (see the discussion on combining heterogeneous data below, under "Sharing and Accessing Data"). While comparisons across datasets should be made carefully, completely discarding information is rarely appropriate in historical studies of marine systems, where data are difficult and expensive to collect. Possible consequences of ignoring available information include poor fisheries management, flawed extinction-risk assessments, and, consequently, inadequate conservation planning (McClenachan et al. 2012; also see chapter 10, this volume). Learning how to use every bit of relevant and available information can produce more robust assessments of current population and ecosystem status and, therefore, better management and conservation plans.

Fomenting a Philosophical Shift

It is generally accepted that historical data can be used qualitatively to contextualize more recent quantitative information (Wolff 2000, Sandin et al. 2008). In practice, authors of ecological papers often cite historical sources in describing baseline conditions of altered ecosystems. For example, Sandin et al. (2008) described historical shark abundance in the Line Islands by quoting eighteenth-century explorers: "On every side of us swam Sharks innumerable, & so voracious that they bit our oars & rudder. . . ." Such qualitative uses of historical data have proved useful to contextualize recent quantitative information but can sometimes produce subjective and ambiguous representations of past conditions.

Conversely, analyses that more formally attempt to incorporate multiple and heterogeneous datasets (e.g., quantitative and qualitative data, continuous and discontinuous data series) are frequently criticized for a lack of precision, for inappropriate comparisons, or for making erroneous inferences from possibly spurious correlations (Burgess et al. 2005, Hampton et al. 2005, Polacheck 2006). As we show in this chapter, when data, analyses, and assumptions are properly defined and clearly presented, integrating heterogeneous datasets can provide new insights and a wide range of opportunities for characterizing baselines of historically depleted species and affected ecosystems. Below, we discuss potential benefits of integrative approaches that draw on a broad range of historical data and present examples of the novel insights derived by these analyses.

Data Are Multifaceted

Data often contain multiple bits of information that can be extracted and analyzed in various ways. For example, written records of species behavior, diet, and taxonomy can also be used to identify the extent of occurrence. Fishes sampled for stomach-content analyses also represent occurrence records of species in a particular time and location, and of the food items found in their stomachs. Link (2004) characterized long-term changes in abundance and distribution of benthic invertebrates in the northeast U.S. continental shelf by using stomach-content data published between 1970 and 2001. Temporal trajectories extracted for several groups of invertebrates were congruent with their vulnerability to fishing perturbations and similar to results of meta-analyses on the effect of trawling on benthic habitats (Link 2004).

Translating qualitative information (e.g., on presence–absence, habitat suitability, range distribution, and even temporal trends in abundance) into quantitative metrics of ecosystem and population status is a practice increasingly used to incorporate historical data into large-scale, long-term baseline research (Pandolfi et al. 2003, Lotze et al. 2006, Kittinger et al. 2013). One of the common criticisms of such research syntheses is that the generated indices of abundance may be biased by the subjective perceptions of the researchers who are interpreting historical data (Beyth-Marom 1982, McBride and Burgman 2012). However, Al-Abdulrazzak et al. (2012) found general agreement among multiple individuals in their rankings of anecdotal terms of abundance, showing that irrespective of different individuals' characteristics (e.g., age and ethnicity), people tend to interpret historical information similarly.

This problem can also be addressed by calibrating qualitative data with contemporary quantitative information. With the objective of converting two centuries (1800–2000) of qualitative indices of fish species abundance in the northern Adriatic Sea, Fortibuoni et al. (2010) extracted time series of perceived abundance (very rare, rare, common, very common, etc.) from historical publications up to 1950 and compared them to a time series of landings from a major fish market in the same region. These time series were used to construct frequency distributions of qualitative and quantitative abundances, calibrate the qualitative indices to the quantitative ones, and then produce quantitative abundance estimates for the categorical classes (perceived abundance; Fortibuoni et al. 2010). The authors were thus able to hindcast the quantitative indices of species abundance to periods not covered by the fish landings data and analyze long-term temporal variation in focal species and catch composition.

Creating Order out of Scattered Data

Identifying, combining, and quantifying historical data to understand ecological change requires an approach spanning multiple spheres of knowledge (Figure 4.1). As applied to ecology, this process can benefit from information technology science to gather and organize data, mathematics and statistics to identify the right tools for analysis, and history to understand and interpret data. The diverse skills and perspectives needed to obtain, analyze, and interpret historical data call for interdisciplinary training and collaboration.

In many cases, data are not scarce but are merely scattered, poorly described, or not organized in readily accessible archives. The Mediterranean Sea, for example, has been traditionally regarded as deficient in fishery statistics, and lack of management for many exploited marine species was often excused on these grounds. Yet this region has one of the highest densities of research facilities in the world and, consequently, one of the highest intensities of scientific monitoring (Coll et al. 2010), arguably the longest exploitation history (Roberts 2007), and the longest history of marine observation (Aristotle 350 BC). Thus, a considerable amount of historical information is dispersed in local publications and in institutional and personal archives (Ferretti et al. 2005, 2008, 2013). Such bodies of written records should be preserved, catalogued, and eventually converted to electronic records to avoid temporal degradation and loss of information (Michener 2006, Ray 2009). Describing, cataloguing, and making data accessible once published is increasingly encouraged and

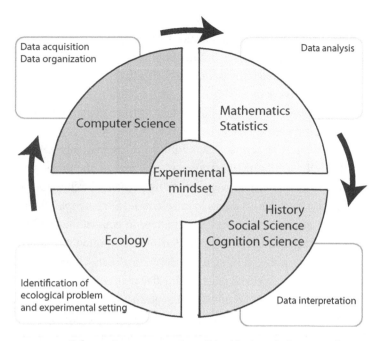

FIGURE 4.1 Schematic representation of the ideal analytic approach to problems of historical data integration. Pie slices represent the different disciplines that can contribute to the process of data integration and analysis. Corresponding boxes are the activities to which the disciplines can efficiently contribute, but contributions are not strictly limited to these. In an ideal historical and integrative analytical exercise, the analyst begins with the identification of the ecological problem and the identification of the system. Then s/he searches for the data and analytical tools required for addressing such a problem. Computer science is instrumental for extracting data from unstructured information and collating and organizing multiple datasets. Mathematics and statistics can help in extracting new information from data and drawing inferences from this information. History, social science, cognition science, and other nonecological disciplines can be useful in interpretation of results, but also in data identification and collection.

rewarded (Michener 2006, Reichman et al. 2011). Producing an inventory of available data and published information (Evans and Foster 2011) allows data tracking, improves understanding of data history and caveats, and facilitates combining data in research syntheses such as meta-analyses (Hafley and Lewis 1963).

Sharing and Accessing Data

In many cases, large datasets are unavailable to the scientific community despite urgent conservation and management needs. The governments of the United Kingdom, European Union, and United States are working toward mandating that all publicly funded research be published in open-access journals (Noorden 2012, 2013). However, this perspective frequently meets strong opposition from institutions when it pertains to raw data. Examples of datasets relevant for historical analyses but not publicly available include data on distribution, abundance, and population demographics of fish and invertebrates sampled by national and international trawl surveys in the Mediterranean (Relini 1998, Anonymous 2007); and European Commission data on fishing effort (e.g., real-time vessel monitoring system data on location, course, and speed of fishing boats) in European waters.

In some cases, data may be accessed through formal agreements with the institutions mandated to perform the sampling operations. In other cases, the process can be tortuous and less transparent and requires an investment of time and research money to sort through the intricacies of policies regulating access to information. For example, researchers seeking access to trawl-survey data to assess long-term changes to populations of sharks and rays in the Adriatic Sea were required to resort to the Aarhus Convention, an international agreement regulating access to publicly funded environmental data (Rodenhoff 2002, Ferretti et al. 2013). Hampering data access in practice shields useful information from a potentially huge creative analytical capacity represented by the international scientific community, which might be capable of tackling the issues that the collecting agencies are mandated to resolve.

A similar problem exists for heterogeneous datasets, which many managers or researchers holding key decision-making positions do not believe can or should be integrated, and data are therefore not made accessible. The case of trawl surveys in the Mediterranean is emblematic. GRUND (National Group for Demersal Resource Evaluation) is a publicly funded scientific trawl-survey program developed to evaluate the status of demersal marine animals in Italian territorial waters (Relini 1998). As part of this program, data on the biology, distribution, and abundance of many commercial and incidentally caught marine species (i.e., bycatch) were systematically collected between 1985 and 2002. Over time, several changes were made to survey design and sampling gear; consequently, the data were often not directly comparable among and within survey sectors. Thirty years after the trawl surveys were initiated, the collected raw data are still not organized into a single database and are not available to the public. Yet they represent the only source of information for many noncommercial species for which fishery-dependent data are unavailable (e.g., many species of sharks and rays). This lack of information caused by the lack of availability of unique—albeit problematic— datasets provides barriers to researchers seeking to use this information to inform management. Making these and similar datasets broadly available for integration and analysis is crucial, and cultural shifts in how this type of data is viewed and utilized are urgently needed.

Extracting Data: Learning from Other Disciplines

Ecoinformatics

Creating data from information and combining disparate datasets benefit from technical capabilities that are often beyond the conventional sphere of ecological training (Michener and Jones 2012). Ecoinformatics and the similar field of bioinformatics are the study of ecological information structure and resulting development of computer technology for its management and analysis. Ecoinformatics is emerging as an important field in ecology as expertise in computer science is needed for mining, manipulating, and visualizing the growing amount of data available through digital publications, websites, online databases, and social networks. The Internet stores a vast amount of information, historical and otherwise, from which it is not always trivial to extract data of interest. It is speculated that ≥80% of the information online is contained in free-form text (Grimes 2008). To be analyzable, this information must be identified and converted to datasets.

Literature analysis employs computational linguistics and statistics to mine the growing body of text available online. Databases can be built automatically through the use of software that extracts the data of interest from structured web pages. The software package rfishbase (an extension for the R programming environment), for example, accesses the FishBase database (www.fishbase.org) through its web pages and encapsulates data in a form that can be readily used for analyses (Boettiger et al. 2012). Text data mining is particularly promising for historical ecological research now that a large body of historical literature is being digitized and made available by academic libraries and Internet-related services (Crane 2006). Notable is the Google Book Library Project, which in April 2013 comprised 30 million digitized volumes (Darnton 2013). Books dating to the sixteenth or seventeenth century, traditionally very hard to access and consult (e.g., Rondelet 1554, Aldrovandi 1613), are now available in searchable format online. Finally, the availability of translation software is also facilitating the access to online literature previously obstructed by linguistic barriers (Crane 2006).

Social Science

Interview surveys are a means of capturing historical ecological information where observations of marine ecosystems or fisheries were not recorded (Johannes et al. 2000, Shackeroff et al. 2011). Interviewing resource users can be a valuable way of gathering historical information about spawning grounds, seasonal migrations, patterns of fishing for many exploited stocks, and other aspects of biology relevant to fishery management (Neis et al. 1999). However, planning and evaluating interview studies is an interdisciplinary process that requires expertise from social science, economics, statistics, and biology (Neis and Felt 2000, Drew 2005). For example, recollection uncertainty and cultural and cognition biases must be taken into account when planning interviews and analyzing their results (Neis and Felt 2000, Daw 2010), and aspects of memory with the least associated uncertainty (i.e., some events are more memorable than others) can be exploited. Saenz-Arroyo et al. (2005), for example, asked 108 fishermen from 11 fishing communities in Baja California, Mexico, simple questions about very memorable moments of their fishing career in relation to their target species, the Gulf grouper (*Mycteroperca jordani*). Questions included: How many fish did you catch on your best day ever? What was the size of the largest fish you ever caught? In what year were these catches made? By posing these questions to subjects in different age classes, they reconstructed a trajectory of change in maximum size and maximum catch of this species from the 1940s to the present day. In general, this process of eliciting information from resource users or experts can be structured so that the resulting data can be applied to particular analyses (Kuhnert et al. 2010). For example, estimates of recollection or expert-opinion uncertainty can be used to build informative priors for Bayesian analytical approaches and then incorporated into more formal stock assessments (more on Bayesian analysis below; Mäntyniemi et al. 2013).

Interpreting Historical Data

Once extracted and organized, data have to be interpreted. Marine historical ecology borrows approaches and expertise from historians, psychologists, archaeologists, and cognitive scientists for interpreting pieces of biological information. Sawfishes (*Pristis* spp.) provide an illustrative case. In the Mediterranean, there is an animated debate on whether locally reproducing sawfish populations have ever existed in the region, in part because the region is considered seasonally too cold to host stable populations (Ferretti 2014). In the absence of any direct physical evidence of these fishes having occurred there (e.g., via museum collections; cf. Box 4.1), Ferretti (2014) conducted an extensive historical search and used a collection of historical publications spanning from 350 BC to the present time to document sawfish occurrence and eventual extinction from the region.

A major challenge in this study was the interpretation of historical records coming from the classical and medieval periods. Aristotle, Pliny, and Oppian described the sawfish in their contemporary treatises of natural history (Aristotle 350 BC, Diaper and Jones 1722, Bostock and Riley 1855), though their quantitative, taxonomic, and geographic detail was vague or absent. The authors' historical context, their biographies, and the aims of their publications suggest that most of their natural descriptions referred to the Mediterranean Sea (Diaper and Jones 1722, Romero 2012). Yet these authors were also exposed to knowledge and information coming from other known ocean basins such as the Red Sea and Indian Ocean. Similarly, in the Middle Ages, sawfish were consistently included in bestiaries produced across Europe and the Mediterranean. However, extracting relevant zoological information from these descriptions is nearly impossible, because in most cases the animals described are mythological and religious allegories (White 2002). Working with experts who have a thorough understanding of how human knowledge is handed down, interpreted, and influenced by culture and religious faith (White 2002) is essential for selecting zoologically relevant facts from a literature that otherwise is obscure to an ecological readership.

Combining Data

Selecting an Analytical Approach

Combining disparate datasets requires identifying the information they contain, understanding their limitations, and implementing approaches for integrating data of diverse type and quality. Every piece of information has a certain degree of credibility, quality, and level of detail. This makes integrative analysis inferential in nature and dependent on a solid base of probability and statistical theory. To combine independent pieces of information appropriately, their individual uncertainty has to be taken into account and propagated. As Ben Halpern relates in his discussion of the Ocean Health Index (Box 4.2), integrative analyses cannot be based on the ideal of a study designed around one's research question and a

BOX 4.1 Viewpoint from a Practitioner: The Role of Natural History Collections in the Field of Historical Ecology

Louise K. Blight

In 2013, researchers at the Smithsonian National Museum of Natural History confirmed the discovery of a new raccoon-like mammal they called the "olinguito" (pronounced "oh-lin-GEE-toh"), the first new carnivore discovered in the Americas in nearly four decades (Stromberg 2013). For biologists and lovers of natural history, it is always exciting to hear about the discovery of a new species of animal, but such events grow increasingly rare as fewer places on our planet remain to be explored. The existence of the olinguito, however, was suspected not because some intrepid twenty-first-century scientist scoured the treetops of the Andean cloud forests with a remotely operated drone, but because a researcher opened a drawer of poorly described specimens at the Field Museum in Chicago.

While extensive use has been made of archived written records in uncovering past and present states of animal populations—as described by Francesco Ferretti and colleagues in this chapter—researchers in the field of marine historical ecology may also make use of the more traditional items found in museum collections. Skins, bones, and even whole animals have long been collected by the Western naturalists and biologists who traveled the world and returned with mementos of the natural wonders found on their journeys. And like the written descriptions of early natural riches (McClenachan et al. 2012), these physical remnants have hidden tales to tell.

Sophisticated modern techniques such as stable isotope analysis (e.g., Schell 2000; also see Box 3.1 in chapter 3, this volume) and DNA tests provide the tools to decipher these stories, as do simpler, more old-fashioned physical measurements. For my PhD research, I searched museum collections from across Canada, the United States, and the United Kingdom, looking for clues about what might be causing long-term population declines in the glaucous-winged gull (*Larus glaucescens*), a common and widespread species of the Pacific coast of North America. Rather than written records, I was seeking gull eggs and study skins (the preserved skin and feathers of a bird), which had been widely collected by early naturalists. I located hundreds of specimens carefully stowed in museums at Cambridge University, the Smithsonian Institution, and elsewhere, and compared them with data collected at my field site in 2008 and 2009. Both eggs and feathers told a detailed tale of changes to local environments over time: stable isotope analysis of feathers indicated that gull diet has changed over the past 150 years, with birds eating less fish over time after the advent of commercial fishing (Blight 2012). More interestingly, I also found that gulls now lay smaller eggs than they did 100 years ago and that they are laying fewer of them (Blight 2011). Overall, these results point to stressed seabird populations and an ecosystem that is likely less productive than it was before Europeans began their commercial extraction of marine resources.

Written museum records and old articles also featured in this research. These provided complementary information showing seabird nesting colonies were eradicated by egg harvesters in the late 1800s. However, the most surprising discoveries, that gull populations of the past had different diet and reproductive output, were derived from the physical remnants of the long-ago ancestors of the birds we presently see along the coast. It has become commonplace to dismiss museum collections as artifacts of a bygone era (Winker 2005), but the emergence of novel applications to new historical ecology questions shows that they represent a valuable resource for this field, with many stories left to tell.

Louise K. Blight is Senior Scientist with Procellaria Research & Consulting, Victoria, Canada.

BOX 4.2   Viewpoint from a Practitioner: Drawing Management Insights
from Disparate Data

Ben Halpern

Gaps in data, information, and understanding will always be a vexing problem for science and conservation, whether one is trying to piece together historical patterns and abundances of species, as described in this chapter, or trying to describe and manage ecosystems. All of the challenges, opportunities, and methods the authors describe are equally relevant for efforts to assess and protect the complex social-ecological systems that exist today.

Two messages from this chapter resonate particularly strongly with me, given the type of research I do: the challenges and opportunities that exist when combining disparate data sources, and the urgency to capture and make available any and all data. The world is replete with data—in historical texts, data servers, and people's personal collections—and we risk making poorly informed decisions and potentially losing that information if we neglect to compile and synthesize it all.

Over the past decade, I have focused my work on pulling together data about how humans interact with, influence, and benefit from marine ecosystems, from regional to global scales (e.g., Halpern et al. 2008, 2009, 2012). Doing this work has required an inclusive approach to the data types and sources used in the analyses, innovative solutions to filling key gaps, and substantial efforts tracking down and synthesizing disparate data sources.

A recent example of this kind of work is the Ocean Health Index (Halpern et al. 2012), which synthesizes qualitative and quantitative data from current and past sources across ecologi-

cal, social, institutional, and economic domains to produce a single assessment of the health of the ocean. Compiling all these data, and then making them freely available, serves as a great resource for scientists and managers, while combining them into a single index allows one to understand the "whole picture" in a way that is nearly impossible from just looking at the individual data layers.

Modern-day management can learn a huge amount from data reconstruction and synthesis efforts such as the ones described by these authors, even if the historical information lacks precision. By assessing what once existed, we can gain insight into what could be. Put another way, our path forward benefits from understanding where we came from. For example, imagine setting fisheries stock-rebuilding targets, species recovery plans, or habitat restoration goals based only on current abundance information. In all these cases, we need to know past abundances and extents of stocks, species, and habitats in order to set appropriate future targets.

The relevance of this chapter to current ecosystem assessment efforts highlights the broader utility of the work described here, while also reinforcing the validity of, and need for, efforts to reconstruct past ecological patterns and processes. The bottom line is that managers must make decisions in spite of missing data. The more that science can inform those decisions, even if the science is only able to paint broad-brushstroke pictures of how things were (or are), the better those decisions will be.

Ben Halpern is Professor in the Bren School of the Environment, University of California, Santa Barbara, and Chair in Marine Conservation, Imperial College London.

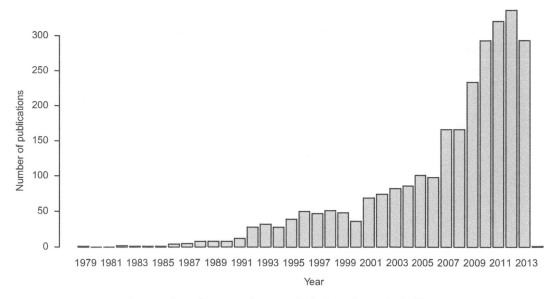

FIGURE 4.2 Increasing number of meta-analyses available from the ecological literature. Bars represent the number of primary publications containing the keywords "ecology" and "meta-analysis" published each year and available through the ISI Web of Knowledge database (papers published before 14 December 2013).

predetermined best analytical approach. Rather, they need to identify analytical approaches that fit data already collected. One way to address this is to consider gradients of environmental or human-induced conditions as treatment levels (Baum and Worm 2009) and select the analytical approach that can most efficiently exploit the identified experimental setting and available data. Below, we describe three methodological approaches that are particularly instrumental in integrative analyses.

Meta-analysis

The increasing availability of data described above has promoted a growing number of research syntheses (Figure 4.2). Meta-analysis is particularly suited for integrative historical ecological studies, including incorporation of data characterized by high uncertainty or limited information. In meta-analysis, values from individual studies are pooled in a larger experimental framework and treated as single data points. A study's uncertainty is used to weight the study's influence on the overall pattern so that data from studies with higher uncertainty have a lower weighting than those from more comprehensive ones, yet all data are exploited (Cooper and Hedges 1994, Normand 1999).

In such a framework, a test of a given hypothesis can reach statistical significance even though the composing studies are inconclusive or contradictory. This is because a meta-analysis increases statistical power by reducing the error of weighted average effect sizes (Cohn and Becker 2003). In ecology, where phenomena are multifactorial and often appear

BOX 4.3   Viewpoint from a Practitioner: Historical Data Revealed Massive Declines of Large Sharks in the Mediterranean

Francesco Ferretti

Mediterranean elasmobranch populations are among the most depleted in the world. This has been evident for decades in comparisons of FAO landings across world regions (Ferretti and Myers 2006), with multiple local analyses showing overexploitation of many shark and ray populations (Aldebert 1997, Jukić-Peladic et al. 2001, Ferretti et al. 2005, Cavanagh and Gibson 2007). However, a formal status assessment of all the species occurring in the basin came only in 2007, when the IUCN produced the first regional assessment of Mediterranean elasmobranchs. Of the 71 species assessed, 42% were Threatened, 18% Near Threatened, 14% of Least Concern, and 26% Data Deficient (Cavanagh and Gibson 2007). Most of the endangered species were assessed by using local aggregated or semiquantitative information such as changes in frequency of occurrence, sightings, and comparisons with qualitative descriptions, and by using a precautionary approach. However, few were based on periods >50 years (Cavanagh and Gibson 2007), and none relied on direct analyses of population change (McClenachan et al. 2012).

Following this, colleagues and I identified and assembled new data and qualitative information from additional regions and periods (Ferretti et al. 2008). These included sightings records from newspapers, museum and library records, commercial trawl catch data, landings from tuna traps, landings and observer data from pelagic longline fisheries, and catch statistics from recreational fishing clubs. Most of these data were available prior to the analyses in Ferretti et al. (2008) but had never been combined quantitatively to produce a regional synthesis.

We were able to select comparable data for five shark species or species groups—blue shark (*Prionace glauca*), common thresher (*Alopias vulpinus*), shortfin mako (*Isurus oxyrinchus*), porbeagle (*Lamna nasus*), and hammerheads (*Sphyrna* spp.)—estimating local trend analyses of catch per unit effort data (CPUEs) or other indices of abundance for each species. Instantaneous rates of change were also produced for each area and species and, as a common currency, were combined in a meta-analytical framework. The analyses revealed that over periods of 50–200 years, populations of these sharks had declined by 96% to 99%, implying a near extinction of these species in the Mediterranean Sea and,

contradictory and system dependent (Lawton 1996), meta-analysis has been effective in evaluating a broad array of questions, from testing the occurrence of top-down regulation in oceanic food webs (Worm and Myers 2003) and characterizing their dynamics (Micheli 1999) to assessing the nature and relative importance of interspecific interactions in animal communities (Gurevitch et al. 2000). Meta-analysis is also useful for assessing the status and extinction risk of species: inconsistent information on population trends, when evaluated meta-analytically, can reveal misclassification errors, such as classifying species to some lower risk category when the actual extinction risk is higher (Fernandez-Duque and Valeggia 1994, Ferretti et al. 2008). Box 4.3 provides a case study of these applications, using the example of sharks in the Mediterranean Sea.

according to IUCN criteria, a status of "Critically Endangered" in the region (Ferretti et al. 2008).

The results clearly warranted immediate conservation action and were communicated to the public through a coordinated outreach effort directed at translating the study's technical aspects and main findings into a language suitable for the greater public and policymakers. Consequently, the study received global media coverage, successfully bringing its finding to international attention, including that of interested political institutions.

In July 2008, the European Commission (EC) requested that the Scientific, Technical and Economic Committee for Fisheries (STECF) review these results. The STECF determined that combining different kinds of information was the only feasible strategy for addressing shark conservation status assessments, given a lack of long-term monitoring programs. The STECF further encouraged additional analyses of available time series in an attempt to reduce uncertainty (Anonymous 2008) and, importantly, advised that the EC implement a European Union (EU) Action Plan for Sharks. This included the establishment of bycatch reduc-

tion programs for Critically Endangered or Endangered elasmobranchs in instances where a zero Total Allowable Catch or prohibited status were not already in force for these species (Anonymous 2008).

On 5 February 2009, the EC adopted the first ever EU Plan of Action for the Conservation and Management of Sharks (European Commission 2009). Although the adoption of this policy framework was the result of the work of multiple nongovernmental organizations, international conservation bodies, and academic scientists, the empirical quantitative evidence provided in Ferretti et al. (2008) materially contributed to accelerating the EC's formal commitment to elasmobranch conservation (S. Maso, Shark Alliance, personal communication). The Action Plan did not impose fishery regulations for European sharks and rays, but it did create a new legal platform for developing specific legislation and management actions—for example, developing bycatch reduction programs for elasmobranchs in the northeast Atlantic and the Mediterranean and for four species caught by European fleets in the open ocean (Cavanagh and Gibson 2007, Gibson et al. 2008, Camhi et al. 2009).

Francesco Ferretti is a postdoctoral scholar at Hopkins Marine Station, Stanford University.

Bayesian Analysis

While meta-analysis relies on null hypothesis testing and thus evaluates the data while ignoring any previous information on the hypotheses being tested (McCarthy 2007), Bayesian analysts recognize that there is nearly always some previous knowledge about the process under study (Punt and Hilborn 1997, Wade 2000, Ellison 2004, McCarthy 2007; for an introduction to Bayesian statistics, see Kruschke 2010). In Bayesian frameworks, the likelihood or credibility of a given hypothesis (i.e., the posterior) is evaluated in light of previous information about the hypothesis (the prior) and the probability of the available data given such a hypothesis (the likelihood function). Existing information, such as historical semiquantitative or qualitative data, can be used in the construction of a prior probability distribution for the parameters under study, and in formulating the likelihood of the data given these hypotheses (Myers et al. 1995, Punt and Hilborn 1997). When a research question is clearly articulated and there is a clear definition of the analytical framework used to address it, historical and heterogeneous data can be incorporated into Bayesian analyses by translating information into probability distributions acting as priors of the Bayes rule.

For example, Gayeski et al. (2011) reconstructed historical populations of winter steelhead (*Oncorhynchus mykiss*) occurring in Puget Sound, Washington, from the nineteenth and twentieth centuries. They used multiple sources of information, including historical commercial catch reports from the late nineteenth century, data on the pace and extent of human settlement in the region, information on historical habitat extent, and other historical sources on the impact of aboriginal and European inhabitants. The authors used this information to build priors of total catch, catch rate, and unreported catch, and integrated these in a Bayesian binomial likelihood function to estimate posteriors of abundance of the overall local steelhead population and of subpopulations occupying different Puget Sound rivers. The authors integrated quantitative and qualitative historical information into quantitative stock assessment models. They estimated a discrepancy of population size between now and the end of the nineteenth century of about 25-fold. Their range of plausible population sizes (485,000–930,000) differed substantially from a previous official estimate (327,522–545,997) that failed to incorporate historical information on levels of steelhead catch by aboriginal peoples and early European settlers (Gayeski et al. 2011).

Newton (2010) provides another example relevant to Bayesian analysis and historical data: development of a quantitative framework for incorporating expert knowledge into assessments of species' conservation status. When the conservation status of a species is being assessed by the World Conservation Union (IUCN) and other similar bodies at the national level (e.g., the U.S. Fish and Wildlife Service, or the Committee on the Status of Endangered Wildlife in Canada), expert knowledge, including perceptions of historical change, is integrated into the assessment process, usually qualitatively and by consensus. However, this process is seldom conducted in a structured and objective way. Bayesian methods can be used to frame this process more objectively by constructing prior probability distributions from expert knowledge to be used in more quantitative assessments for data-poor species (Newton 2010).

Hierarchical Modeling

Multilevel hierarchical models can combine meta-analyses and Bayesian inference in a single statistical framework. Hierarchical modeling allows the analyst to incorporate multiple layers of information, including uncertainty on how the system works (process error) and on the measure of its state (observation errors; Parent and Rivot 2013). Probability models for the estimation of coefficients or the estimation of uncertainty can be plugged into higher-level models (Gelman and Hill 2006); hence, a variety of sparse and heterogeneous historical sources can be used to estimate different aspects of the ecological processes under investigation. Cornulier et al. (2011), for example, used a state-space Bayesian additive model to estimate change in hedgerow availability in the UK countryside to explain the loss of a farmland bird (the yellowhammer, *Emberiza citrinella*) previously detected by other studies. The authors collected multisource point estimates of hedgerow length and rates of change over time and developed a model of decline from these sparse historical data; original data were collected from several regions and periods, using variable protocols and levels of reporting detail. These authors stated that their modeling framework was highly generalizable and could be applied to the reconstruction of a time series of variables from "a variety of sparse and heterogenous historical sources" (Cornulier et al. 2011).

Information, when missing, can be borrowed from elsewhere. Gerber (2006) estimated historical trends in abundance of sharks and rays in the Gulf of Lion, southern France, by using a mix of Bayesian analysis and meta-analysis on aggregated trawl catch data (presence–absence). The author borrowed the dispersion parameters needed to characterize the species-specific negative binomial distributions of the catches from more detailed trawl-survey data recording similar species on the eastern U.S. continental shelf, and combined these dispersion parameters meta-analytically to construct informative priors necessary for the Bayesian trend analyses (Gerber 2006).

Integrative Analyses Stimulate Discussion and Further Work

Overall, attempts to integrate multiple datasets have led to useful debates and productive discussions that have advanced the field by making datasets available, contextualizing previously published data, and generating new hypotheses and research programs. Myers and Worm (2003), for example, analyzed historical time series of pelagic longline logbook catch data and fishery independent surveys to reconstruct steep declines of large predatory fishes from the onset of industrial fishing. This paper received worldwide media coverage and stimulated political and scientific discussions, including critiques of the authors' analytical approach and conclusions. These disputes had the positive effect of stimulating analysis and publication of large regional datasets refining the global patterns highlighted by this initial study (Hampton et al. 2005, Magnuson et al. 2006, Polacheck 2006, Sibert et al. 2006, Juan-Jordá et al. 2011). These and other analyses have shown that fish population trends differ regionally, with some species declining rapidly as a result of fishing, and others perhaps benefiting from the declines in predators or competitors (Myers et al. 2007).

Similarly, Worm et al. (2006) used historical FAO catch data to estimate the number of collapsed fish stocks and make projections about the future state of world fisheries, an approach that was heavily criticized (Murawski et al. 2007). The debate over the paper's conclusions ultimately promoted a larger collaboration between the authors and critics of the original study, which resulted in a further synthesis of the status of world fisheries (Worm et al. 2009). This collaboration partly resolved the disagreement and stimulated the assessment of unmanaged fisheries (Costello et al. 2012) and the development of the most comprehensive stock-assessment database available today (Ricard et al. 2011).

Uncertainty also has bounds, even in historical data. While researchers have disagreed about the details of estimates of historical population change, the overall magnitude and direction of change have generally emerged as points of agreement. Overfishing in New England, Canada, and the North Sea caused the collapse of multiple stocks of cod (*Gadus morhua*) in recent decades. Despite the variability of virgin biomass estimates across areas (Myers et al. 2001, Rosenberg et al. 2005) and disputes over the relative contributions of anthropogenic and species interactions in determining these trajectories (Hutchings and Myers 1995, Yodzis 2001), it became evident that fishing exploitation had to be reduced by at least half to preserve these stocks (Rosenberg 2007).

IMPLICATIONS FOR CONSERVATION AND MANAGEMENT

Moving from knowledge to actual conservation efforts requires effectively communicating such knowledge to policymakers and motivating them to take action. Public support is a crucial ingredient in this process. Integrative historical analyses can offer important insights into long-term population and ecosystem trends, patterns of decline or recovery, and baselines for conservation and management efforts. They can also play a key role in motivating and guiding conservation actions by rescaling our perception of what might constitute a natural level of population abundance for many animal populations, and helping document the magnitude of long-term population depletion (e.g., McClenachan and Cooper 2008).

Integrative historical analyses have the capability to engage and stimulate the imagination of the public through use of unconventional data (McClenachan et al. 2012). For example, by examining archaeozoological remains and the illustrations of fish in ancient Roman and Greek mosaics, Guidetti and Micheli (2011) found that the Mediterranean groupers (e.g., dusky grouper, *Epinephelus marginatus*) represented in these works of art were much larger than most of the animals seen today, and occupied shallower habitats than documented even at today's best-protected sites. These artifacts were not exaggerations, as the grouper sizes inferred from the mosaics in general were consistent with those estimated from bones found in archaeological sites. The peculiarity of the data source attracted worldwide media attention to the depleted state of these populations and to overexploitation of marine resources in general. Historical studies may also inspire policy changes, as has resulted from historical analyses of large pelagic sharks in the Mediterranean (Box 4.3).

CONCLUSIONS

Understanding long-term ecological processes and population trends requires an expanded approach to the synthesis and integration of data. However, several challenges to such approaches must be overcome; these issues range from a lack of inventories, organization, and standardization of the data available to the technical difficulties of making inferences from complex and heterogeneous datasets. Nearly all data types are amenable to being integrated into historical meta-analytical frameworks, but this process requires collaborative interdisciplinary approaches or requires ecologists to develop skills in areas outside their familiar bounds.

A key cultural shift will facilitate future scientific progress in using all available information in ecological analyses. Specifically, it will be important to move from seeking *precise* science to seeking *necessary* science. Integrative analyses are seldom precise but often offer broader perspectives on data that could rarely be achieved with conventional analytical approaches. While such analyses frequently attract criticism, they can also motivate action in the research, policy, and even political arenas. Movement on conservation issues is usually achieved by stimulating the imagination, interest, and engagement of the public, which consequently prompts political action for conservation.

Even though we recognize that the restoration of species or ecosystems to a relatively pristine state (defined through historical baselines) might not be feasible at present, if ever, providing evidence of the one-time existence of such baselines can enhance people's imagination about what might be possible. And once people understand such possibilities based on the past, they can make more enlightened decisions as to how to proceed into the future.

ACKNOWLEDGMENTS

The authors wish to acknowledge Jack Kittinger, Louise Blight, and Loren McClenachan for their editorial assistance and their comments on the manuscript, and the financial support of the Lenfest Ocean Program.

REFERENCES

Aires-da-Silva, A. M., Hoey, J. J., and Gallucci, V. F. (2008) A historical index of abundance for the blue shark *Prionace glauca* in the western North Atlantic. *Fisheries Research* 92, 41–52.

Al-Abdulrazzak, D., Naidoo, R., Palomares, M. L. D., and Pauly, D. (2012) Gaining perspective on what we've lost: the reliability of encoded anecdotes in historical ecology. *PLoS ONE* 7, e43386.

Aldebert, Y. (1997) Demersal resources of the Gulf of Lions (NW Mediterranean). Impact of exploitation on fish diversity. *Vie Milieu* 47, 275–285.

Aldrovandi, U. (1613) *De piscibus libri V. et de cetis lib. unus.* apud Bellagambam.

Anonymous (2007) *International Bottom Trawl Survey in the Mediterranean (Medits). Instruction Manual, version 5.* IFREMER, Nantes, France.

Anonymous (2008) 28th plenary meeting report of the scientific, technical and economic committee for fisheries. Plen-08–02. Commission staff working document, Commission of the European Communities, Helsinki, Finland.

Aristotle (350 BC) *The History of Animals.* University of Adelaide Library, Adelaide, Australia.

Baum, J., and Blanchard, W. (2010) Inferring shark population trends from generalized linear mixed models of pelagic longline catch and effort data. *Fisheries Research* 102, 229–239.

Baum, J. K., Myers, R. A., Kehler, D. G., et al. (2003) Collapse and conservation of shark populations in the northwest Atlantic. *Science* 299, 389–392.

Baum, J. K., and Worm, B. (2009) Cascading top-down effects of changing oceanic predator abundances. *Journal of Animal Ecology* 78, 699–714.

Beyth-Marom, R. (1982) How probable is probable? A numerical translation of verbal probability expressions. *Journal of Forecasting* 1, 257–269.

Blight, L. K. (2011) Egg production in a coastal seabird, the glaucous-winged gull (*Larus glaucescens*), declines during the last century. *PLoS ONE* 6, e22027.

Blight, L. K. (2012) Glaucous-winged gulls *Larus glaucescens* as sentinels for a century of ecosystem change: long-term trends in population, diet, and egg production in North America's Salish Sea. PhD dissertation, University of British Columbia, Vancouver, BC.

Block, B. A., Jonsen, I. D., Joregensen, S. J., et al. (2011) Tracking apex marine predator movements in a dynamic ocean. *Nature* 475, 86–90.

Boettiger, C., Lang, D. T., and Wainwright, P. C. (2012) Rfishbase: exploring, manipulating and visualizing FishBase data from R. *Journal of Fish Biology* 81, 2030–2039.

Bonebrake, T. C., Christensen, J., Boggs, C. L., and Ehrlich, P. R. (2010) Population decline assessment, historical baselines, and conservation. *Conservation Letters* 3, 371–378.

Bostock, J., and Riley, H. T. (1855) *The Natural History of Pliny. Translated, with copious notes and illustrations by the late John Bostock and H. T. Riley*. Bohn's Classical Library, H. G. Bohn, London, England.

Burgess, G., Beerkircher, L., Calliet, G. M., et al. (2005) Is the collapse of shark populations in the Northwest Atlantic Ocean and Gulf of Mexico real? *Fisheries* 30, 19–26.

Camhi, M. D., Valenti, S. V., Fordham, S. V., et al. (2009) The conservation status of pelagic sharks and rays. *Report of the IUCN Shark Specialist Group Pelagic Shark Red List Workshop*. IUCN Species Survival Commission Shark Specialist Group, Newbury, UK.

Cavanagh, R., and Gibson, C. (2007) Overview of the conservation status of cartilaginous fishes (Chondrichthyans) in the Mediterranean Sea. *World Conservation Union, Gland, Switzerland, and Malaga, Spain*. http://cmsdata.iucn.org/downloads/med_shark_rep_en_1.pdf.

Clarke, S. C., McAllister, M. K., Milner-Gulland, E. J., et al. (2006) Global estimates of shark catches using trade records from commercial markets. *Ecology Letters* 9, 1115–1126.

Cohn, L. D., and Becker, B. J. (2003). How meta-analysis increases statistical power. *Psychological Methods* 8, 243.

Coll, M., Piroddi, C., Steenbeek, J., et al. (2010) The biodiversity of the Mediterranean Sea: estimates, patterns, and threats. *PLoS ONE* 5, e11842.

Collie, J., Richardson, K., and Steele, J. (2004) Regime shifts: can ecological theory illuminate the mechanisms? *Progress in Oceanography* 60, 281–302.

Cooper, H., and Hedges, L. V. (1994) *The Handbook of Research Synthesis*. Russell Sage Foundation, New York, NY.

Cornulier, T., Robinson, R., Elston, D., et al. (2011) Bayesian reconstitution of environmental change from disparate historical records: hedgerow loss and farmland bird declines. *Methods in Ecology and Evolution* 2, 86–94.

Costello, C., Ovando, D., Hilborn, R., et al. (2012) Status and solutions for the world's unassessed fisheries. *Science* 338, 517–520.

Crane, G. (2006) What do you do with a million books? *D-Lib Magazine* 12, 1.

Darnton, R. (2013) The national digital public library is launched! *New York Review*, April 25.

Daw, T. M. (2010) Shifting baselines and memory illusions: what should we worry about when inferring trends from resource user interviews? *Animal Conservation* 13, 534–535.

Diaper, W., and Jones, J. (1722) *Oppian's Halieuticks of the Nature of Fishes and Fishing of the Ancients*. Printed at the Theater, Oxford, England.

Drew, J. A. (2005) Use of traditional ecological knowledge in marine conservation. *Conservation Biology* 19, 1286–1293.

Ellison, A. (2004) Bayesian inference in ecology. *Ecology Letters* 7, 509–520.

Estes, J., Terborgh, J., Brashares, J. S., et al. (2011) Trophic downgrading of planet Earth. *Science* 333, 301.

European Commission (2009) Communication from the Commission to the European Parliament and the Council on a European Community Action Plan for the Conservation and Management of Sharks. Communication, Brussels, February 5.

Evans, J., and Foster, J. (2011) Metaknowledge. *Science* 331, 721.

Fernandez-Duque, E., and Valeggia, C. (1994) Meta-analysis: a valuable tool in conservation research. *Conservation Biology* 8, 555–561.

Ferretti, F. (2014) Geographical distribution and status. 7.2.1 Mediterranean Sea. In *Sawfish: A Global Strategy for Conservation* (L. H. Harrison and N. K. Dulvy, Eds.). IUCN Shark Specialist Group, Vancouver, BC. pp. 50–51.

Ferretti, F., and Myers, R. A. (2006) By-catch of sharks in the Mediterranean Sea: available mitigation tools. In *Proceedings of the Workshop on Mediterranean Cartilaginous Fish with Emphasis on Southern and Eastern Mediterranean* (N. Basuşta, C. Keskin, F. Serena, and B. Serét, Eds.). Turkish Marine Research Foundation, Instanbul. pp. 149–161.

Ferretti, F., Myers, R. A., Sartor, P., and Serena, F. (2005) Long term dynamics of the chondrichthyan fish community in the upper Tyrrhenian Sea. ICES CM 2005/N 25.

Ferretti, F., Myers, R. A., Serena, F., and Lotze, H. K. (2008) Loss of large predatory sharks from the Mediterranean Sea. *Conservation Biology* 22, 952–964.

Ferretti, F., Osio, G. C., Jenkins, C. J., et al. (2013) Long-term change in a meso-predator community in response to prolonged and heterogeneous human impact. *Scientific Reports* 3, article 1057.

Ferretti, F., Worm, B., Britten, G. L., et al. (2010) Patterns and ecosystem consequences of shark declines in the ocean. *Ecology Letters* 13, 1055–1071.

Fortibuoni, T., Libralato, S., Raicevich, S., et al. (2010) Coding early naturalists' accounts into long-term fish community changes in the Adriatic Sea (1800–2000). *PLoS ONE* 5, e15502.

Gayeski, N., McMillan, B., and Trotter, P. (2011) Historical abundance of Puget Sound steelhead, *Oncorhynchus mykiss*, estimated from catch record data. *Canadian Journal of Fisheries and Aquatic Sciences* 68, 498–510.

Gelman, A., and Hill, J. (2006) *Data Analysis Using Regression and Multilevel/Hierarchical Models.* Cambridge University Press, Cambridge, UK.

Gerber, L. (2006) A Bayesian method for determining trends in abundance from negative binomial count data. Master's thesis, Dalhousie University, Halifax, NS.

Gibson, C., Valenti, S. V., Fordham, S. V., and Fowler, S. L. (2008) The conservation status of northeast Atlantic chondrichthyans. *Report of the IUCN Shark Specialist Group Northeast Atlantic Red List Workshop.* World Conservation Union, Gland, Switzerland.

Grimes, S. (2008) Unstructured data and the 80 percent rule: investigating the 80%. Technical report. *Clarabridge Bridgepoints.*

Guidetti, P., and Micheli, F. (2011) Ancient art serving marine conservation. *Frontiers in Ecology and the Environment* 9, 374–375.

Gurevitch, J., Morrison, J. A., and Hedges, L. V. (2000) The interaction between competition and predation: a meta-analysis of field experiments. *American Naturalist* 155, 435–453.

Hafley, W. L., and Lewis, J. S. (1963) Analyzing messy data. *Industrial & Engineering Chemistry* 55, 37–39.

Halpern, B. S., Kappel, C. V., Selkoe, K. A., et al. (2009) Mapping cumulative human impacts to California Current marine ecosystems. *Conservation Letters* 2, 138–148.

Halpern, B. S., Longo, C., Hardy, D., et al. (2012) An index to assess the health and benefits of the global ocean. *Nature* 488, 615–620.

Halpern, B. S., Walbridge, S., Selkoe, K. A., et al. (2008) A global map of human impact on marine ecosystems. *Science* 319, 948–952.

Hampton, J., Sibert, J. R., Kleiber, P., et al. (2005) Fisheries decline of Pacific tuna populations exaggerated? *Nature* 434, E1–E2.

Hutchings, J., and Myers, R. (1995) The biological collapse of Atlantic cod off Newfoundland and Labrador. In *The North Atlantic Fisheries: Successes, Failures, and Challenges* (L. Felt, Ed.). Island Institute Studies, Charlottetown, Prince Edward Island, Canada. pp. 37–94.

Johannes, R., Freeman, M., and Hamilton, R. (2000) Ignore fishers' knowledge and miss the boat. *Fish and Fisheries* 1, 257–271.

Juan-Jordá, M., Mosqueira, I., Cooper, A., et al. (2011) Global population trajectories of tunas and their relatives. *Proceedings of the National Academy of Sciences USA* 108, 20650–20655.

Jukić-Peladic, S., Vrgoc, N., Krustulovic-Sifner, S., et al. (2001) Long-term changes in demersal resources of the Adriatic Sea: comparison between trawl surveys carried out in 1948 and 1998. *Fisheries Research* 53, 95–104.

Kittinger, J. N., Houtan, K. S. V., McClenachan, L. E., and Lawrence, A. L. (2013) Using historical data to assess the biogeography of population recovery. *Ecography* 36, 868–872.

Kruschke, J. (2010) *Doing Bayesian Data Analysis: A Tutorial Introduction with R and BUGS*. Academic Press, Boston, MA.

Kuhnert, P. M., Martin, T. G., and Griffiths, S. P. (2010) A guide to eliciting and using expert knowledge in Bayesian ecological models. *Ecology Letters* 13, 900–914.

Lawton, J. (1996) Patterns in ecology. *Oikos* 75, 145–147.

Lee, T. E., McCarthy, M. A., Wintle, B. A., et al. (2013) Inferring extinctions from sighting records of variable reliability. *Journal of Applied Ecology* 51, 251–258.

Link, J. S. (2004) Using fish stomachs as samplers of the benthos: integrating long-term and broad scales. *Marine Ecology Progress Series* 269, 265–275.

Lotze, H. K., Lenihan, H. S., Bourque, B. J., et al. (2006) Depletion, degradation, and recovery potential of estuaries and coastal seas. *Science* 312, 1806–1809.

Magnuson, J., Cowan, L., Jr., Crowder, L. B., et al. (2006) *Dynamic Changes in Marine Ecosystems: Fishing, Food Webs, and Future Options*. Ocean Studies Board, National Research Council, National Academy Press, Washington, DC.

Mäntyniemi, S., Haapasaari, P., Kuikka, S., et al. (2013) Incorporating stakeholders' knowledge to stock assessment: central Baltic herring. *Canadian Journal of Fisheries and Aquatic Sciences* 70, 591–599.

McBride, M. F., and Burgman, M. A. (2012) What is expert knowledge, how is such knowledge gathered, and how do we use it to address questions in landscape ecology? In *Expert Knowledge and Its Application in Landscape Ecology* (A. H. Perera, C. A. Drew, and C. J. Johnson, Eds.). Springer, New York, NY. pp. 11–38.

McCarthy, M. A. (2007) *Bayesian Methods for Ecology*. Cambridge University Press, Cambridge, UK.

McClenachan, L., and Cooper, A. (2008) Extinction rate, historical population structure and ecological role of the Caribbean monk seal. *Proceedings of the Royal Society of London Series B* 275, 1351–1358.

McClenachan, L., Ferretti, F., and Baum, J. K. (2012) From archives to conservation: why historical data are needed to set baselines for marine animals and ecosystems. *Conservation Letters* 5, 349–359.

Micheli, F. (1999) Eutrophication, fisheries, and consumer-resource dynamics in marine pelagic ecosystems. *Science* 285, 1396–1398.

Michener, W. K. (2006) Meta-information concepts for ecological data management. *Ecological Informatics* 1, 3—7.

Michener, W. K., and Jones, M. B. (2012) Ecoinformatics: supporting ecology as a data-intensive science. *Trends in Ecology & Evolution* 27, 85–93.

Murawski, S., Methot, R., and Tromble, G. (2007) Biodiversity loss in the ocean: how bad is it? *Science* 316, 1281–1284.

Musick, J. A. (Ed.) (1999) *Life in the Slow Lane: Ecology and Conservation of Long-lived Marine Animals*. Proceedings of the Symposium Conservation of Long-lived Marine Animals, Monterey, CA, USA, 24 August 1997. American Fisheries Society, Bethesda, MD.

Myers, R. A., Barrowman, N. J., Hutchings, J. A., and Rosenberg, A. A. (1995) Population dynamics of exploited fish stocks at low population levels. *Science* 269, 1106.

Myers, R. A., Baum, J. K., Shepherd, T., et al. (2007) Cascading effects of the loss of apex predatory sharks from a coastal ocean. *Science* 315, 1846–1850.

Myers, R. A., MacKenzie, B., Bowen, K., and Barrowman, N. (2001) What is the carrying capacity for fish in the ocean? A meta-analysis of population dynamics of North Atlantic cod. *Canadian Journal of Fisheries and Aquatic Sciences* 58, 1464–1476.

Myers, R. A., and Worm, B. (2003) Rapid worldwide depletion of predatory fish communities. *Nature* 423, 280 283.

Neis, B., and Felt, L. (2000) *Finding Our Sea Legs: Linking Fishery People and Their Knowledge with Science and Management*. ISER Books, St. John's, NL.

Neis, B., Schneider, D. C., Lawrence, F., et al. (1999) Fisheries assessment: what can be learned from interviewing resource users? *Canadian Journal of Fisheries and Aquatic Sciences* 56, 1949–1963.

Newton, A. (2010) Use of a Bayesian network for Red Listing under uncertainty. *Environmental Modelling & Software* 25, 15 23.

Noorden, R. (2012) Europe joins UK open-access bid. Britain plans to dip in to research funding to pay for results to be freely available. *Nature* 487, 285.

Noorden, R. (2013) US science to be open to all. *Nature* 494, 414–415.

Normand, S. T. (1999) Tutorial in biostatistics. Meta-analysis: formulating, evaluating, combining, and reporting. *Statistics in Medicine* 18, 321–359.

Pandolfi, J., Bradbury, R., Sala, E., et al. (2003) Global trajectories of the long-term decline of coral reef ecosystems. *Science* 301, 955 958.

Parent, E., and Rivot, E. (2013) *Introduction to Hierarchical Bayesian Modeling for Ecological Data*, vol. 8. Chapman & Hall/CRC, Boca Raton, FL.

Pauly, D. (1995) Anecdotes and the shifting baseline syndrome of fisheries. *Trends in Ecology & Evolution* 10, 430–430.

Pauly, D., Watson, R., and Alder, J. (2005) Global trends in world fisheries: impacts on marine ecosystems and food security. *Philosophical Transactions of the Royal Society of London Series B* 360, 5–12.

Polacheck, T. (2006) Tuna longline catch rates in the Indian Ocean: did industrial fishing result in a 90% rapid decline in the abundance of large predatory species? *Marine Policy* 30, 470–482.

Pulliam, H. (2000) On the relationship between niche and distribution. *Ecology Letters* 3, 349–361.

Punt, A. E., and Hilborn, R. (1997) Fisheries stock assessment and decision analysis: the Bayesian approach. *Reviews in Fish Biology and Fisheries* 7, 35–63.

Ray, J. (2009) Sharks, digital curation, and the education of information professionals. *Museum Management and Curatorship* 24, 357–368.

Reichman, O., Jones, M., and Schildhauer, M. (2011) Challenges and opportunities of open data in ecology. *Science* 331, 703.

Relini, G. (1998) Valutazione delle risorse demersali in Italia. *Biologia Marina Mediterranea* 5, 3–19.

Ricard, D., Minto, C., Jensen, O., and Baum, J. (2011) Examining the knowledge base and status of commercially exploited marine species with the RAM legacy stock assessment database. *Fish and Fisheries* 13, 380–398.

Roberts, C. (2007) *The Unnatural History of the Sea*. Island Press, Washington, DC.

Roberts, D., and Solow, A. (2003) When did the dodo become extinct. *Nature* 426, 245.

Rodenhoff, V. (2002) The Aarhus convention and its implications for the "institutions" of the European Community. *Review of European Community & International Environmental Law* 11, 343–357.

Romero, A. (2012) When whales became mammals: the scientific journey of cetaceans from fish to mammals in the history of science. In *New Approaches to the Study of Marine Mammals* (A. Romero and E. O. Keith, Eds.). www.intechopen.com/books/new-approaches-to-the-study-of-marine-mammals/when-whales-became-mammals-the-scientific-journey-of-cetaceans-from-fish-to-mammals-in-the-history-0.

Rondelet, G. (1554) *Libri de Piscibus Marinis: in Quibus Verae Piscium Effigies Expressae Sunt*. Apud Mathiam Bonhomme, Lugduni.

Rosenberg, A. A. (2007) Fishing for certainity. *Nature* 449, 989.

Rosenberg, A. A., Bolster, W. J., Alexander, K., et al. (2005) The history of ocean resources: modeling cod biomass using historical records. *Frontiers in Ecology and the Environment* 3, 84–90.

Saenz-Arroyo, A., Roberts, C. M., Torre, J., et al. (2005) Rapidly shifting environmental baselines among fishers of the Gulf of California. *Proceedings of the Royal Society of London Series B* 272, 1957–1962.

Sandin, S. A., Smith, J. E., DeMartini, E. E., et al. (2008) Baselines and degradation of coral reefs in the Northern Line Islands. *PLoS ONE* 3, e1548.

Schell, D. (2000) Declining carrying capacity in the Bering Sea: isotopic evidence from whale baleen. *Limnology and Oceanography* 45, 459–462.

Shackeroff, J., Campbell, L., and Crowder, L. (2011) Social–ecological guilds: putting people into marine historical ecology. *Ecology and Society* 16, 52.

Sibert, J., Hampton, J., Kleiber, P., and Maunder, M. (2006) Biomass, size, and trophic status of top predators in the Pacific Ocean. *Science* 314, 1773.

Solow, A. R. (2005) Inferring extinction from a sighting record. *Mathematical Biosciences* 195, 47–55.

Stromberg, J. (2013) For the first time in 35 years, a new carnivorous mammal species is discovered in the Americas. Smithsonianmag.com. www.smithsonianmag.com/science-nature/for-the-first-time-in-35-years-a-new-carnivorous-mammal-species-is-discovered-in-the-americas-48047/#New-Mammal-Olinguito-1.png.

Wade, P. (2000) Bayesian methods in conservation biology. *Conservation Biology* 14, 1308–1316.

White, T. (2002) *The Book of Beasts: Being a Translation from a Latin Bestiary of the Twelfth Century*. Parallel Press, Madison, WI.

Winker, K. (2005) Bird collections: development and use of a scientific resource. *Auk* 122, 966–971.

Wolff, W. (2000) The south-eastern North Sea-losses of vertebrate fauna during the past 2000 years. *Biological Conservation* 95, 209–217.

Worm, B., Barbier, E. B., Beaumont, N., et al. (2006) Impacts of biodiversity loss on ocean ecosystem services. *Science* 314, 787.

Worm, B., Hilborn, R., Baum, J.K., et al. (2009) Rebuilding global fisheries. *Science* 325, 578–585.

Worm, B., and Myers, R.A. (2003) Meta-analysis of cod–shrimp interactions reveals top-down control in oceanic food webs. *Ecology* 84, 162–173.

Yodzis, P. (2001) Must top predators be culled for the sake of fisheries? *Trends in Ecology & Evolution* 16, 78–84.

PART TWO

CONSERVING FISHERIES

LEAD SECTION EDITOR: JOHN N. KITTINGER

Fisheries have been one of the primary focal areas for marine historical ecology research. Daniel Pauly's now famous "shifting baselines syndrome" was first described in the context of fisheries (Pauly 1995), and some of the first work in this field was directed toward understanding how fisheries ecosystems have changed over time. Jeremy B.C. Jackson, inspired by Pauly's work, convened a working group from 1999 to 2002 at the National Center for Ecological Analysis and Synthesis in Santa Barbara, California. Their first paper was published in 2001, documenting the long-term effects of fishing in coastal ecosystems, and showing that exploitation pressures in these environments preceded other impacts. Three other papers would follow—all published in the journal *Science*—that would build on this seminal work.

Jackson's and Pauly's work on historical changes in fisheries has inspired similar efforts around the globe to understand long-term changes in fisheries. It also helped redefine the nature of data validity, or, as Jackson put it, "The addition of a deep historical dimension to analyze and interpret ecological problems requires that we sacrifice some of the apparent precision and analytical elegance prized by ecologists" (Jackson et al. 2001:630). Perhaps more importantly, Jackson and colleagues showed unequivocally that Daniel Pauly's shifting baselines syndrome—if ignored—could have dangerous implications for fisheries management.

Researchers in marine historical ecology have increasingly focused not only on the ecological history of fisheries, but also on the corresponding social dynamics of these systems. Fished species are, of course, embedded in ocean ecosystems, but they also function as critical resources for human communities, influencing coastal cultures through resource dependence, food security, and cultural values. Just as fishers have affected the status and condition of fisheries through time, so have fisheries resources determined the character and culture of coastal communities. Cod fisheries in the northwest Atlantic, for example, were highly influential in the history and development of coastal communities and economies throughout the Atlantic (Rosenberg et al. 2005). Similarly, coastal fisheries have featured centrally in the historical population of the Americas and other regions (Erlandson et al. 2007, Rick and Erlandson 2009) and continue to play a central cultural role for indigenous communities worldwide (Johannes 2002, Garibaldi and Turner 2004).

Marine historical ecology can tell us more than just the magnitude or timeline of fisheries declines and what actions might be taken to make fisheries sustainable in the future. We define "sustainable fisheries" as those where human communities fish in a way that preserves critical ecosystem processes, while still maintaining socioeconomic and cultural benefits associated with fishing industries and practices. From stock assessments to reconstructed fisheries yields and the use of historically situated management methods—all tools presented in this section—there are many potential applications to fisheries sustainability of past knowledge, information, and practice.

Several crosscutting themes emerge from this section on conserving fisheries. First, these chapters focus primarily on small-scale fisheries, which are typically more data-poor than their industrialized equivalents, and account for most fishery livelihoods at the global scale. It is perhaps appropriate that the history of small-scale fisheries receive more attention, since the solutions that flow from historical analyses have much potential to improve the assessment, management, and solutions for these fisheries. For example, Friedlander, Nowlis, and Koike provide insights on how to incorporate nontraditional data sources into stock assessments, providing insights on how to better manage small-scale fisheries. Similarly, Al-Abdulrazzak, Zeller, and Pauly show how historical data can be used to reconstruct trends in catch for data-poor fisheries, and discuss what these reconstructions offer in terms of management prescriptions.

Second, there has been substantial investment around the globe in conventional, top-down fisheries management approaches, but the returns and outcomes from these approaches have been poor, with mounting evidence of fishery collapses and declines globally. By contrast, there is promise and progress in alternative approaches that engage fishers and communities. This is a growing area of research that is increasingly informed by understanding historical and customary management methods that were successful in the past. Kittinger, Cinner, Aswani, and White in this section focus on how customary management approaches that are based on historical practices can be implemented into fisheries management.

To conclude, all chapters in this section are explicitly focused on applied approaches to fisheries sustainability—or managing fisheries for continued use and human benefits. The impetus to understand the historical and contemporary roles of fisheries in global food security, sustainable livelihoods, and coastal cultures has never been greater, and historical ecology—as this section illustrates—has much to offer in illuminating pathways to sustainability.

REFERENCES

Erlandson, J. M., Graham, M. H., Bourque, B. J., et al. (2007) The kelp highway hypothesis: marine ecology, the coastal migration theory, and the peopling of the Americas. *Journal of Island and Coastal Archaeology* 2, 161–174.

Garibaldi, A., and Turner, N. (2004) Cultural keystone species: implications for ecological conservation and restoration. *Ecology and Society* 9, article 1. www.ecologyandsociety.org/vol9/iss3/art1.

Jackson, J. B. C., Kirby, M. X., Berger, W. H., et al. (2001) Historical overfishing and the recent collapse of coastal ecosystems. *Science* 293, 629–637.

Johannes, R. E. (2002) The renaissance of community-based marine resource management in Oceania. *Annual Reviews in Ecology and Systematics* 33, 317–340.

Pauly, D. (1995) Anecdotes and the shifting baseline syndrome of fisheries. *Trends in Ecology & Evolution* 10, 430.

Rick, T. C., and Erlandson, J. M. (2009) Coastal exploitation. *Science* 325, 952–953.

Rosenberg, A. A., Bolster, W. J., Alexander, K. E., et al. (2005) The history of ocean resources: modeling cod biomass using historical records. *Frontiers in Ecology and the Environment* 3, 78–84.

FIVE

Improving Fisheries Assessments Using Historical Data

Stock Status and Catch Limits

ALAN M. FRIEDLANDER, JOSHUA NOWLIS,
and HARUKO KOIKE

The health of marine fish stocks is inherently difficult to assess because catches are only partially recorded and abundance cannot be directly observed. Understanding the current status of stocks requires an estimate of what the stock is capable of producing in the absence of fishing, yet fisheries data almost never extend back to pre-exploitation states. Without catch and abundance estimates across a range of fish densities, it can be exceedingly difficult to estimate the productive capacity of a fishery, or to develop reference points to approximate this capacity. Historical data (e.g., historical records, archaeological information, geological records, ecological reconstructions, local ecological knowledge, and traditional ecological knowledge) provide unconventional opportunities to develop more realistic reference points and examine stock status prior to large-scale intensive fishing. These data can help managers avoid the pitfalls of the "shifting baseline syndrome," in which conventional stock rebuilding programs are heavily influenced by the most recent peak in productivity. Such approaches can also be productively employed in small-scale fisheries, where standard stock-assessment techniques and assumptions are not applicable and where historically based analyses can provide valid scientific advice to guide management decisions. This chapter focuses on how historical data can inform nontraditional fishery-assessment methods, using case studies from small-scale tropical fisheries, which present particularly complex assessment challenges due to the large number of species exploited, the wide variety of gear employed, and the diffuse nature of fishing locations and landing sites. A growing number of communities around the world are combining historical data and locally situated knowledge systems such as local ecological knowledge and traditional ecological knowledge to assess and manage fish stock status. By incorporating these data into population assessment models and management practices, we gain insight into the yield of these ecosystems in the past and provide guidance for future management actions.

INTRODUCTION TO STOCK ASSESSMENT MODELS AND METHODS

The main tool for gauging the health of a fish stock is the statistical stock assessment model. It fits biological parameters, including life-history traits such as size-at-age, age, sex, recruitment rates, and natural mortality, to a predefined model structure using measures of the relative abundance of fish and the quantity of fish caught over time. Models identify the biological growth and mortality required of a population to explain observed changes in abundance, given the quantity of fish removed by fishing. These growth patterns are then compared across different stock abundances to estimate the effects of fish density on productivity. Together, the growth and mortality patterns, as modulated by density, allow predictions of how the stock will respond to fishing pressure and, thus, form the basis for policy guidance, such as total allowable catch limits (Clark 1990, Quinn and Deriso 1999).

Time plays a crucial role in stock assessment models. Catch data are often used to calculate abundance parameters, and a long history of data provides insight over a greater range of conditions and potentially reduces variance in models of a stock's productive capacity and response to fishing. However, data spanning the history of a fishery are rare. Usually, data-collection programs are not enacted until a fishery has already matured and been fished well below its pristine (pre-exploitation) abundance level. This nearly universal data gap causes problems, which historical information can help address.

At first approximation, growth and mortality patterns can be described using three key reference points, including (1) the pristine abundance (B_0), typically measured as biomass that the stock may have achieved in the absence of fishing; (2) the abundance of the stock that would sustain maximum sustainable yields (B_{MSY}); and (3) the associated fishing mortality rate (F_{MSY}). Maximum sustainable yields are associated with intermediate abundance levels, often less than half of B_0. Low abundance can affect a stock's productivity through a lack of individuals to reproduce. High stock abundance can also affect overall productivity, because of density-dependent factors such as competition, disease, and predation. To understand the challenges in establishing these reference points, it is helpful to look in greater detail at how they are estimated from catch measures and abundance indices.

Data Sources and Limitations

Fisheries data are fraught with gaps and uncertainties (Haltuch et al. 2009, Gårdmark et al. 2011). Although catches are arguably the easiest component of fisheries to measure, such data carry many challenges. In fact, it is exceptionally rare that catches are fully recorded. Some fleets carry observers who record all fish caught under their watch. However, observer programs cover only a few large-boat industrial fisheries, and even in these fisheries, observers typically cover only a small percentage of all trips conducted. Large-boat industrial fisheries are often characterized by a few easily monitored landing sites that allow for dockside measurements of fish brought to port. If there are no observers, we also have to estimate at-sea discards, which can be substantial and are often underestimated (Hall and Mainprize 2005, Harrington et al. 2005). When it comes to small-boat commercial, artisanal, and recreational fisheries, even landings are difficult to monitor because of the diffuse nature of fishing and landing sites (Galluci et al. 1996, Pauly 2006). For these fisheries, we often rely on self reporting, and it is common for these reports to lump multiple species into market categories because of the difficulty of identification or the burden of reporting when many species are caught simultaneously. Even some large-boat fisheries contribute to missing catch data via illegal, unreported, and unregulated (IUU) fishing (Bray 2000).

If catch seems difficult to measure accurately, fish abundance is even harder to estimate. In some fisheries, scientists perform random scientific sampling, which provides a fishery-independent index of abundance. These data can be highly variable as a result of patchy habitats and factors such as seasonal variability in fish populations. Surveys provide better guidance when performed regularly over wide areas and with consistent methods. While many large-scale industrial fisheries have such indices that stretch back many years, they are rare in small-scale fisheries, particularly those in diverse tropical regions. Every index we are aware of started after the fisheries were already well developed or, in some cases, overexploited. When fishery-independent indices are absent or weak, we typically rely on fishery-dependent measures of abundance. To do so, we use catch per unit effort (CPUE), under the assumption that a more abundant stock will yield a higher CPUE. After standardizing effort, we typically assume that CPUE is directly proportional to abundance. Thus, measuring and standardizing effort are essential for fisheries that lack well-developed fishery-independent indices.

If ships carry observers, effort can be directly measured and categorized. Few do, but there are some techniques to estimate effort in the absence of observers. Dockside monitoring can provide data on the number of days fishing, the crew size, and the type of gear used. However, these measures are crude approximations of the actual effort expended and are particularly lacking in information to standardize effort. Visual surveys (e.g., aerial, satellite, land, and boat-based observers) can monitor fishing activity close to shore, but these are rare and expensive. An emerging technique, available only to some fisheries (typically large-scale industrial fleets), is the use of vessel-monitoring systems, which provide real-time location and movement of registered fishing vessels and sometimes include video monitoring of the catch.

However, for most fisheries, particularly small-scale and data-poor ones, we must depend on self reporting. Often, these reports are limited to days at sea, crew size, and gear employed. These limits make it challenging to standardize effort. CPUE will vary, sometimes dramatically, depending on the location and timing of fishing, and on the detailed configuration and deployment of fishing gear. Standardizing CPUE indices requires taking into account spatiotemporal patterns, which range from broad-scale (e.g., latitude and season) to fine-scale (e.g., depth, habitat, and time of day), and accounting for differences in gear type, configuration, and deployment. Without information to guide standardization, the quantification of effective effort may contain large errors and lead to inappropriate conclusions.

Recall that the ultimate goal of a stock assessment is the statistical estimation of the growth potential of the fishery. Poor data can make such estimations impossible. For example, undocumented increases in the efficiency of a fleet can lead to seemingly higher CPUE at the same time that catches are increasing. In this case, we may not be able to fit sensible parameters to a model built on the assumption that increasing catches should lead to lower stock abundance and CPUE. With all these data limitations and challenges, we often cannot directly estimate fisheries productivity and must rely on proxies for the above-mentioned reference points: unfished biomass (B_o), maximally productive biomass (B_{MSY}), and maximally productive fishing mortality rate (F_{MSY}).

Current State of Knowledge and Practice: Ongoing Uncertainty

Given the complexities of fisheries and data limitations, it is not surprising that the global status of fisheries is poorly understood. What is most striking about the state of world fisheries is how little we know. In the United States, the overfished condition, overfishing status, or both remain unknown for over half of all stocks under federal management (National Marine Fisheries Service 2012). For nonpelagic (primarily reef) vertebrate fisheries under federal management in the tropical United States, 75% are unknown (National Marine Fisheries Service 2012). Globally, the problem is almost certainly far greater because of resource and data limitations associated with fisheries (e.g., Beddington et al. 2007, Costello et al. 2012).

There are a number of reasons why stocks go unassessed. In the United States, it is common for formal assessments to be conducted only every 3–6 years because of limited resources, although most stocks are reviewed annually (National Marine Fisheries Service 2012). Elsewhere around the world, the capacity for formal assessments is more limiting. A frequent cause of unassessed stocks is a paucity of adequate conventional fisheries data. In some cases, data exist but are inadequate to inform a stock assessment model, because of either data gaps or contradictory trends. In other cases, the data simply do not exist. For example, when catches are reported by market categories rather than by species, species-level assessments are not possible. Scientists are making laudable efforts to assess mixed stock complexes (e.g., Hutchinson 2008, McClanahan et al. 2011), but these efforts are fraught with problems. In particular, multispecies assessments allow the relatively weaker stocks within a complex to suffer the brunt of the effects of fishing pressure while their decline goes undetected (Hilborn et al. 2004).

In summary, stock assessment efforts are plagued by a scarcity of conventional data and constrained by methods that were developed primarily for large-scale and relatively data-rich industrial fisheries. Most fisheries worldwide do not fit the assumptions embedded in these methods (Ruddle and Hickey 2008, Fenner 2012). To adequately assess more stocks, we need to explore the promise of unconventional data sources and develop and implement techniques to make better use of them in fisheries management practice.

ALTERNATIVES TO CURRENT PRACTICES

Many promising alternatives, based in part on greater incorporation and use of historical data, have emerged as viable alternatives to conventional stock assessment. Below, we first describe a set of complex modeling efforts, which aim to use ecological relationships as a way of inferring details of important fish populations. Some of this work has specifically focused on recreating past ecosystems by following energy flows and species interactions. Next, we review key studies that used historical information, rather than models (and assumptions), to gain a perspective on the productive potential of fish stocks. Third, we explore the information that can be gleaned from traditional ecological knowledge and local ecological knowledge, because fishing communities often know a great deal about the ecosystems that sustain them. We then look at geological and archaeological evidence for estimating historical abundance, showing how these records allow us to assess the development of fisheries and characterize natural cycles in fishery populations. Next, we discuss the option of using the biomass of pristine unfished areas as a reference point, using a space-for-time substitution approach. Finally, we examine data-limited management and review historical management practices, illustrating how fisheries were sustained for centuries using some simple techniques.

Simulation Modeling

Rebuilding fish stocks implies reconstructing elements of past ecosystems. A trophic mass-balance model, ECOPATH (Polovina 1984), and two derived dynamic simulations, ECOSIM (Walters et al. 1997) and ECOSPACE (Walters et al. 1998, 1999), are some of the most widely used tools to model past ecosystems. Using data on fisheries catch by sector, production-to-biomass ratios, consumption rates, and a diet matrix for up to 50 defined components of an ecosystem, ECOPATH tallies the flows of matter within the components of a system, defines trophic levels, and can be used to estimate biomass per trophic level, given diet, mortality, and consumption rates (Christensen and Pauly 1992, 1993). ECOSIM evaluates the impact of changes in fishing rates selectively across gear types, and investigators can tune the model to defined time series for biomass estimates (Christensen and Walters 2004). ECOSPACE allows investigators to engage in spatial ecosystem modeling by replicating the ECOSIM simulations across a grid of habitat cells (Walters et al. 1999).

An interesting reconstruction process that utilizes this EcoPath simulation package is "Back to the Future" (BTF), which employs traditional ecological knowledge and local ecological knowledge, historical documentation, and archaeology to facilitate ecological modeling of past systems and uses these states to help provide policy goals for the future (Pitcher 2001, 2005). For example, a reconstruction of the Strait of Georgia, British Columbia, marine ecosystem was conducted for (1) the present day, (2) 100 yr BP, and (3) 500 yr BP (Pitcher 1998). Results from this work highlight the enormous changes in abundance, size, and composition of fisheries populations over a 500-year period and show how a long-term approach is essential for determining the natural productivity of an ecosystem (Dalsgaard et al. 1998). This methodology has been applied to a variety of ecosystems in locations ranging from Hong Kong (Buchary et al. 2003) to Newfoundland (Pitcher et al. 2002) and northern British Columbia (Ainsworth et al. 2002), all with similar outcomes, revealing evidence of dramatic declines in biomass and shifts to lower trophic levels across these diverse systems.

Although complex models are tempting as a means to generate reference points, particularly historical ones, we nevertheless must use caution. ECOPATH and other complex models rely on data to inform them, just like conventional fisheries models. Unlike conventional models, ECOPATH incorporates ecological interactions and thus adds complexity, which requires additional data needs. Although we sometimes have some information about the type, strength, and direction of ecological interactions within a food web, this information is frequently missing. Researchers can nevertheless fill in data gaps via assumptions. Typically, a modeler will examine the behavior of their modeled ecosystem and then tweak these assumptions until the model's behavior seems plausible, a process referred to as "tuning." As a reality check, we must remember that tuning is subjective and driven by assumptions rather than observation. Thus, we recommend using complex models as a tool for generating testable hypotheses and focusing on empirical evidence to learn about the historical capacity of ecosystems. Fortunately, we have a growing toolbox for doing so.

Historical Catch Records

Historical records give us a rare opportunity to look into the past without complicated methods or dangerous assumptions. For some fisheries, it is possible to use historical sources to examine trends over long periods (e.g., decades to centuries) and estimate biomass in the early stages of fisheries exploitation. By taking a long-term view, we can provide more realistic insights into the past productivity of these ecosystems, not just recently observed catch levels (Rosenberg et al. 2005). A shortcoming of this approach is that it can be difficult to find records that date back to the early stages of a fishery, let alone enough reliable data to reconstruct catches. Despite this limitation, there are numerous examples of the utility of historical catch records in estimating past stock abundances.

Case 1: Cod Fishery in the Western North Atlantic

Commercial fishing for cod in the western north Atlantic dates back to the 1500s, when Basque fishermen discovered the Grand Banks off Newfoundland (Kurlansky 1997). Over the next 300 years, the fishing industry thrived, with many periods of low and high production. Using detailed catch logs from cod fishing schooners based in Beverly, Massachusetts, from 1852 to 1859, Rosenberg et al. (2005) were able to reconstruct biomass of cod for the Scotian Shelf, Canada, and compare these results to present-day estimates of cod standing stock for the same area.

Estimates of adult biomass of cod from 1852 (~1.3 million mt) are 96% higher than biomass estimates by Canada's Department of Fisheries and Oceans in the 1980s, and three orders of magnitude greater than biomass estimates from 2002 (Rosenberg et al. 2005). The Scotian Shelf was heavily fished in the 1850s by well over a thousand fishing schooners, so the estimates of biomass are far from pristine, or pre-exploitation. Current fisheries management policies for the region use the 1980s biomass estimate as a target for stock rebuilding despite the fact that these values represent 4% of the biomass estimates from the 1850s, which in itself is likely very distant from unfished biomass (Rosenberg et al. 2005). Using these historical logs provides a more accurate assessment of cod stocks in the past and offers a more realistic benchmark for future management.

Case 2: Hawaiian Archipelago

Ancient Hawai'i had a long history of sustainable subsistence fishing (Kirch 1982, Kittinger et al. 2011), followed by a shift to commercial fishery after Western contact (after AD 1778; Schug 2001). By 1900, commercial fisheries had become a dominant feature in local island economies, with fish markets established on each of the main Hawaiian Islands (McClenachan and Kittinger 2012).

As a result of their importance to the Hawaiian economy, major quantitative surveys of the commercial fisheries were conducted in 1900 and 1903 by the U.S. Fish Commission (Cobb 1902, 1905a, 1905b), followed by data collection by the Territory of Hawai'i in the 1920s and 1930s, with continuous data collection by the territory, and then by the state of Hawaii, since 1948. Records of landings by trip began in 1966, enabling calculations of CPUE starting at this point.

Commercial landings for a number of important species have shown dramatic declines since the early 1900s, with most recent landings at <1% of maximum landings (Figure 5.1). The character of Hawaii's commercial fisheries has changed dramatically over the past 100 years, so trends in total landings may be driven by a number of factors other than fish abundance (e.g., transition to a tourism-based economy, other economic opportunities). Therefore, it is illustrative to look at trends in catch rates, which are more closely correlated with stock abundance than are total landings. Improvements in boats, engines, and fishing technology over time have resulted in large increases in fishing-gear efficiency. We used a conservative estimate of a 2% increase in fishing efficiency per year and examined $CPUE_{now}/CPUE_{max}$ to determine current catch rates compared with maximum catch rates. Because data by trip began in 1966, decades after the onset of commercial fishing, our ratio is a conservative estimate, given that $CPUE_{max}$ does not likely represent an unfished, or even lightly fished, condition. For nearly all species and all gear types examined, current CPUE values are generally well below 1% of historical highs (Table 5.1). A few exceptions include the handline fishery for soldierfishes (Holocentridae), where CPUE is currently 15% of $CPUE_{max}$; and the lay gill net fishery for rudderfishes (Kyphosidae), with current CPUE at 9.5% of $CPUE_{max}$. These ratios are still extremely low compared to historical values and suggest substantial declines in stock size.

Case 3: Florida and Cuba Grouper Fisheries

The histories of the fisheries of Cuba and Florida are intertwined and date back to pre-Columbian times. Prior to 1955, insular fisheries on the Cuban shelf consisted mostly of small-scale artisanal fisheries with catch never exceeding 10,000 t annually (Claro et al. 2001). By the mid-1980s, however, increases in fishing effort and efficiency resulted in landings of 79,000 t, but signs of overfishing were evident for a number of valuable target

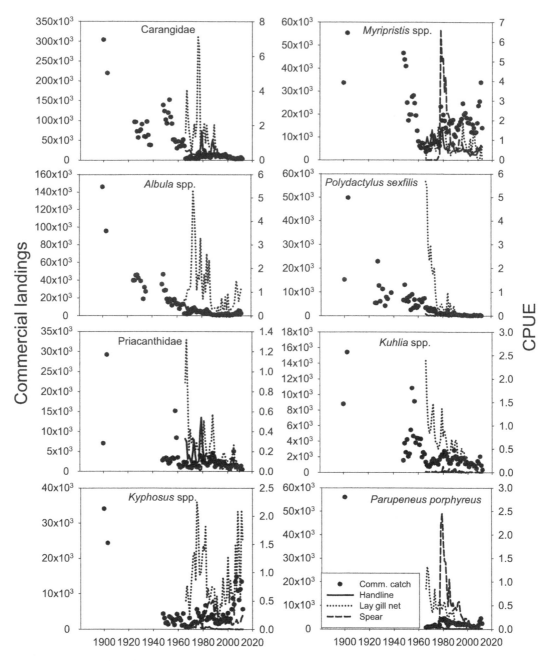

FIGURE 5.1 Total landings (kg) and catch per unit effort (CPUE; kg trip⁻¹) by gear type from Hawaii commercial fisheries catch records. Major quantitative surveys of the commercial fisheries were conducted in 1900 and 1903 by the U.S. Fish Commission (Cobb 1902, 1905a, 1905b), followed by data collection by the Territory of Hawai'i in the 1920s and 1930s, with continuous data collection by the territory, and then the state of Hawaii, since 1948. Records of landings by trip began in 1966 (Hawaii Division of Aquatic Resources data), thus enabling calculations of CPUE starting at this point.

TABLE 5.1 Hawaii Commercial Fish Landing and Catch Rates
CPUE = catch per unit effort (kg trip⁻¹)

Taxa	Statistics	Handline CPUE	Lay Gill Net CPUE	Spear CPUE	Commercial Catch (kg)
Jacks (Carangidae)	Maximum	1.7	7.1	1.2	303,848
	Current % of maximum	<1.0	<1.0	<1.0	<1.0
Bonefishes (*Albula* spp.)	Maximum	0.50	5.33	0.04	145,886
	Current % of maximum	<1.0	2.1	1.5	<1.0
Soldierfishes (Holocentridae)	Maximum	1.65	1.89	6.64	55,316
	Current % of maximum	15.0	1.6	1.9	6.3
Threadfin (*Polydactylus sexfilis*)	Maximum	0.08	5.67	0.02	49,768
	Current % of maximum	<1.0	<1.0	2.5	<1.0
Whitesaddle goatfish (*Parupeneus porphyreus*)	Maximum	0.11	1.32	2.45	43,820
	Current % of maximum	<1.0	<1.0	1.6	1.4
Rudderfishes (Kyphosidae)	Maximum	0.15	2.25	0.51	34,047
	Current % of maximum	<1.0	9.5	3.3	1.3
Bigeyes (Priacanthidae)	Maximum	0.54	1.32	0.18	29,211
	Current % of maximum	3.5	<1.0	3.6	1.6
Flagtail (*Kuhlia* spp.)	Maximum	0.04	2.39	0.12	15,403
	Current % of maximum	<1.0	<1.0	1.6	2.2

Notes: Landings data begin in 1900. Catch rates begin in 1966, when data began being reported by individual trip.

species despite the introduction of significant management measures (Claro et al. 2001, 2009). The government-owned and -managed fishing industry in Cuba provides a uniquely detailed multidecadal database of fisheries landings that is a valuable tool for assessing trends in a wide variety of species and gear types (Claro et al. 2009).

Following European contact, commercial fisheries developed in the Florida Keys because of their proximity to the mainland North American settlements and northern Caribbean islands. Recreational fisheries commenced in the mid-1800s, and over the past 4 decades there has been a dramatic increase in the amount of recreational fishing pressure, with the number of registered vessels quadrupling during this period (Ault et al. 1998).

Nassau grouper (*Epinephelus striatus*) was once an important component of Cuba's near-shore fisheries, reaching 1,728 t in 1963, but has declined to <2% of that peak in recent years (Claro et al. 2009; Figure 5.2). Examination of the headboat and commercial catch of Nassau grouper in south Florida since the 1980s shows a similar proportional decline to that of the Cuba data, irrespective of the two-orders-of-magnitude overall difference in catch (Bohnsack 2003, NOAA unpublished data). These similarities in more recent trends suggest that Nassau grouper populations may have been much larger in south Florida in the past. This

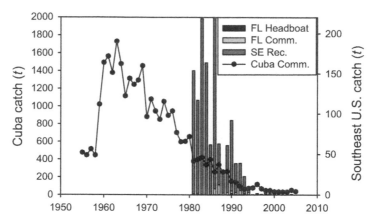

FIGURE 5.2 Historical Nassau grouper (*Epinephelus striatus*) landings for Cuba and south Florida. Commercial landings from Cuba date back to 1958 (Claro et al. 2001, 2009). Data from Florida come from commercial landing data (Fish and Wildlife Research Institute; http://myfwc.com/research/saltwater/fishstats/commercial-fisheries/landings-in-florida/) and recreational headboats (Bohnsack 2003, National Marine Fisheries Service 2012) starting in the mid-1980s.

case, as with the cod fisheries in the northeast Atlantic and Hawaiian coral reef fisheries, highlights the "shifting baseline syndrome" (Pauly 1995), which results when there is inadequate long-term data prior to full exploitation, or overexploitation, of a fishery (Bohnsack 2003).

Catch Reconstructions

Nontraditional sources (e.g., fishers' interviews, maritime records, unpublished reports, etc.) can be used to reconstruct past landings in the absence of catch data (Pauly et al. 1998, Zeller et al. 2007; also see chapter 6, this volume). These diverse data sets can be used to estimate historical catch rates, which, combined with demographic information, can be used to estimate total catch (Zeller et al. 2006, 2008). For example, reconstruction of the fishery in American Samoa showed a nearly 80% decline in overall coral reef catches since the 1950s, revealing long-term overfishing of nearshore resources, which can go undocumented in the absence of formal fisheries catch data (Zeller et al. 2006). These reconstructions have been performed for numerous locations around the world (see chapter 6, this volume) and highlight the limitations of fisheries catch statistics and the need to take a long-term perspective when it comes to estimating potential stock productivity. However, catch reconstructions can also pose challenges for fisheries managers. As noted in Box 5.1, catch reconstructions can be less important to managers than more pressing challenges, such as protecting and restoring critical fisheries habitats.

Catch reconstructions can help establish reference points to estimate the full productivity of unfished marine ecosystems. For example, a comparison of reconstructed yields from Hawaiʻi and Florida provides insight into precontact productivity for these two regions.

The "shifting baseline syndrome" may influence the frame of reference in rebuilding targets for depleted fisheries, but how is this significant? While historical evidence of an abundant "unfished" biomass of a presently smaller stock may speak volumes about failures of management and long-term changes in fishery ecosystems, it is not necessarily an appropriate reference point for present-day managers working to balance stock rebuilding with fisheries yield. Fishing mortality, loss or damage to habitat, shifts in trophic dynamics, and impacts of a changing climate might individually or collectively preclude rebuilding a stock to its unfished level. Even when there are adequate data to support a historical reconstruction of stock size, the indelible impacts of human activity and natural shifts within the ecosystem over decades, and in some cases centuries, often place the past and present into differing contexts.

In recommending harvest strategies that are compliant with the 10 national standards of U.S. federal fisheries law (the Magnuson-Stevens Fishery Conservation and Management Act) and the implementation guidelines for those standards, I am much more interested in current estimates of maximum sustainable yield, and in the trajectory of the stock in relation to that metric. That said, there are many data-poor scenarios in which traditional and local ecological knowledge (TEK and LEK) may prove to be valuable management inputs. This is particularly the case when such data can inform the identification of sustainable fishery mortality rates, which may, in many cases, prove more effective than biomass reference points. Understanding the role and vulnerability of the species within a dynamic ecosystem enables managers to be responsive and proactive within current natural states, and TEK and LEK may provide valuable insight into these dynamics.

TEK and LEK can, and should, play a significant role in achieving these goals. Where conventional assessment data are lacking, TEK and LEK can be invaluable tools for estimating present abundance of stocks as well as current trends within the ecosystem. Even in data-rich scenarios, TEK and LEK may assist scientists and managers in validating conclusions and in responding to observed changes within the ecosystem. Generally speaking, I do not consider the best use of TEK and LEK to be the reconstruction of historical baselines, but rather to assist in understanding the current ecosystem and stock dynamics to support the achievement of sustainable yield going forward. Looking beyond the potential value of TEK and LEK in establishing biological reference points, these tools can also inform managers about sustainable fishing practices and allocation methods in contemporary small-scale fisheries.

John Henderschedt is Executive Director of the Fisheries Leadership and Sustainability Forum and Vice Chair of the North Pacific Fishery Management Council.

Despite markedly different exploitation histories, the modern levels of extraction on Florida (12–13 t km^{-2}) and Hawaiian (10–12 t km^{-2}) coral reefs are similar (McClenachan and Kittinger 2012). Based on population data and potential per capita catch rates, the total reconstructed yield for wild-caught coral reef fisheries in Hawai'i achieved a maximum in the mid-1400s (>17 mt km^{-2}; McClenachan and Kittinger 2012). Precontact fisheries' reconstructed yields ranged from 12 to 17 mt km^{-2} for nearly 400 years, suggesting sustainable rates of fishing mortality over that period. In Florida, by contrast, total catch remained <5 t km^{-2} until 1930 (McClenachan and Kittinger 2012). Before European settlement, catch was well below 1 t km^{-2} and slowly increased until the second half of the 20th century, when rates increased rapidly, exceeding 20 t km^{-2} for much of the 1980s and 1990s. Reductions in overall landings occurred since the mid-1990s, with a decline of 50% between 1996 and 2008. Historical reconstructions help establish more accurate reference points and need to be considered when developing contemporary fisheries policies.

Traditional and Local Ecological Knowledge

Traditional ecological knowledge (TEK) is defined as "a cumulative body of knowledge, practice and belief, evolving by adaptive processes and handed down through generations by cultural transmission, about the relationships of living beings (including humans) with one another and with their environments" (Berkes 2008). In the Pacific, island cultures have depended on the sea as their primary source of food for millennia, and along the way they invented nearly all of the basic fisheries conservation measures that we have in place today (e.g., closed areas, closed seasons, size limits, and restricted entry; Johannes 1998). One key management measure was the recognition of property rights by local communities, with local chiefs able to enact fishery management policies knowing that the future benefits of present sacrifices would benefit their own community. Area and seasonal closures were common, particularly when the chief felt that a stock was overfished (Johannes 1978, 1981). Areas were also left in reserve in anticipation of future needs, such as closing a particular area to accumulate fish to be caught for upcoming ritual feasts, as well as unanticipated ones, such as closing calm inshore areas to be used only during extended periods of rough weather (Titcomb 1972). Additionally, some fish were purposely allowed to escape, with the intention that they would serve to repopulate the stock.

In Hawai'i, Polynesian cultures developed a lunar calendar, which encompassed a detailed understanding of the marine environment and was used to help regulate fishing effort and timing (Poepoe et al. 2007). The moon calendar emphasized certain repetitive biological and ecological processes (e.g., fish spawning, aggregation, and feeding habits), which function at different time scales (e.g., seasonal, monthly, and daily; Friedlander et al. 2013). These practices were based on detailed TEK of these marine systems and allowed these cultures to maintain resilient and responsive management practices that reduced pressures at signs of overfishing and allowed heavier fishing for healthy stocks.

Similarly, the native inhabitants of the Lower Klamath River basin, in coastal northern California, relied on salmon for the bulk of their dietary protein and developed a complex system of legal rights and religious observations to maintain these stocks (McEvoy 1986). Fish were primarily caught in large communal weirs (funneling fish traps) that had strict requirements for construction and dismantling based on ritual and religious beliefs (Swezey and Heizer 1977). Construction and blessing of the weir took 10 days, during which time salmon escaped upstream. The weir was dismantled 10 days later, allowing additional escapement. In this way, escaped salmon could swim to upstream spawning grounds before and after the use of the weir. In good years, catches would be moderated by escapements, which indirectly limited human population growth. In bad years, catches would be low but fish would still be allowed to escape and maintain the viability of the fishery.

In addition to TEK, contemporary local ecological knowledge (LEK) can be used to help inform the status of fish stocks, particularly when other forms of data are limited or absent (Neis et al. 1999). For example, Australia's southeast region is one of the most important fishing areas in the country (Bax and Williams 2001). With fishers' input, scientists were able to identify fisheries-independent survey sites that are important for these fisheries without the intense effort usually associated with mapping projects (Williams and Bax 2007). Another example where critical habitat was identified through fishers' knowledge was the cod spawning grounds in New England. Prior to a fisher-based spawning-ground study, very few spawning locations were known and researchers had difficulty determining the basic life history of these local stocks without the knowledge of fishers (E. P. Ames 2004, T. Ames 2007).

One recently developed approach is Participatory Fisheries Stock Assessment (ParFish), which aims to obtain information on stock condition in situations where data are inadequate for a conventional assessment (Walmsley et al. 2005). This multicriterion decision-making methodology uses interviews of fishers to identify stakeholder preference among various management outcomes and to create a preliminary estimate of stock status. It utilizes Bayesian decision analysis, with uncertainty in the results shown as probability density functions that can be broken down into simpler components, thus making multispecies assessments more viable. ParFish methodology has been developed and tested through a number of pilot studies conducted on various fisheries throughout the world, including in the Caribbean, East Africa, and India.

Research on TEK and LEK systems has increased dramatically over the past several decades. Customary fishery management practices based on TEK and LEK are being implemented in policy in many places worldwide and are increasingly integrated with conventional management approaches (see chapter 7, this volume). Managers are also increasingly engaging with TEK and LEK systems, particularly when conventional fishery data are lacking (Box 5.1). Such approaches allow researchers and practitioners to engage productively with fishers, who interact with the resources on a daily basis and are intimate with the status of many fish stocks. Inclusion of TEK and LEK provides further insight into effective management systems, which in many cases were successful at maintaining sustainable fisheries. Additionally, it is important to match the scales of management to those of the community that engages with fishery resources. (See Box 5.2, which discusses the importance of matching scales and how incorporation of LEK and TEK can help aid decision-making processes.)

BOX 5.2   Viewpoint from a Practitioner: The Importance of Matching Scales
in Fisheries Comanagement

Dean Wendt

It is my opinion that any sustainable management system has to be at a scale that matches
people's inherent connection with the ecosystem and the services it provides them. People
will only conserve and protect that with which
they identify. Fisheries management in the
United States occurs on a geographic scale that
is so large that it misses the inherent social and
ecological heterogeneity of fishing communities and fish stocks, thereby weakening the critical connection between people and the ecological resources on which they rely. Recognition of
the mismatch in social and ecological scales and
our fisheries management system is creating
momentum in the United States for a more bottom-up, comanagement approach whereby
communities engage with authority in the management of their local resource.

In addition, the policy change in 1998 of the
Magnuson-Stevens Fishery Conservation and
Management Act states that federal fisheries
management councils should adopt a precautionary approach to specifying optimum yield
of a stock. The federal government has essentially reversed the burden of proof, and we now

implement significant conservation measures
even in the absence of scientific evidence that a
stock is being overexploited. Put another way,
without good data on the status of a fish stock,
the "Restrepo rule" dictates that historical
catches be reduced significantly (Restrepo
et al. 1998).

As Friedlander and colleagues show in this
chapter, such shifts in management and policy
necessarily require the development and use of
new tools and methods in the assessment, allocation, and utilization of a broad diversity of data
sources. They also incentivize fishermen and
fishing communities to engage in formal data
collection for management and to contribute
their knowledge to the management process.
Management at smaller, more appropriate social
and ecological scales using a precautionary
approach will require incorporation of local and
traditional ecological knowledge into decision-
making processes. If communities are given a
viable way to incorporate their knowledge and
collaborate with government, I'm confident that
decisions will be made that yield the best solu-
tion for sustained social and ecological benefits.

Dean Wendt is Dean of Research, California Polytechnic State University.

Paleoecological and Archaeological Evidence

Paleoecological evidence has also been used to examine historical changes in fishery species. Since most fishery species have some form of calcified body parts, changes in catch can
be inferred through shifts in abundance and sizes in archaeological deposits and sediment
samples, which may date back thousands of years. This paleoecological evidence offers
important insight into both human harvesting behaviors and natural cycles of abundance
and can sometimes be used to elucidate whether recent changes in resource abundance
might be natural, rather than human-induced, cycles. For example, a fish population may
show random variation among years but may demonstrate large oscillations that occur at
scales of decades to centuries (Finney et al. 2010, Valdés et al. 2008). If we step back even

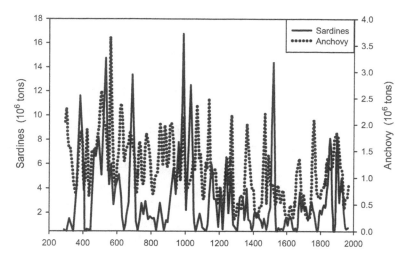

FIGURE 5.3 Two-thousand-year reconstruction of sardine and anchovy populations based on scale deposits in the Santa Barbara Channel, California (modified from Baumgartner et al. 1992).

farther and look at evolutionary time scales of millions of years, we may see population changes due to shifts in local, regional, and even global climatic conditions (Jackson 2010).

Furthermore, by unearthing paleoecological evidence, we can place recent population changes into a longer-term historical context. For example, Baumgartner et al. (1992) counted the sardine- and anchovy-scale deposits in sediment from the fishing ground off the coast of California and calculated the biomass of these two fished species for the past 2,000 years (Figure 5.3). The reconstructed sardine and anchovy stocks showed repeated fluctuation over the interval of several decades or longer, associated with cycles in oceanic conditions (Jacobson and MacCall 1995). Sardines were the subject of intense fishing pressure throughout the early 1900s, but the fishery collapsed in the 1950s. The collapse led to a moratorium on sardine fishing in 1967, by which time stocks along the west coast of the United States had already collapsed (Radovich 1982). Both the intensive pressure and the moratorium were policies that did not reflect the cyclical nature of this fishery. Starting in 1980, sardine populations started to show recovery, and the fishery is currently under much stricter regulations, which limit catches to 25% of estimated maximum sustainable yields. This policy was crafted with a desire to moderate annual fluctuations in allowable catch levels while considering production cycles and the potential for a future stock collapse. However, these estimates continue to generate controversy (e.g., Sugihara et al. 2012).

Zooarchaeological remains can also tell us a great deal about past fish stocks over extremely long time series, although the temporal resolution is rather coarse, usually on the order of decades to centuries (Erlandson and Rick 2010). Using species remains from archaeological deposits, a number of researchers have demonstrated reductions in the size of fish caught through time (Amorosi et al. 1994, Jackson et al. 2001), changes in fish growth rates (Van Neer et al. 2002), changes in the genetic diversity of populations (Larson et al. 2002), and changes in species composition (Butler 2001), all of which can yield information about fishing intensity. Prior to European contact, faunal assemblages recovered from midden deposits in a number of Pacific Islands show overall declines in both catch and effort through time, suggesting overfishing, changes in agriculture or other subsistence practices, or a combination of the two (Erlandson and Rick 2010, Nagaoka 2001).

Wing and Wing (2001) studied the faunal remains from archaeological sites on five Caribbean islands, each with an early (1,850–1,280 yr BP) and late (1,415–560 yr BP) occupation. On each island, the mean size of reef-obligate species (e.g., parrotfishes, surgeonfishes, snappers, and groupers) showed large declines while facultative species such as jacks and herring showed little change in size. The authors also found a sharp decline in total reef fish biomass and mean trophic level from early to late occupation, suggesting heavy exploitation even in prehistoric times. These results are consistent with modern patterns of overexploitation and suggest that growth overfishing and fishing down the food web occurred long before European contact.

In several well-studied midden sites in Hawai'i, fish remains shift over time from a predominance of carnivorous to herbivorous reef fishes (Kittinger et al. 2011), suggesting early examples of fishing down the food web (Pauly et al. 1998). However, a decreased reliance on marine protein as a result of increased animal husbandry and sophisticated resource-management systems resulted in several hundred years of stable harvest levels. For example, modest increases in the size of parrotfish bones and limpet shells suggest release of these populations from exploitation pressure during the development of an agrarian society (about AD 1400–1778+; Kittinger et al. 2011). Zooarchaeological remains can therefore tell us a great deal about long-term historical changes in catch composition and mortality rates and can serve as proxies for estimates of abundance.

Another way to study archaeological materials is through ancient DNA. Recent improvements in molecular techniques to recover genetic material have allowed scientists to compare genetic diversity and population structures of current fish stocks to their historical state. Studies of herring in British Columbia (Speller et al. 2012), North Sea cod (Hutchinson et al. 2003), and snappers in New Zealand (Hauser et al. 2002) have shown large reductions in genetic diversity since the onset of each of these fisheries, while examination of Plaice DNA from the 1920s for the North Sea and Iceland show an effective population size five orders of magnitude smaller than the estimated population size today, with significant heterozygote deficiencies that coincide with increased fishing mortality after World War II (Hoarau et al. 2005).

Using Unfished Reference Areas

"Space-for-time" substitution has been used in many instances as an alternative to long-term studies to assess the impact of human-induced changes where pre-impact records are sparse or nonexistent (Pickett 1989). Surveys of remote coral reefs in the Pacific (Friedlander and DeMartini 2002, Sandin et al. 2008, Williams et al. 2008) support historical reports of high fish abundance and predator domination that characterized coral reefs before extensive fishing occurred. These areas therefore give us a window into the past as to what reefs looked like prior to human extraction and provide baselines for comparisons with more exploited locations (Knowlton and Jackson 2008).

Biomass estimates from unfished areas have recently been used as a substitute for pristine, unfished estimates in fishery stock-assessment approaches (Babcock and MacCall 2011, McClanahan et al. 2011). The ratio of fish density outside versus inside unfished areas can be used as a proxy for biomass depletion (B/B_{target}) in fisheries, thus eliminating the need for a stock assessment to estimate depletion. This approach is also advantageous in that it requires no historical data, and because in some places older, large marine reserves exist, which can be used as rough approximations of unfished biomass (Babcock and MacCall 2011). For example, in the western Indian Ocean, McClanahan et al. (2011) used unfished reference areas and the oldest no-take marine parks in the region, estimating the unfished reef fish biomass (B_0) at ≈1,200 kg ha^{-1}.

There are two things to note when conducting or interpreting these types of analyses. The first is that no-fishing areas (e.g., marine reserves) used for analyses must be large when approximating unfished conditions. Most no-fishing reserves are small and likely inadequate, at least for mobile predators. The second point of caution is that a large unfished area may show strong biogeographic gradients among species, therefore obscuring potential responses due to fishing. Large unfished areas with the potential to serve as robust reference sites include the Chagos Marine Protected Area–Indian Ocean (640,000 km²), the Phoenix Islands Protected Area–Kiribati (408,250 km²), the Papahānaumokuākea Marine National Monument (362,073 km²), and no-take areas within the Great Barrier Reef Marine Park (113,982 km²).

A total of 57 fish species in Hawai'i were assessed by comparing biomass within the populated main Hawaiian Islands to the remote and virtually unfished northwestern Hawaiian Islands (Papahānaumokuākea Marine National Monument). Based on this assessment technique, one-quarter of the species examined in the main Hawaiian Islands were depleted below 10% of unfished abundance, while close to half were below 25% of unfished abundance. Large predators were especially affected, but many other target and nontarget species also appeared to be depleted. This study highlighted the value of large unfished areas as reference points for fisheries management and contrasted with previous works, which identified no-fishing areas as impediments to assessing stocks because their effects can complicate the interpretation of conventional fisheries data (Punt and Methot 2004, Field et al. 2006). However, this study also showed that small and sparse no-fishing areas in the MHI were inadequate to reestablish the full biological potential of many species and did not represent adequate reference areas.

A number of stock assessment parameters can be estimated with the help of unfished reference areas. For example, because the northwestern Hawaiian Island (NWHI) fish populations experience little or no fishing pressure, all mortality is considered to be natural mortality (M), whereas the MHI populations experience both natural and fishing mortality (F). Size frequency analysis of the blue trevally (*Caranx melampygus*), a highly prized recreational species, in the unfished NWHI produced an estimate of $M = 0.27$ and an estimate of total mortality ($Z = M + F$) in the MHI of 0.69, which can be used to calculate $F = 0.42$ (Figure 5.4A, B; Fried-

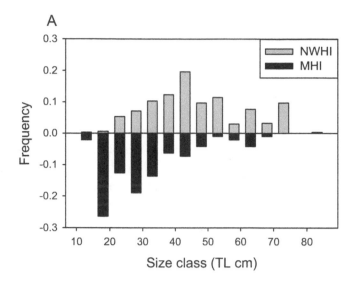

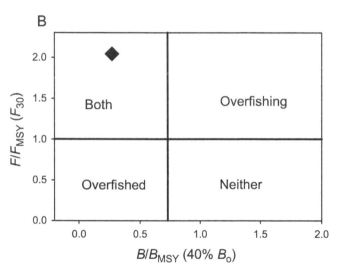

FIGURE 5.4 Length-based assessments of blue trevally (*Caranx melampygus*) in the main Hawaiian Islands (MHI) using the northwestern Hawaiian Islands (NWHI) as an unfished reference area. (A) Comparison of length frequencies of blue trevally in the MHI and NWHI (TL = total length). (B) Estimates of actual biomass (*B*), $B_{MSY} = 40\%\ B_o$, fishing mortality rate (*F*), and $F_{MSY} = F_{30}$ in the MHI. See text for details.

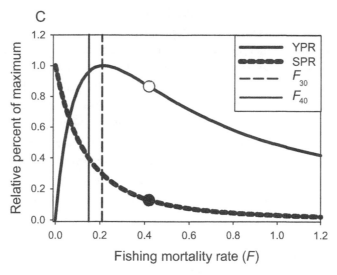

FIGURE 5.4 *(continued)* (C) Spawning potential ratio (SPR) and yield per recruit (YPR) models for *C. melampygus* in the MHI. See text for details.

lander et al. 2008). It is common to set fishing mortality-rate limits that allow individuals within the population to produce 30–40% of its reproductive potential in the absence of fishing (F_{30} to F_{40}; National Marine Fisheries Service 2012). For this species, F_{30} was estimated at 0.21 and F_{40} at 0.15, which suggests that recent fishing rates were 2 to 3 times higher than a reasonable fishing limit. The spawning-potential ratio calculation in the study indicated that blue trevally in MHI are currently producing only 13% of their reproductive potential (Figure 5.4C). These results are consistent with analyses of the relative biomass densities of this species in the MHI and NWHI that indicated that the MHI population may have dropped to 2% of its unfished abundance (Friedlander and DeMartini 2002).

APPLICATION OF REFERENCE POINTS USING UNCONVENTIONAL MANAGEMENT TECHNIQUES

Effective fisheries management systems can be developed even if historical analyses provide only a few reference points (see Box 5.2). Fisheries management systems perform well, even with limited data, if they identify a target abundance level, allow reasonable levels of fishing when stocks are healthy, and reduce fishing effort decisively when stocks drop below target levels (e.g., Restrepo et al. 1998, Sladek Nowlis and Bollermann 2002, MacCall 2009). Simulations have demonstrated the potential of these techniques to yield fairly accurate catch limits (MacCall 2009), sustain high average catches (Sladek Nowlis and Bollermann 2002), and achieve a desired balance among several competing objectives (Sladek Nowlis 2004), even when the techniques rely solely on roughly estimated reference points.

To illustrate these capabilities, let's consider the dynamics of a fish population in terms of additions and removals as functions of population size. For additions, let's focus on natural productive capacity, which includes growth and reproduction as mediated by natural mortality. We expect this natural productive capacity to be zero when a population is extinct, for lack of seedstock; and when a population is highly abundant, for lack of resources (Figure 5.5, parabolic curves). For removals, we have fish catches, which are influenced by the fishery management system. A given system may result in a point where additions balance subtractions (Figure 5.5A), associated with a target catch and target abundance. These are the sorts of reference points that can be informed by historical studies but also need to reflect societal objectives. We can choose targets (Figure 5.5B), recognizing that higher catches lead to lower abundance. We can also choose how responsive our systems will be to changes in abundance both below (Figure 5.5C) and above target levels (Figure 5.5D). Responsiveness increases sustainability (Sladek Nowlis and Bollermann 2002), which may be especially important for data-poor fisheries, but decreases the predictability of catches and, consequently, economic returns (Sladek Nowlis 2004). Annual catch-limit rules are commonly used as part of such a system, but we can also use size limits and closed areas (Sladek Nowlis and Bollermann 2002). Ultimately, the best policy should reflect a balance among competing objectives based on societal needs (Sladek Nowlis 2004). Viewing fishery management systems in this manner highlights ways in which approaches can be developed for data-poor stocks, which are often overlooked. When only catch data exist, Restrepo et al. (1998) suggested estimating sustainable catch levels using average recent catches for fisheries with a suitably long history of catch records, as long as there is no evidence of stock decline (also see Box 5.2). MacCall (2009) recommended accounting for the stock depletion expected with the development of a new fishery with a fairly simple formula that relies on catch history, a proxy for $B_{MSY}/B_0 = 0.4$ and a proxy for $F_{MSY} = cM$, where c is a tuning parameter whose value would be chosen by expert judgment, often somewhere in the range of 0.6 to 0.8; M would also be estimated by expert judgment, aided by records of oldest observed fish where available. The final variable necessary for this technique is an estimate of the degree to which a stock has been depleted over the history of the fishery. Absent a fishery-independent or CPUE index of abundance, such information might be collected anecdotally from experienced fishermen.

Interestingly, the principles behind these data-limited management systems are consistent with evidence we have from TEK. Not all indigenous groups managed their fishery resources sustainably (e.g., Diamond 2005). Examples where they did (e.g., the previously discussed Hawaiian Island and Lower Klamath River societies), though, show strong evidence for adaptive management approaches where adjustments were made to catches at early signs of depletion. These systems are being described with increasing intensity by researchers (see chapter 7, this volume). Some related concepts have recently been proposed or even implemented, partly in response to tightening standards that address the need for annual catch limits in federally managed U.S. fisheries.

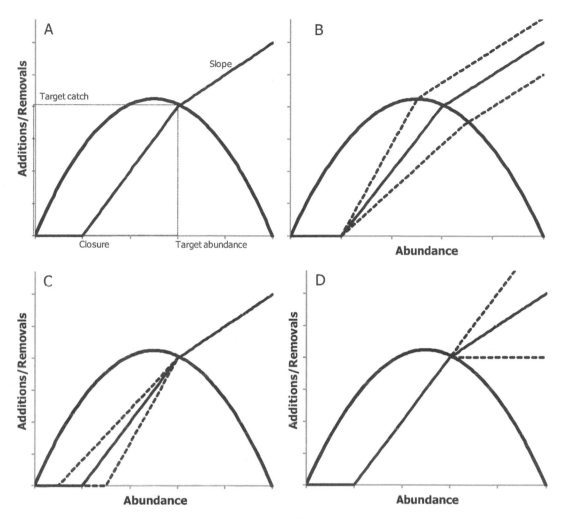

FIGURE 5.5 Data-poor management in theory. Fisheries can be characterized by abundance-based additions (in the form of natural productive capacity; the parabolas) and removals (in the form of catch limits; kinked lines, both dotted and solid), the intersection of which represents target catch and abundance (A). Fisheries productivity and catch limits can be paired to achieve desired results, subject to fundamental trade-offs. We can vary the targets (B) and the responsiveness of the management system to abundance changes below (C) or above (D) target levels. (Details adapted from Sladek Nowlis and Bollermann 2002, Sladek Nowlis 2004.)

CONCLUSIONS

The global fisheries crisis, combined with ever increasing exploitation pressures on these resources, requires us to dramatically change the way we interact with and manage our marine ecosystems. This includes taking into account the important ecological roles of fishery species in maintaining ecosystem resilience (for an example from the Caribbean, see

BOX 5.3   Viewpoint from Practitioners: Historical Baselines and Fishery
Management—Parrotfish and Caribbean Corals

Steve Roady and Andrea Treece

Elkhorn coral (*Acropora palmata*) and staghorn coral (*A. cervicornis*) were once the major reef builders in the U.S. Caribbean. Both have declined precipitously, with most populations losing 80–98% of their abundance as measured against a 1970s baseline. The U.S. Endangered Species Act protects these two coral species in recognition of their precarious condition.

Along with declines in such key species, many Caribbean reefs have also undergone a "phase shift" during the past three decades; as a result, macroalgae now dominate these systems. The regional shift from high coral cover to high macroalgal cover is attributed to a variety of factors, including the overfishing that has depleted herbivorous grazing species such as parrotfish (Scaridae), which control macroalgal growth.

Following the catastrophic near extinction of the sea urchin *Diadema antillarum* due to disease, parrotfish are now the only group of species remaining in the U.S. Caribbean that are capable of removing significant amounts of fleshy macroalgae from coral reefs. Parrotfish and other herbivorous fish help preserve coral reef habitat by grazing on algae that would otherwise crowd out these corals. Abundant and diverse herbivorous fish populations are necessary to remove sufficient algae to prevent algal overgrowth and create open space for new corals. Algae-dominated reefs provide far less productive habitat than coral-dominated ones, leading to a negative feedback loop in which fish populations decline as coral cover declines, and vice versa.

The struggle to protect Caribbean parrotfish and to foster their vital role in preserving critically depleted corals graphically demonstrates the importance of including historical baselines in fisheries management. A failure by the U.S. government to consider such baselines in Puerto Rico and the U.S. Virgin Islands threatens to degrade these corals further by allowing continued fishing of parrotfish. Despite the acknowledged importance of parrotfish in protecting the critically depleted elkhorn and staghorn corals (NOAA Biological Opinion 2011:116, Mumby et al. 2007), the U.S. government failed to consider history when establishing catch limits for those fish in 2011. The government neither endeavored to reconstruct historical baselines of parrotfish populations nor analyzed the numbers necessary to restore their previous role in reef ecosystems. Indeed, rather than relying on historical data or generating estimates of the parrotfish abundance required to restore healthy elkhorn and staghorn populations, the government downplayed the relative importance of grazing by the current (depleted) parrotfish assemblage (NOAA Biological Opinion 2011:50).

This failure to establish a historical baseline incorporating a robust and functional Caribbean reef ecosystem deprives managers of the context required to consider the true effects of fishing. Resulting management decisions allow continued fishing pressure on parrotfish—vital herbivores essential to coral protection—and risk irreversible damage to the coral reef ecosystem that sustains the fishery and, thus, to the fishery itself.

Steve Roady is Managing Attorney for Oceans at Earthjustice, Washington, D.C., and an adjunct faculty member in the Nicholas School of the Environment, Duke University. Andrea Treece is Staff Attorney for Earthjustice, San Francisco.

Box 5.3). Fish stocks that have been exploited close to their maximum capacity have lower resilience to stochastic events due to truncated size structure and decreased recruitment potential (Berkeley et al. 2004, Musick 1999). As our climate changes, the number of these extreme climatic events is expected to increase. Historical fisheries data allow us to examine past variability, and it is critical to have this historical knowledge if we are to efficiently manage fisheries stocks into the future with a rapidly changing climate.

A long-term perspective is critical to avoid the pitfalls of the shifting baseline and to help us better understand the past productivity of many marine ecosystems around the world. Assessing the status of most stocks has been hindered by the lack of adequate information. In order to assess a greater number of stocks, we must utilize a wider range of data sets that are often overlooked by conventional stock assessment. Historical data are typically underutilized in fisheries stock assessments because of a lack of standardization and difficulties in incorporating these data into standard stock assessment models. A better understanding of past ecosystem productivity is critical if we are to effectively manage these ecosystems into the future. Unfished biomass is almost never known, and this nearly universal data gap causes problems, ones that historical information offers promise for addressing. By incorporating historical data, TEK, and LEK into population assessment models and management practices, we bring insight into the yield of these ecosystems in the past and provide guidance for future management actions.

ACKNOWLEDGMENTS

The authors would like to thank the editors for giving us the opportunity to contribute to this important work. We thank Dr. Michael Larkin of NOAA Fisheries, Southeast Regional Office, for providing south Florida recreational fisheries data for Nassau grouper; Reginald Kokubun of the Hawai'i Department of Land and Natural Resources, Division of Aquatic Resources, for providing State of Hawaii commercial catch data; and Dr. Jack Kittinger for providing historical commercial catch data from Hawai'i. Our thanks go out to the members of the Fisheries Ecology Research Laboratory at the University of Hawai'i for providing comments on drafts of this chapter.

REFERENCES

Ainsworth, C., Heymans, J. J., Pitcher, T. J., and Vasconcellos, M. (2002) *Ecosystem Models of Northern British Columbia for the Time Periods 2000, 1950, 1900 and 1750. Fisheries Centre Research Reports* 10(4).

Al-Abdularazzak, D., Zeller, D., and Pauly, D. (2014) Understanding fisheries through historical reconstructions: implications for fishery management and policy. Chapter 6, this volume.

Ames, E. P. (2004) Atlantic cod stock structure in the Gulf of Maine. *Fisheries* 29, 10–28.

Ames, T. (2007) Putting fishers' knowledge to work: reconstructing the Gulf of Maine cod spawning grounds on the basis of local ecological knowledge. In *Fishers' Knowledge in Fisheries Science and Management* (N. Haggan, B. Neis, and I. G. Baird, Eds.). Coastal Management Sourcebooks 4. UNESCO, Paris, France. pp. 353–363.

Amorosi, T., McGovern, T. H., and Perdikaris, S. (1994) Bioarchaeology and cod fisheries: a new source of evidence. *ICES Marine Science Symposia* 198, 31–48.

Ault, J. S., Bohnsack, J. A., and Meester, G. A. (1998). A retrospective (1979–1996) multispecies assessment of coral reef fish stocks in the Florida Keys. *Fishery Bulletin* 96, 395–414.

Babcock, E. A., and MacCall, A. D. (2011) How useful is the ratio of fish density outside versus inside no-take marine reserves as a metric for fishery management control rules? *Canadian Journal of Fisheries and Aquatic Sciences* 68, 343–359.

Baumgartner, T., Soutar, A., and Ferreira-Bartrina, V. (1992) Reconstruction of the history of Pacific sardine and northern anchovy populations over the past two millennia from sediments of the Santa Barbara Basin, California. *California Cooperative Oceanic Fisheries Investigations Reports* 33, 24–40.

Bax, N. J., and Williams, A. (2001) Seabed habitat on the south-eastern Australian continental shelf: context, vulnerability and monitoring. *Marine Freshwater Research* 52, 491–512.

Beddington, J. R., Agnew, D. J., and Clark, C. W. (2007) Current problems in the management of marine fisheries. *Science* 316, 1713–1716.

Berkeley, S. A., Hixon, M. A., Larson, R. J., and Love, M. S. (2004) Fisheries sustainability via protection of age structure and spatial distribution of fish populations. *Fisheries* 29, 23–32.

Berkes, F. (2008) *Sacred Ecology*, 2nd ed. Routledge, New York, NY.

Bohnsack, J. A. (2003) Shifting baselines, marine reserves, and Leopold's biotic ethic. *Gulf and Caribbean Science* 14, 1–7.

Bray, K. (2000) A global review of illegal, unreported, and unregulated (IUU) fishing. AUS:IUU/2000/6. FAO, Rome.

Buchary, E. A., Cheung, W. L., Sumaila, U. R., and Pitcher, T. J. (2003) Back to the future: a paradigm shift for restoring Hong Kong's marine ecosystem. *American Fisheries Society Symposium* 38, 727–746.

Butler, V. L. (2001) Changing fish use on Mangaia, Southern Cook Islands: resource depression and the prey choice model. *International Journal of Osteoarchaeology* 11, 88–100.

Christensen, V., and Pauly, D. (1992) Ecopath II—a software for balancing steady-state ecosystem models and calculating network characteristics. *Ecological Modelling* 61, 169–185.

Christensen, V., and Pauly, D. (1993) Flow characteristics of aquatic ecosystems. In *Trophic Models in Aquatic Systems, vol. 26* (V. Christensen and D. Pauly, Eds.). ICLARM, Manila, Philippines. pp. 338–352.

Christensen, V., and Walters, C. J. (2004) Ecopath with Ecosim: methods, capabilities and limitations. *Ecological Modelling* 172, 109–139.

Clark, C. (1990) *Mathematical Bioeconomics: The Optimal Management of Renewable Resources.* John Wiley and Sons, New York, NY.

Claro, R., Baisre, J. A., Lindeman, K. C., and García-Arteaga, J. P. (2001) Cuban fisheries: historical trends and current status. In *Ecology of the Marine Fishes of Cuba* (R. Claro, K. C. Lindeman, and L. R. Parenti, Eds.). Smithsonian Institution Press, Washington, DC. pp. 194–219.

Claro, R., Sadovy de Mitcheson, Y. S., Lindeman, K. C., and Garca-Cagde, A. (2009) Historical analysis of commercial Cuban fishing effort and the effects of management interventions on important reef fishes: 1960–2005. *Fisheries Research* 99, 7–16.

Cobb, J. N. (1902) Commercial fisheries of the Hawaiian Islands. U.S. Fish Commission Report for 1901, plates 21–27 (pp. 353–499). Government Printing Office, Washington, DC.

Cobb, J. N. (1905a) The commercial fisheries. In *Bulletin of the U.S. Fish Commission, vol. XXIII for 1903: The Aquatic Resources of the Hawaiian Islands* (D. S. Jordan and B. W. Evermann, Eds.). Government Printing Office, Washington, DC. pp. 715–765.

Cobb, J. N. (1905b) The commercial fisheries of the Hawaiian Islands in 1903. Appendix to the Report of the Commissioner of Fisheries to the Secretary of Commerce and Labor for the year ending June 30, 1904 (pp. 433–512). U.S. Bureau of Fisheries. Government Printing Office, Washington, DC.

Costello, C., Ovando, D., Hilborn, R., et al. (2012) Status and solutions for the world's unassessed fisheries. *Science* 26, 517–520.

Dalsgaard, J., Wallace, S. S., Salas, S., and Preikshot, D. (1998) Mass–balance model reconstructions of the Strait of Georgia: the present, one hundred, and five hundred years ago. *Back to the Future: Reconstructing the Strait of Georgia Ecosystem. Fisheries Centre Research Reports* 6(5), 72–91.

Diamond, J. M. (2005) *Collapse: How Societies Choose to Fail or Succeed.* Viking, New York, NY.

Erlandson, J. M., and Rick, T. C. (2010) Archaeology meets marine ecology: the antiquity of maritime cultures and human impacts on marine fisheries and ecosystems. *Annual Review of Marine Science* 2, 165–185.

Fenner, D. (2012) Challenges for managing fisheries on diverse coral reefs. *Diversity* 4, 105–160.

Field, J. C., Punt, A. E., Methot, R. D., and Thomson, C. J. (2006) Does MPA mean 'Major Problem for Assessments'? Considering the consequences of place-based management systems. *Fish and Fisheries* 7, 284–302.

Finney, B., Alheit, P., Emeis, J., et al. (2010) Paleoecological studies on variability in marine fish populations: a long-term perspective on the impacts of climate change on marine ecosystems. *Journal of Marine Systems* 79, 316–326.

Friedlander, A., Aeby, G., Brainard, R., et al. (2008) The state of coral reef ecosystems of the main Hawaiian Islands. In *The State of Coral Reef Ecosystems of the United States and Pacific Freely Associated States* (J. E. Waddell and A. M. Clarke, Eds.). NOAA Technical Memorandum NOS NCCOS 73. pp. 158–199.

Friedlander, A. M., and DeMartini, E. E. (2002) Contrasts in density, size, and biomass of reef fishes between the northwestern and the main Hawaiian Islands: the effects of fishing down apex predators. *Marine Ecology Progress Series* 230, 253–264.

Friedlander, A. M., Shackeroff, J. M., and Kittinger, J. N. (2013) Customary marine resource knowledge and use in contemporary Hawai'i. *Pacific Science* 67, 441–460.

Gallucci, V. F., Saila, S. B., Gustafson, D. J., and Rothschild, B. J. (1996) *Stock Assessment, Quantitative Methods and Applications for Small-Scale Fisheries.* CRC Press, Boca Raton, FL.

Gårdmark, A., Nielsen, A., Floeter, J., and Möllmann, C. (2011) Depleted marine fish stocks and ecosystem-based management: on the road to recovery, we need to be precautionary. *ICES Journal of Marine Science* 68, 212–220.

Haggan, N., Neis, B., and Baird, I. G. (Eds.) (2007) *Fishers' Knowledge in Fisheries Science and Management.* UNESCO Publishing, Paris.

Hall, S. J., and Mainprize, B. M. (2005) Managing by-catch and discards: how much progress are we making and how can we do better? *Fish and Fisheries* 6, 134–155.

Haltuch, M. A., Punt, A. E., and Dorn, M. W. (2009) Evaluating the estimation of fishery management reference points in a variable environment. *Fisheries Research* 100, 42–56.

Harrington, J. M., Myers, R. A., and Rosenberg, A. A. (2005) Wasted fishery resources: discarded by-catch in the USA. *Fish and Fisheries* 6, 350–361.

Hauser, L., Adcock, G. J., Smith P. J., et al. (2002) Loss of microsatellite diversity and low effective population size in an overexploited population of New Zealand snapper (*Pagrus auratus*). *Proceedings of the National Academy of Sciences USA* 99, 11742–11747.

Hilborn, R., Punt, A. E., and Orensanz, J. (2004) Beyond band-aids in fisheries management: fixing world fisheries. *Bulletin of Marine Science* 74, 493–507.

Hoarau, G., Boon, E., Jongma, D. N., et al. (2005) Low effective population size and evidence for inbreeding in an overexploited flatfish, plaice (*Pleuronectes platessa* L.). *Proceedings of the Royal Society of London Series B* 272, 497–503.

Hutchinson, W. F. (2008) The dangers of ignoring stock complexity in fishery management: the case of the North Sea cod. *Biology Letters* 4, 693–695.

Hutchinson, W. F., van Oosterhout C., Rogers S. I., and Carvalho G. R. (2003) Temporal analysis of archived samples indicates marked genetic changes in declining North Sea cod (*Gadus morhua*). *Proceedings of the Royal Society of London Series B* 270, 2125–2132.

Jacobson, L. D., and MacCall, A. D. (1995) Stock-recruitment models for Pacific Sardine (*Sardinops sagax*). *Canadian Journal of Fisheries and Aquatic Sciences* 52, 566–577.

Jackson, J. B. C. (2010) The future of the oceans past. *Philosophical Transactions of the Royal Society of London Series B* 365, 3765–3778.

Jackson, J. B. C., Kirby, M., Berger, W. H., et al. (2001) Historical overfishing and the recent collapse of coastal ecosystems. *Science* 293, 629–638.

Johannes, R. E. (1978) Traditional marine conservation methods in Oceania and their demise. *Annual Review of Ecology and Systematics* 9, 349–364.

Johannes, R. E. (1981) *Words of the Lagoon: Fishing and Marine Lore in the Palau District of Micronesia*. University of California Press, Berkeley, CA.

Johannes, R. E. (1998) The case for data-less marine resource management: examples from tropical nearshore finfisheries. *Trends in Ecology & Evolution* 13, 243–246.

Kirch, P. V. (1982) The ecology of marine exploitation in prehistoric Hawaii. *Human Ecology* 10, 455–476.

Kittinger, J. N., Pandolfi, J. M., Blodgett, J. H., et al. (2011) Historical reconstruction reveals recovery in Hawaiian coral reefs. *PLoS ONE* 6, e25460.

Knowlton, N., and Jackson, J. B. C. (2008) Shifting baselines, local impacts, and global change on coral reefs. *PLoS Biology* 6, 54.

Kurlansky, M. (1997) *Cod: A Biography of the Fish That Changed the World*. Walker, New York, NY.

Larson, S., Jameson, R., Etnier, M., et al. (2002) Loss of genetic diversity in sea otters (*Enhydra lutris*) associated with the fur trade of the 18th and 19th centuries. *Molecular Ecology* 11, 1899–1903.

MacCall, A. (2009) Depletion-corrected average catch: a simple formula for estimating sustainable yields in data-poor situations. *ICES Journal of Marine Science* 66 2267–2271.

McClanahan, T. R., Graham, N. A. J., MacNeil, M. A., et al. (2011) Critical thresholds and tangible targets for ecosystem-based management of coral reef fisheries. *Proceedings of the National Academy of Sciences USA* 108, 17230–17233.

McClenachan, L. E., and J. N. Kittinger. (2012) Multicentury trends and the sustainability of coral reef fisheries in Hawai'i and Florida. *Fish and Fisheries* 14, 239–255.

McEvoy, A. F. (1986) *The Fisherman's Problem: Ecology and Law in the California Fisheries, 1850–1980*. Cambridge University Press, Cambridge, UK.

Mumby, P. J., Hastings, A., and Edwards, H. J. (2007) Thresholds and the resilience of Caribbean coral reefs. *Nature* 450, 98–101.

Musick, J. A. (1999). Criteria to define extinction risk in marine fishes: the American Fisheries Society initiative. *Fisheries* 24, 6–14.

Nagaoka, L. (2001) Using diversity indices to measure changes in prey choice at the Shag River Mouth site, southern New Zealand. *International Journal of Osteoarchaeology* 11, 101–111.

National Marine Fisheries Service (2012) Annual Report to Congress on the Status of U.S. Fisheries—2011. U.S. Department of Commerce, National Oceanic and Atmospheric Administration, National Marine Fisheries Service, Silver Spring, MD.

National Research Council (1998) *Improving Fish Stock Assessments*. National Academy Press, Washington, DC.

Neis, B., Schneider, D. C., Felt, L., et al. (1999) Fisheries assessment: what can be learned from interviewing resource users? *Canadian Journal of Fisheries and Aquatic Sciences* 56, 1949–1963.

NOAA Biological Opinion (2011) Endangered Species Act Section 7 Consultation Biological Opinion for the Continued Authorization of Reef Fish Managed under the Reef Fish Fishery Management Plan of Puerto Rico and the U.S. Virgin Islands, dated October 4, 2011. National Oceanic and Atmospheric Administration (NOAA), Silver Spring, MD.

North Pacific Fishery Management Council (2011) Stock assessment and fishery evaluation report for the groundfish resources of the Bering Sea/Aleutian Islands regions. North Pacific Fishery Management Council, Anchorage, AK.

Pauly, D. (1995) Anecdotes and the shifting baseline syndrome of fisheries. *Trends in Ecology & Evolution* 10, 430.

Pauly, D. (2006) Major trends in small-scale marine fisheries, with emphasis on developing countries, and some implications for the social sciences. *Maritime Studies* 4, 7–22.

Pauly, D., Christensen, V., Dalsgaard, J., et al. (1998) Fishing down marine food webs. *Science* 279, 860–863.

Pickett, S. T. A. (1989) Space-for-time substitution as an alternative to long-term studies. In *Long-term Studies in Ecology: Approaches and Alternatives* (G. E. Likens, Ed.). Springer-Verlag, New York, NY. pp. 110–135.

Pitcher, T. J. (1998) "Back to the future": a novel methodology and policy goal in fisheries. *Back to the Future: Reconstructing the Strait of Georgia Ecosystem. Fisheries Centre Research Reports* 6(5), 4–7.

Pitcher T. J. (2001) Fisheries managed to rebuild ecosystems: reconstructing the past to salvage the future. *Ecological Applications* 11, 601–617.

Pitcher, T. J. (2005) Back-to-the-future: a fresh policy initiative for fisheries and a restorationecology for ocean ecosystems. *Philosophical Transactions of the Royal Society of London Series B* 360, 107–121.

Pitcher, T. J., Heymans, J. J., and Vasconcellos, M. (Eds.) (2002) *Ecosystem Models of Newfoundland for the Time Periods 1995, 1985, 1900 and 1450. Fisheries Centre Research Reports* 10(5).

Poepoe, K., Bartram, P., and Friedlander, A. (2007) The use of traditional Hawaiian knowledge in the contemporary management of marine resources. In *Fishers' Knowledge in Fisheries Science and Management* (N. Haggan, B. Neis, and I. Baird, Eds.). UNESCO, Paris. pp. 117–141.

Polovina, J. J. (1984) Model of a coral reef ecosystem, part I: ECOPATH and its application to French Frigate Shoals. *Coral Reefs* 3, 1–11.

Punt, A. E., and Methot, R. D. (2004) Effects of marine protected areas on the assessment of marine fisheries. *American Fisheries Society Symposium* 42, 133–154.

Quinn, T. J., and Deriso, R. B. (1999) *Quantitative Fish Dynamics*. Oxford University Press, Oxford, UK.

Radovich, J. (1982) The collapse of the California sardine fishery: what have we learned? *CalCOFI Report* 23.

Restrepo, V. R., Thompson, G. G., Mace, P. M., et al. (1998) Technical guidance on the use of precautionary approaches to implementing National Standard 1 of the Magnuson-Stevens Fishery Conservation and Management Act. NOAA Technical Memorandum NMFS-F/SPO-31.

Rosenberg, A. A., Bolster, W., Alexander, K. E., et al. (2005) The history of ocean resources: modeling cod biomass using historical records. *Frontiers in Ecology and the Environment* 3, 78–84.

Ruddle, K., and Hickey, F. R. (2008) Accounting for the mismanagement of tropical nearshore fisheries. *Environment, Development and Sustainability* 10, 565–589.

Sandin, S. A., Smith, J. E., DeMartini, E. E., et al. (2008) Degradation of coral reef communities across a gradient of human disturbance. *PLoS ONE* 3, e1548.

Schug, D. (2001) Hawaii's commercial fishing industry: 1820–1945. *Hawaiian Journal of History* 35, 15–34.

Sladek Nowlis, J. (2004) Performance indices to facilitate informed, value-driven decision making in fisheries management. *Bulletin of Marine Science* 74, 709–726.

Sladek Nowlis, J., and Bollermann, B. (2002) Methods for increasing the likelihood of restoring and maintaining productive fisheries. *Bulletin of Marine Science* 70, 715–731.

Speller, C. F., Hauser, L., Lepofsky, D., et al. (2012) High potential for using DNA from ancient herring bones to inform modern fisheries management and conservation. *PLoS ONE* 7, e51122.

Sugihara, G., May, R., Ye, H., et al. (2012) Detecting causality in complex ecosystems. *Science* 338, 496–500.

Swezey, S. L., and Heizer, R. F. (1977) Ritual management of salmonid fish resources in California. *Journal of California Anthropology* 4, 5.

Titcomb, M. (1972) *Native Use of Fish in Hawaii*. University of Hawaii Press, Honolulu.

Valdés, J., Ortlieb, L., Gutiérrez, D., et al. (2008) 250 years of sardine and anchovy scale deposition record in Mejillones Bay, northern Chile. *Progress in Oceanography* 79, 198–207.

Van Neer, W., Ervynck, A., Bolle, L., et al. (2002) Fish otoliths and their relevance to archaeology: an analysis of medieval, post-medieval and recent material of plaice, cod and haddock from the North Sea. *Environmental Archaeology* 7, 61–76.

Walmsley, S., Howard, C., and Medley, P. (2005) *Participatory Fisheries Stock Assessment (ParFish) Guidelines*. Marine Resources Assessment Group, London, UK.

Walters, C., Christensen, V., and Pauly, D. (1997) Structuring dynamic models of exploited ecosystems from trophic mass-balance assessments. *Reviews in Fish Biology and Fisheries* 7, 139–172.

Walters, C., Pauly, D., and Christensen, V. (1999) Ecospace: prediction of mesoscale spatial patterns in trophic relationships of exploited ecosystems, with emphasis on the impacts of marine protected areas. *Ecosystems* 2, 539–554.

Williams, A., and Bax, N. (2007) Integrating fishers' knowledge with survey data to understand the structure, ecology and use of a sea scape off south-eastern Australia. In *Fishers' Knowledge in Fisheries Science and Management* (N. Haggan, B. Neis, and I. G. Baird, Eds.). UNESCO Publishing, Paris. pp. 365–379.

Williams, I. D., Walsh, W. J., Schroeder, R. E., et al. (2008) Assessing the relative importance of fishing impacts on Hawaiian coral reef fish assemblages along regional-scale human population gradients. *Environmental Conservation* 35, 261–272.

Wing, S., and Wing, E. (2001) Prehistoric fisheries in the Caribbean. *Coral Reefs* 20, 1–8.

Zeller, D., Booth, S., Craig, P., and Pauly, D. (2006) Reconstruction of coral reef fisheries catches in American Samoa, 1950—2002. *Coral Reefs* 25, 144–152.

Zeller, D., Booth, S., Davis, G., and Pauly, D. (2007) Re-estimation of small-scale fisheries catches for U.S. flag island areas in the western Pacific: the last 50 years. *Fishery Bulletin* 105, 266–277.

Zeller, D., Darcy, M., Booth, S., et al. (2008) What about recreational catch? Potential impact on stock assessment for Hawaii's bottomfish fisheries. *Fisheries Research* 91, 88–97.

SIX

Understanding Fisheries through Historical Reconstructions

Implications for Fishery Management and Policy

DALAL AL-ABDULRAZZAK, DIRK ZELLER, and DANIEL PAULY

Global fisheries catch statistics are often incomplete. The contribution of many sectors, including small-scale fisheries, illegal catches, and discards are frequently absent from or underreported in statistics submitted annually by member countries of the United Nations Food and Agriculture Organization (FAO). This incomplete accounting in official statistics, and the resulting distorted historical trends, impairs our understanding of the management and policy prescriptions necessary for fisheries sustainability. This chapter describes an approach to retroactively estimate catches where comprehensive time series data are lacking. Data are gathered from nontraditional sources, such as unpublished studies, gray literature, published studies, and surveys; or from sources unrelated to fisheries, such as satellite imagery. We present examples of the discrepancy between reported and reconstructed catches and discuss the implications of such misreporting for management and fisheries policy on national, regional, and global scales.

INTRODUCTION

Indisputably, the single most important piece of information about a fishery is the total catch, by species, over time (Froese et al. 2012). Catch data increase our knowledge of a fishery by determining two essential things: (1) the scale of the fishery (Zeller et al. 2006, 2011a) and (2) the status of the stock over time (Froese et al. 2012, Kleisner et al. 2012). Understanding long-term trends (decades to centuries) in catch is critical to managing fisheries sustainably because it allows us to assess the "health" of a fishery within the context of species life spans, human impacts, environmental disturbances such as El Niño events, as well as long-term changes in oceanographic regimes. Temporal trends also allow us to establish viable restoration targets based on past abundances of species.

Since its creation, the FAO has been tasked to collect, analyze, and disseminate information related to food, nutrition, and agriculture, including fisheries (Ward 2004). Since 1950, the FAO has assembled, standardized, and distributed annual marine fishery landings reported by its member countries, broken down by taxa as well as by 19 large statistical areas. This global-capture fishery database, known as FishStat (www.fishstat.org), provides the only global time series of national fishery landings (Garibaldi 2012), and a detailed analysis of its contents is published every 2 years (see, e.g., FAO 2011).

FishStat has served as a major source of data in many global and regional fisheries studies that analyze and interpret fisheries trends. These global studies include, among others, the first estimation of the primary production required to sustain global fisheries (Pauly and Christensen 1995), evidence that humans are "fishing down" marine food webs (Pauly et al. 1998), the first global estimation of fuel usage by fishing fleets (Tyedmers et al. 2005), a prediction that fisheries will collapse in 2048 (Worm et al. 2006), and fish biomass trends since 1950 (Tremblay-Boyer et al. 2011, Watson et al. 2013). These data have also formed the core of global catch-mapping efforts (Watson et al. 2004), which, in turn, enabled a greater understanding of the spatial expansion of global fisheries (Swartz et al. 2010a) and, conversely, of the manner in which they supply fish markets (Swartz et al. 2010b).

Besides scientific studies, FAO data have been used to inform fisheries management policy. For example, organizations such as the World Resources Institute often use FAO catch data to estimate national earnings from fish exports, production trends, and per capita fish consumption rates. Additionally, the Marine Trophic Index (which was derived from the "fishing down" concept of Pauly et al. 1998) is partly based on FAO catch data and is one of the indicators used to measure the biodiversity of large fishes by the countries party to the Convention on Biological Diversity (Pauly and Watson 2005).

Despite their wide use, numerous studies have called into question how closely FAO data resemble reality (Zeller et al. 2006, 2007a; Clarke et al. 2006; Watson and Pauly 2001). Because catch data and other fishery statistics are generally submitted to FAO by national entities, the quality of FAO data is dependent on the accuracy and reliability of statistical data collection within these member countries (Garibaldi 2012). Thus, perverse incentives (for the case of China, see Watson and Pauly 2001), the historical legacies of governmental or institutional change (e.g., in Tanzania; see Jacquet et al. 2010), or political interference (e.g., in India; Bhathal and Pauly 2008) can cause such statistical reporting systems to underperform or decay.

In general, FAO data are recognized as incomplete in many regions; more than half of FAO member countries failed to report their annual fishery statistics over the past decade (Garibaldi 2012). Additionally, FAO does not account for discards in its database and, thus, implicitly requires countries to report fisheries "landings" rather than "catches." By contrast, at least one regional fisheries management organization, the Commission for the Conservation of Antarctic Marine Living Resources (CCAMLR), includes discarded catches obtained via onboard observer programs into their catch database (although they do not label these as discards). Overall, however, the contribution of many sectors, including small-scale fisheries, recreational fishing, and illegal catches are frequently absent or substantially underreported (e.g., Zeller et al. 2007a, 2007b, 2011a, 2011b; Jacquet et al. 2010; Le Manach et al. 2012).

The chronic misreporting of fisheries data is not trivial; the consequences extend far beyond resource conservation and can lead to inequitable policy decisions that jeopardize food security (Jacquet et al. 2010, Le Manach et al. 2013), underestimate the contribution of small-scale fisheries to rural food security and GDP (Pauly 2006, Zeller et al. 2006), and incorrectly assume that global fish catches are increasing (Watson and Pauly 2001). This may lead to negative policy outcomes, such as when misreported catches undermine stock assessments and management plans (Zeller et al. 2008), potentially leading to overoptimistic investments or between-country access agreements, which consequently subvert the sustainability of fisheries.

Breaking this cycle requires estimating overall national catches from independent data if possible, or by complementing the data provided by countries to the FAO with estimates for all missing or underreported components. Pauly (1998) suggested that a complex, multifaceted sector of a given country's or territory's economy cannot operate without casting a "shadow" on the other sectors of that economy (ranging from boat building, repairs, and supplies to domestic fish consumption). Therefore, if catch data are unavailable, they can be indirectly inferred from other related sectors.

These ideas were operationalized by Zeller et al. (2007a), who devised an approach for retroactively estimating or "reconstructing" the catch of neglected fisheries, in order to enable a more accurate picture of historical catch patterns and trends. Often, catch reconstructions rely on information and data sources originally intended for other purposes. Such information can be found in a number of sources, including unpublished studies, gray literature, maritime records, aerial photos, and interviews with fishers, or obtained from published studies and surveys whose primary focus is other than catch reporting. (See Box 6.1, where Loren McClenachan relates her stories on how archival research can yield important—and sometimes unexpected—sources for marine historical ecologists.) These nontraditional sources can be used to derive estimates of catch and of catch rates per unit area, per fisher, or per capita for a given period. Using a series of such data "anchor points" (Zeller et al. 2007a), total-catch time-series estimates can then be derived for years with missing data using interpolations.

Fishing predates all other human activities in the ocean (Jackson et al. 2001), and data sets compiled over longer periods (decades to centuries) can be used to detect greater declines in large fishes (McClenachan 2009). However, the reconstruction method developed by Zeller and colleagues typically starts at 1950 for several reasons, including the following: (1) this is the year when FAO began to publish its annual "yearbook" of global fisheries statistics; (2) in many parts of the world, this period marks the start of industrialized fishing; and (3) greater data availability for this period allows for the development of more robust catch time series. As discussed in Box 6.2, these catch reconstructions can provide important information on fishery trends for conservation practitioners.

Marine fisheries catches have been successfully reconstructed in many regions of the world, including for Pacific Island countries and territories (Zeller et al. 2006, 2007a), East Africa and the Western Indian Ocean (Jacquet et al. 2010, Le Manach et al. 2012), South America (Wielgus et al. 2010), Europe (Zeller et al. 2011b), and the High Arctic (Zeller et al. 2011a). As part of the research of the Sea Around Us (www.seaaroundus.org), we now have complete global coverage of reconstructions for all maritime countries in the world. Here, we use the basic conceptual framework and approach outlined by these studies to discuss reconstructions and their policy implications at three different scales: national, regional, and global.

THE NATIONAL SCALE

To illustrate the need and scope for reconstructing catches at the national level, we use the nation of Kuwait. Like many countries, Kuwait possesses a diversity of coastal fishery types and faces challenges in data collection, accuracy, and lack of understanding of historical

Public Library in Key West, Florida—I found a box of photographs of tourists posing near fish caught on recreational fishing boats in the 1950s, '60s, '70s, and '80s. To most people, the photos would be interesting for their local flair, so they had not immediately come to mind when I initially described my interest in the history of Keys fisheries to the library's archivist. Finding the photographs—which contained a nearly unbroken record of data on the size structure and species composition of reef fisheries decades before traditional ecological or fisheries data existed (McClenachan 2009)—took weeks of following leads and conversing with librarians, local people, and other researchers. Similarly, after several archival visits and conversations with local residents, Kyle Van Houtan realized that decorative menus from Hawaiian seafood restaurants could be used to describe the historical harvest of reef fish, at a critical period at the beginning of the twentieth century for which no other data existed (Van Houtan et al. 2013).

As technology advances, the detective process of historical research becomes easier. Increasingly, we can search electronically the contents of digitized books, digital archives, and photographs (also see chapters 4 and 11, this volume). For example, the Monroe County Library has now digitized hundreds of their "Dead Fish" pictures and archived these online (Florida Keys Public Libraries n.d). Advances like this should encourage more ecologists to undertake historical research and expedite the use of the results for conservation and management. However, there is great value to elements of the detective process that cannot be replicated online: searching through a variety of sources to determine the best ones and following clues toward sources that may be unexpectedly rich. This process is enhanced through conversations with librarians, local people, and researchers sitting across the table at the archives. As historical ecology advances, ecologists should make full use of new technologies, but these cannot fully replace the detective process of historical research.

Loren McClenachan is Assistant Professor of Environmental Studies at Colby College.

trends in fisheries. Despite their predominantly small-scale nature, Kuwait's fisheries remain its second-most-important natural resource after oil (Carpenter et al. 1997). For centuries prior to the discovery of oil in 1938, pearl diving dominated the country's economy. Indeed, at the height of pearling, 80% of the world's natural pearls came from the Persian Gulf. However, the production of Japanese cultured pearls in the 1920s, the global recession in the 1930s, World War II, and the production of crude oil in the late 1940s dealt consecutive blows to the Gulf's pearl industry, leading to its rapid demise.

In general, Kuwait's fisheries are poorly managed, and officially reported catch data are haphazard at best. Substantial sectors are missing from reported data, including recreational catches, discards by the shrimp fishery, and illegal catches. Using catch reconstruction methods, we can see that the reported catches for Kuwait's artisanal and industrial fisheries potentially underestimate total catches by a factor of 7 over the 1950–2010 period (Figure 6.1).

Rod Fujita

The global fisheries catch is a key indicator of the status of one of the planet's most important ecosystem services. Conservation practitioners and funders interested in food security, ocean conservation, and sustainable development focus on global catch and its trend to set agendas for action and to generate urgency. Catch is the most salient attribute of fisheries to most people; as such, it has a strong effect on the dominant worldview regarding the status of fisheries. And of course, the dominant view determines demand for change and the amount of funding and political will for making change.

Knowing how many fish are being caught is just as important at other scales, for many different reasons. Catch statistics are often used to estimate the status of fish stocks, the very basis of scientific fisheries management—essential for setting sustainable catch limits, a primary fisheries conservation tool. And once a catch limit is set, of course, catch statistics are necessary for ensuring that the limit is not exceeded. Catch records are often used to allocate catch privileges, helping to establish rules for member states, regions, groups, or even individual fishermen that can alleviate the competition to maximize catch. This competition tends to deplete fish populations and make fisheries unprofitable in the absence of governance or rules.

Clearly, catch statistics are extremely important for fisheries management. But unfortunately, the reliability of these statistics is very uneven. Some countries collect high-quality catch information, while others do not. One way to remedy this is to reconstruct the catch, using whatever information is available to piece together the missing history of a fishery.

Catch reconstruction is easy to criticize, and there is an active debate among scientists about it (Pauly et al. 2013). But the bottom line is that the alternative—a massive new data-collection program—is infeasible in the short term. Moreover, it would not yield information on catches early in the history of fisheries, which is critically important for practitioners because it provides a baseline against which to measure the performance of a fishery.

The FAO's global fisheries database includes >19,000 unique entries, but only a few hundred fisheries have been assessed. The available evidence (stock status of unassessed fisheries extrapolated from catch statistics) suggests that most unassessed stocks are overfished and are becoming more overfished as time goes on. If we are to reverse this trend, one of the first things we need to do is assess as many stocks as possible, and that means reducing data requirements and developing

Such substantial differences between reported landings and reconstructed total catches illustrate the magnitude of the data-reporting problems faced by countries with small-scale, data-poor fisheries (e.g., Zeller et al. 2007a, 2011a, 2011b). It also points at a fundamental problem of fisheries catch data being viewed purely from a commercial, market perspective, which accounts only for what is landed and utilized for commercial sale or export (Zeller and Pauly 2004, Pauly and Zeller 2003). By contrast, fisheries data collection—and, hence, catch accounting—needs to account for total catches, particularly given the global move toward assessment methods that are accessible to a much larger number of people.

One of my priorities is to remedy this situation, and I think that catch reconstructions will help. Most fisheries collect limited data; catch is virtually the only fishery statistic that is collected around the globe, however spotty that record may be. Fortunately, simple and rapid methods are now available for analyzing these kinds of stocks. However, they depend on estimates of unfished biomass—hard to come by in most fisheries, because few fisheries bother to collect data from day one. Catch reconstructions that yield estimates of historical catch may provide valuable insight into what unfished biomass might have been, facilitating the kinds of data-limited analytical methods that I and many others have been promoting.

While many fisheries collect catch data, and catch reconstruction can fill in many gaps, I often encounter fisheries that have no data whatsoever—particularly in developing countries. As the authors of this chapter assert, new data-collection efforts should focus on using assessment methods that require less time, fewer resources, and less expertise than conventional methods. We practitioners should also mine the available data more effectively. For example, my team is trying to generate abundance indices using visual census data, which are often quite rich in coral reef countries that lack traditional fisheries data. Moreover, if every fishery included an unfished reference area, there would be less need for long historical catch records, as the unfished area provides a good baseline for assessing stock status. We are using ratios of abundance in fishing grounds to abundance in no-take marine reserves as an index of stock status, which we hope can serve as a performance indicator in combination with empirical thresholds for coral reef tipping points related to fish biomass in adaptive management. If the ratio stays above the threshold associated with coral-dominated reef systems, no change in fishing mortality is required. But if the ratio falls below thresholds associated with transitions to less desirable ecosystem states, a harvest control rule is triggered in order to reduce fishing mortality to levels that are likely to increase the ratio. Similarly, information on the distribution of size classes in a fish population can provide a snapshot of stock status useful for management without the need for many years of data. Methods that utilize unfished areas and size data are much simpler to use than conventional methods and offer hope that many more stocks will be assessed and managed, improving fishery outcomes for stocks, ecosystems, and society in the future.

Rod Fujita is Director of Research and Development, Environmental Defense Fund.

viewing and managing fisheries on an ecosystem scale (Pikitch et al. 2004). This requires comprehensive accounting for the entire catch of fish and invertebrates, including the recording (or estimation) and reporting of discarded catch, and the estimation and reporting of catches from unregulated sectors such as subsistence, cultural, and recreational fisheries. Given the high costs of monitoring such sectors using traditional catch-monitoring approaches, alternative methods such as utilizing national surveys and census opportunities have been suggested for the more widely dispersed and hard-to-monitor small-scale and

A practical way to reconstruct basic information on small-scale commercial and subsistence fisheries where it is not collected by fisheries agencies, or to verify such information where it is available, is to use the household income and expenditure survey (HIES). These surveys are usually conducted every 5 years by national statistics offices to calculate the consumer price index and other socioeconomic measures.

Provided that the HIES reflects national trends and contains questions on how much fish is purchased, caught, received as gifts, and sold by the household, it provides important information for sustainable management of coastal fish stocks. For example, a well-designed HIES can provide estimates of the total quantities of different types of fish consumed within the country (reef fish, oceanic fish, freshwater fish, etc.) and whether these fish were derived from commercial or subsistence fishing. Each of these categories of fish can be divided into simple classes (e.g., herbivorous and carnivorous reef fish, the main species of tuna) within the HIES to provide catch estimates for important groups of species or to determine the extent of selective fishing where independent data are available on the relative abundance of these groups in the wild.

The information from HIES not only provides useful snapshots of the total catch from small-scale fisheries and per capita fish consumption, but can also be used to monitor the performance of management and conservation measures. Because the HIES is done at regular intervals, trends in fish catch and consumption can be assessed and the results can be used to adapt management as required.

In Pacific Island countries and territories, coastal communities are being encouraged to shift some of their fishing effort from coral reef fish to oceanic fish (especially skipjack and yellowfin tuna) by fishing around fish aggregating devices (FADs) anchored close to shore to attract large pelagic fish. This measure is designed to help provide more fish for food security as human populations grow, and as an adaptation to the declines in coral-reef fish production projected to occur as the climate changes. Fishing around inshore fish aggregating devices also has the conservation benefit of reducing pressure on coral reef fisheries. HIES is a powerful tool for measuring whether this management option changes the percentages of reef fish and tuna in the catches of small-scale commercial and subsistence fisheries.

Johann Bell was previously Principal Fisheries Scientist with the Secretariat of the Pacific Community; he is now Honorary Professorial Fellow at the Australian National Centre for Ocean Resources and Security, University of Wollongong, NSW, Australia.

recreational fisheries sectors (Zeller et al. 2007b). For example, in the Pacific Islands, household income and expenditure surveys are undertaken periodically to assess fisheries' performance and importance to local economies (Box 6.3). Al-Abdulrazzak and Pauly (2014) use Google Earth to "ground truth" weir fish catches, speaking to the potential of satellite technologies for monitoring fisheries remotely, particularly in areas that were once considered too dangerous or expensive for fisheries surveillance and enforcement.

The reconstructed time series also illustrates the magnitude and importance of discards (Figure 6.2). In terms of tonnage, discards amounted to almost 10 times the total amount of

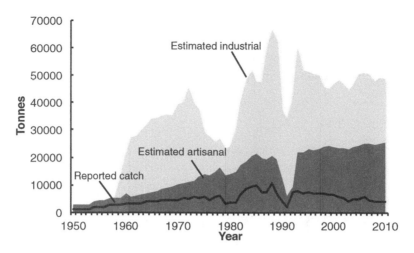

FIGURE 6.1 Reconstructed estimates of small-scale and industrial catches in Kuwait compared to officially reported catch.

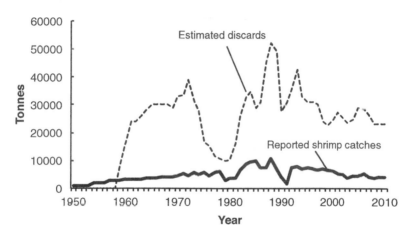

FIGURE 6.2 Estimated discards from Kuwait's industrial shrimp fishery.

landed finfish in Kuwait. Although local fisheries are currently not essential to Kuwait's food security (because the country's food resources are largely imported), recent climate-change studies predict that the Persian Gulf may lose 50% of its fisheries in the coming decades (Huelsenbeck 2012), prompting a greater demand for imports in the near future. To adequately prepare for climate-change impacts, as well as for sound fisheries management, managers should consider policies that reduce discarding practices. Such approaches may also help bolster local food-security policy as fisheries resources shift under global climate change. In general, a more complete estimate of total fisheries extractions is fundamental to effectively manage fisheries resources in order to mediate potential threats to food security and to assess ecological resilience. The future success of many countries relies, in part, on their ability to keep pace with an increasingly global economy while maintaining a healthy supply of resources for domestic purposes.

THE REGIONAL SCALE

The importance of historical reconstructions for food security policy can also be examined from a regional perspective. Historical catches have been reconstructed for almost all small island states and territories of the Caribbean (e.g., see Frotté et al. 2009, Harper et al. 2009), thus enabling some generalizations. The first is that catches are systematically underreported in the region, by 32–35% in Guadeloupe and Martinique and by 430% in Jamaica (with a median of 250% for all islands in the region). Such underreporting—which includes only a small amount of the discarded fish that is reported to FAO—implies that the fisheries of the Caribbean are systematically underreported in terms of their contribution to food security and GDP, just as in other areas of the world (Zeller et al. 2007b).

As in the Caribbean, catches by Pacific Island countries have also been systematically underreported (see, e.g., Zeller and Harper 2009, Harper and Zeller 2011). Most small Pacific Island countries have shifted their focus to lucrative pelagic tuna fisheries. These fisheries are conducted almost exclusively through foreign fishing interests, whether via fishing access agreements, through so-called joint venture or charter operations, or even by reflagging of foreign vessels to permit easier and cheaper access to Pacific Island countries' waters. Fishing access fees often provide a major share of a country's foreign exchange earnings (Gillett et al. 2001; Gillett 2009, 2011) in this region, but such approaches also discount future production from domestic fisheries, which may provide for current and future food security in a region with few other domestic protein sources (Secretariat of the Pacific Community 2008). See Box 6.3 on how regional initiatives in the Pacific are being undertaken to provide local communities more access to pelagic fishery resources and how fisheries experts are helping us better understand the role of fisheries in food security in this region.

Catch reconstructions can also benefit public policy by providing improved catch statistics and a better accounting of the role of fisheries in national, regional, and global economies. For example, reconstructed catches for Mozambique and Tanzania (Jacquet and Zeller 2007a, 2007b; Jacquet et al. 2010) contributed directly to changes in how these two countries estimate and report their data. Mozambique, which was shown to have 6 times higher catches than the reported data suggested at the time (mid-2000s), has subsequently been able to improve its catch sampling program, which is reflected in substantially improved catch data in recent years. Such increased accuracy of reported catch data can assist during foreign fishing-access negotiations, which often focus on a perceived "surplus production." Catch reconstructions can reveal the true catch landed by domestic fisheries, thus preventing excessive allocation to foreign fishing and reducing the likelihood of "selling the same fish twice" to both domestic and foreign fishers operating in the same waters (also see Le Manach et al. 2012, 2013).

Another example comes from Tanzania, where catch statistics from a major region of the country (the island group of Zanzibar), despite being collected by the Zanzibar fisheries agency, were missing entirely from the data reported by the government of Tanzania to FAO (Jacquet and Zeller 2007b). These catch statistics were retroactively included as of 2000 (Jacquet et al. 2010). This reporting oversight was likely brought to light through catch reconstructions, which were shared with local authorities, thus influencing the subsequent data correction.

THE GLOBAL SCALE

Reconstructed time series can also be used to influence policy in the international arena. For example, in recent years the representatives of Caribbean countries sided with Japan in voting for whaling in successive meetings of the International Whaling Commission, citing the need to maintain a "balance in the ecosystem" (Swartz and Pauly 2008). Robust time-series data on catch, which include all withdrawals from fisheries ecosystems, can provide an important first step in provisioning data for ecosystem-based fisheries management, and one that is required for any kind of "balance" to be assessed. This is especially important because it has shown that fisheries have strong impacts on the structure and functioning of Caribbean marine ecosystems (Jackson et al. 2001, Mora 2008). By contrast, marine mammals in general, and whales in particular, have only negligible effects on these ecosystems, even when abundant (Gerber et al. 2009).

One major insight emerging from catch reconstructions is the global footprint of fisheries and the full extent of living-marine-resource extraction. Reconstructions are illustrating the trajectories of fisheries and how they are changing over time, which can contrast with official reported statistics and challenge existing assumptions about global trends. For example, contrary to the FAO's (2011) description of global fisheries being "stable," reconstructed catches, when pooled for the whole world, show a declining trend. This suggests that increased catches from newly exploited areas are no longer replacing those from depleted stocks, as was the case for the past few decades when geographically expanding fisheries masked the overfishing of traditional stocks (Swartz et al. 2010a). These insights, which have critical consequences for the way global fisheries are managed, are not possible without a historical perspective.

CONCLUSIONS

When fisheries managers rely solely on conventional quantitative data (i.e., data collected from mainly commercial fisheries using logbook or landings site/port or market sampling), catch data often suffer from massive underreporting. Such approaches can result in incorrect assumptions that small-scale fishery catches are small or negligible, leading to inappropriate management, often with unintended outcomes and consequences both for fisheries resources and coastal communities. Researchers are increasingly paying more attention to methods that make use of imperfect records of the past. Researchers are accessing and using a wide range of nontraditional sources such as historical anecdotes, which can be "as factual as a temperature record" (Pauly 1995). As we show in the context of fishery reconstructions, marine historical ecology approaches can counter the underestimation of traditionally neglected sectors and the shifting baselines syndrome, which can result in mismanagement of fisheries resources.

Despite data uncertainties and the need to appropriately qualify data assumptions, catch reconstruction approaches are preferable to the alternative, which assumes low levels of catch or—even worse—leaves whole sectors out of reported data, which may be erroneously interpreted as "zero" catch in the policy and management arena (Zeller et al. 2006). Such "zero" interpretations of missing data components will certainly be more misleading than reconstructed estimates (despite uncertainties), especially if reconstructions remain conservative. Some managers and practitioners may be understandably uneasy with the imprecision of catch reconstructions, given the uncertainty of the data and the information sources that underlie them, rather than valuing the trends and improved accuracy they provide (i.e., by accounting for sectors ignored in reported data). However, as the economist John Maynard Keynes is reported to have said, "It is better to be vaguely right than precisely wrong."

With increased availability of catch reconstructions for historical fisheries, managers and policymakers will be able to focus on trends in the available data rather than focusing on the uncertainty surrounding these data (Rosenberg et al. 2005). We hope the increased acceptance of such approaches—and recognition of their validity and value—will help inform much-needed shifts in data collection and accounting toward considering and embedding fisheries in long-term reconstructions of marine ecosystems. As we have shown here, the incorporation of missing fishery sectors often results in distinctly different baselines of past catches, and these can have strong policy and management implications.

Even among fisheries scientists, there remain debates about the global status of fisheries and even the value and need for basic fisheries data such as catch (Pauly et al. 2013). Many believe that rigorous quantification of the uncertainties surrounding stock assessments and the subsequent delivery of results to managers in the form of risk assessments sufficiently address the overfishing crisis (Pauly and Zeller 2003). Yet the major challenges in fisheries are based in the realm of public policy. Issues of equitable resource allocation require public involvement, participatory democratic planning, and provisioning of decision making with the most accurate and reliable data. As we have shown here, fishery reconstructions can provide a critical long-term view on fisheries performance, sectoral contributions, and risks associated with various management strategies (e.g., not managing discards appropriately).

Fishery reconstructions and other data from marine historical ecology can be critical in informing the public, and the public-policy decision makers who represent them, of the true status and trend of fisheries and their associated effects on ocean health (Pauly and Zeller

2003). Findings from marine historical ecology will continue to be challenged, but continuous efforts on behalf of the scientific and practitioner community to effectively communicate this emerging area not only to scientific colleagues, but also to policymakers and the general public, are making headway. Historical reconstruction approaches have much potential to remove "lack of data" from the list of excuses used to maintain the status quo, and to increase public transparency and involvement in fishery policy by the true owners of the marine resources, the present and future global citizens.

ACKNOWLEDGMENTS

This is a contribution from the Sea Around Us, a collaboration between the University of British Columbia and the Pew Charitable Trusts.

REFERENCES

Al-Abdulrazzak, D., and Pauly, D. (2014) Managing fisheries from space: Google Earth improves estimates of distant fish catches. *ICES Journal of Marine Science* 71, 450–454.

Bhathal, B., and Pauly, D. (2008) 'Fishing down marine food webs' and spatial expansion of coastal fisheries in India, 1950–2000. *Fisheries Research* 91, 26–34.

Carpenter, K. E., Krupp, F., Jones, D. A., and Zajonz, U. (1997) *Living Marine Resources of Kuwait, Eastern Saudi Arabia, Bahrain, Qatar, and the United Arab Emirates*. FAO, Rome.

Clarke, S. C., McAllister, M. K., Milner-Gulland, E. J., et al. (2006) Global estimates of shark catches using trade records from commercial markets. *Ecology Letters* 9, 1115–1126.

FAO (2011) *The State of World Fisheries and Aquaculture 2011*. Food and Agriculture Organization of the United Nations (FAO), Rome.

Florida Keys Public Libraries (n.d.) "Dead Fish" Photo Stream. www.flickr.com/photos/keyslibraries /sets/72157607307854787.

Froese, R., Zeller, D., Kleisner, K., and Pauly, D. (2012) What catch data can tell us about the status of global fisheries. *Marine Biology* 159, 1283–1292.

Frotté, L., Harper, S., Veitch, L., et al. (2009) Reconstruction of marine fisheries catches for Guadeloupe from 1950–2007. *Fisheries Catch Reconstructions: Islands, Part I. Fisheries Centre Research Reports* 17(5), 13–19.

Garibaldi, L. (2012) The FAO global capture production database: a six-decade effort to catch the trend. *Marine Policy* 36, 760–768.

Gerber, L., Morissette, L., Kaschner, K., and Pauly, D. (2009) Should whales be culled to increase fishery yields? *Science* 323, 880–881.

Gillett, R. (2009) *Fisheries in the Economies of Pacific Island Countries and Territories*. Asian Development Bank, Mandaluyong City, Philippines.

Gillett, R. (2011) Catch attribution in the Western and Central Pacific Fisheries Commission. WCPFC, Pohnpei, Federated States of Micronesia.

Gillett, R., McCoy, M., Rodwell, L., and Tamate, J. (2001) *Tuna: A Key Economic Resource in the Pacific Islands*. Pacific Studies Series. Asian Development Bank, Manila, Philippines.

Harper, S., Bothwell, J., Bale, S., et al. (2009) Cayman Island fisheries catches: 1950–2007. Fisheries Centre Research Reports, 3–11.

Harper, S., and Zeller, D. (2011) *Fisheries Catch Reconstructions: Islands, Part II. Fisheries Centre Research Reports* 19(4).

Huelsenbeck, M. (2012) Ocean-based food security threatened in a high CO_2 world: a ranking of nations' vulnerability to climate change and ocean acidification. Oceana, Washington, DC.

www.oceanacidification.org.uk/pdf/Ocean-Based_Food_Security_Threatened_in_a_High_CO2_World.pdf.

Jackson, J. B. C., Kirby, M. X., Berger, W. H., et al. (2001) Historical overfishing and the recent collapse of coastal ecosystems. *Science* 293, 629–638.

Jacquet, J. L., Fox, H., Motta, H., et al. (2010) Few data but many fish: marine small-scale fisheries catches for Mozambique and Tanzania. *African Journal of Marine Science* 32, 197–206.

Jacquet, J. L., and Zeller, D. (2007a) National conflict and fisheries: reconstructing marine fisheries catches for Mozambique. *Reconstruction of Marine Fisheries Catches for Key Countries and Regions (1950–2005)*. *Fisheries Centre Research Reports* 15(2), 35–47.

Jacquet, J. L., and Zeller, D. (2007b) Putting the 'United' in the United Republic of Tanzania: reconstructing marine fisheries catches. *Reconstruction of Marine Fisheries Catches for Key Countries and Regions (1950–2005)*. *Fisheries Centre Research Reports* 15(2), 49–60.

Kleisner, K., Zeller, D., Froese, R., and Pauly, D. (2012) Using global catch data for inferences on the world's marine fisheries. *Fish and Fisheries* 14, 293–311.

Le Manach, F., Andriamahefazafy, M., Harper, S., et al. (2013) Who gets what? Developing a more equitable framework for EU fishing agreements. *Marine Policy* 38, 257–266.

Le Manach, F., Gough, C., Harris, A., et al. (2012) Unreported fishing, hungry people and political turmoil: the recipe for a food security crisis in Madagascar? *Marine Policy* 36, 218–225.

McClenachan, L. (2009) Documenting loss of large trophy fish from the Florida Keys with historical photographs. *Conservation Biology* 23, 636–643.

McClenachan, L., and Cooper, A. (2008) Extinction rate, historical population structure and ecological role of the Caribbean monk seal. *Proceedings of the Royal Society of London Series B* 275, 1351–1358.

McClenachan, L., Jackson, J. B. C., and Newman, M. J. H. (2006) Conservation implications of historic sea turtle nesting beach loss. *Frontiers in Ecology and the Environment* 4, 290–296.

Mora, C. (2008) A clear human footprint in the coral reefs of the Caribbean. *Proceedings of the Royal Society of London Series B* 275, 767–773.

Pauly, D. (1995) Anecdotes and the shifting baseline syndrome of fisheries. *Trends in Ecology & Evolution* 10, 430.

Pauly, D. (1998) Rationale for reconstructing catch time series. *EC Fisheries Cooperation Bulletin.* 11(2): 4–7.

Pauly, D. (2006) Major trends in small-scale marine fisheries, with emphasis on developing countries, and some implications for the social sciences. *Maritime Studies* 4, 7–22.

Pauly, D., and Christensen, V. (1995) Primary production required to sustain global fisheries. *Nature* 374, 255–257.

Pauly, D., Christensen, V., Dalsgaard, J., et al. (1998) Fishing down marine food webs. *Science* 279, 860–863.

Pauly, D., Hilborn, R., and Branch, T. A. (2013) Comment: does catch reflect abundance? *Nature* 494, 303–306.

Pauly, D., and Watson, R. (2005) Background and interpretation of the 'marine trophic index' as a measure of biodiversity. *Philosophical Transactions of the Royal Society of London Series B* 360, 415–423.

Pauly, D., and Zeller, D. (2003) The global fisheries crisis as a rationale for improving the FAO's database of fisheries statistics. *From Mexico to Brazil: Central Atlantic Fisheries Catch Trends and Ecosystem Models*. *Fisheries Centre Research Reports* 11(6).

Pikitch, E. K., Santora, C., Babcock, E. A., et al. (2004) Ecosystem-based fishery management. *Science* 305, 346–347.

Rosenberg, A. A., Bolster, W. J., Alexander, K. E., et al. (2005) The history of ocean resources: modeling cod biomass using historical records. *Frontiers in Ecology and the Environment* 3, 84–90.

Secretariat of the Pacific Community (2008) Policy brief: fish and food security. www.spc.int/sppu/images/spc%20policy%20brief%201–2008%20fish%20and%20food%20security.pdf.

Swartz, W., and Pauly, D. (2008) Who's eating all the fish? The food security rationale for culling cetaceans. A report to Humane Society International. www.pewtrusts.org/uploadedFiles/wwwpewtrustsorg/Reports/Protecting_ocean_life/daniel_pauly_paper_iwc_2008_pdf_doc.pdf.

Swartz, W., Sala, E., Tracey, S., et al. (2010a) The spatial expansion and ecological footprint of fisheries (1950 to present). *PLoS ONE* 5, e15143.

Swartz, W., Sumaila, U. R., Watson, R., and Pauly, D. (2010b) Sourcing seafood for the three major markets: the EU, Japan and the USA. *Marine Policy* 34, 1366–1373.

Tremblay-Boyer, L., Gascuel, D., Watson, R., et al. (2011) Modelling the effects of fishing on the biomass of the world's oceans from 1950 to 2006. *Marine Ecology Progress Series* 442, 169–U188.

Tyedmers, P. H., Watson, R., and Pauly, D. (2005) Fueling global fishing fleets. *Ambio* 34, 635–638.

Van Houtan, K. S., McClenachan, L., and Kittinger, J. N. (2013). Seafood menus reflect long-term ocean change. *Frontiers in Ecology and the Environment* 11, 289–290.

Ward, M. (2004) *Quantifying the World: UN Ideas and Statistics*. Indiana University Press, Bloomington, IN.

Watson, R., Cheung, W., Anticamara, J., et al. (2013) Global marine yield halved as fishing intensity redoubles. *Fish and Fisheries* 14, 493–503.

Watson, R., Kitchingman, A., Gelchu, A., and Pauly, D. (2004) Mapping global fisheries: sharpening our focus. *Fish and Fisheries* 5, 168–177.

Watson, R., and Pauly, D. (2001) Systematic distortions in world fisheries catch trends. *Nature* 414, 534–536.

Wielgus, J., Zeller, D., Caicedo-Herrera, D., and Sumaila, R. (2010) Estimation of fisheries removals and primary economic impact of the small-scale and industrial marine fisheries in Colombia. *Marine Policy* 34, 506–513.

Worm, B., Barbier, E. B., Beaumont, N., et al. (2006) Impacts of biodiversity loss on ocean ecosystem services. *Science* 314, 787–790.

Zeller, D., Booth, S., Craig, P., and Pauly, D. (2006) Reconstruction of coral reef fisheries catches in American Samoa, 1950–2002. *Coral Reefs* 25, 144–152.

Zeller, D., Booth, S., Davis, G., and Pauly, D. (2007a) Re-estimation of small-scale fishery catches for US flag-associated island areas in the western Pacific: the last 50 years. *Fishery Bulletin* 105, 266–277.

Zeller, D., Booth, S., Pakhomov, E., et al. (2011a) Arctic fisheries catches in Russia, USA, and Canada: baselines for neglected ecosystems. *Polar Biology* 34, 955–973.

Zeller, D., Booth, S., and Pauly, D. (2007b) Fisheries contribution to the gross domestic product: underestimating small-scale fisheries in the Pacific. *Marine Resource Economics* 21, 355–374.

Zeller, D., Darcy, M., Booth, S., et al. (2008) What about recreational catch? Potential impact on stock assessment for Hawaii's bottomfish fisheries. *Fisheries Research* 91, 88–97.

Zeller, D., and Harper, S. (2009) *Fisheries Catch Reconstructions: Islands, Part I. Fisheries Centre Research Reports* 17(5).

Zeller, D., and Pauly, D. (2004) The future of fisheries: from 'exclusive' resource policy to 'inclusive' public policy. *Marine Ecology Progress Series* 274, 295–303.

Zeller, D., Rossing, P., Harper, S., et al. (2011b) The Baltic Sea: estimates of total fisheries removals 1950–2007. *Fisheries Research* 108, 356–363.

SEVEN

Back to the Future

Integrating Customary Practices and Institutions
into Comanagement of Small-scale Fisheries

JOHN N. KITTINGER, JOSHUA E. CINNER, SHANKAR ASWANI,
and ALAN T. WHITE

In many parts of the world, marine-resource governance systems include aspects of customary marine tenure and traditional sociocultural institutions for resource management. These practices are rooted in historical context and vary by culture and location, with place-specific practices and customs that are based on local knowledge systems. In this chapter, we review the incorporation of customary practices into contemporary management, highlighting the roles of social history, changes in customary practices, and their application in, and influence on, modern legal and policy contexts. Next, we discuss the challenges and opportunities of integrating historical management practices into modern governance systems, exploring the roles of comanagement and participatory approaches in successful "back to the future" management approaches. To conclude, we look to the future of integrated management systems and their potential to address social-ecological challenges in coastal areas that face increasing population densities and growing dependence on coastal and marine resources.

INTRODUCTION

Globally, small-scale fisheries support food security for millions (Berkes et al. 2001, Delgado et al. 2003) and employ 90% of the world's capture fishers (FAO 2012, Teh et al. 2013). By some estimates, these fisheries account for more than half of global landings, albeit with a much lower environmental footprint than industrialized commercial fisheries (Chuenpagdee et al. 2006, Jacquet and Pauly 2008, McClanahan et al. 2009). In regions where growing coastal populations depend on nearshore resources for sustenance and livelihoods (White

2010, Bell et al. 2009), there is a clear need for efficient and effective management systems that effectively protect coastal resources for present and future use (Burke et al. 2011).

Small-scale and artisanal fisheries are often small or localized and use traditional or simpler gear types. Financial investments in such fishing operations are usually made by individuals, families, or small firms rather than by large corporate entities (Kittinger 2013a). Fisheries resources that are locally harvested are often used directly for subsistence or can feed into market systems that reach from local to global scales. Most small-scale fisheries are also characterized as data-poor, presenting a distinct set of challenges for management models that rely on intensive resource assessments (Berkes et al. 2001, Kittinger 2013a, Starr et al. 2010; also see chapter 5, this volume). However, for many small-scale fisheries, collective governance systems were developed historically to manage the interaction between communities and fisheries resources. These customary management systems exhibit tremendous diversity in terms of governance arrangements and the strategies and tools employed but are similar in that most systems reflect the historical context of local cultures and resource systems.

There is mounting evidence that selected customary management systems have achieved positive social and environmental outcomes for fisheries resources and resource-dependent communities (Aswani 2011, Cinner et al. 2005, Friedlander et al. 2013, McClenachan and Kittinger 2012, McClanahan et al. 2006). For this reason, conservation groups, social and ecological scientists, grant-making organizations, and resource managers have become increasingly interested in these customary systems and how they can be effectively integrated into conventional resource-management programs (Aswani and Ruddle 2013, Cinner and Aswani 2007, McClanahan et al. 2009, White et al. 1994, Clarke and Jupiter 2010).

Here, we review recent literature on customary management and explore the integration of historically based practices into modern management efforts. First, we provide brief overviews of customary and conventional fisheries management, highlighting the similarities and differences between these approaches. Next, we present case studies from Hawai'i and the Marquesas Islands to highlight how historical research can shed light on the social and ecological dimensions of customary management systems. We then review evidence for the efficacy of comanagement systems, which often incorporate historically based customary practices into conventional management, with decision making that is shared between local resource users and state management agencies. Finally, we discuss the institutional arrangements conducive to developing and maintaining the integrated approaches that merge customary and conventional management systems. Our overarching purpose is to explore strategic ways of integrating historical practices with modern management approaches to increase local acceptance and enhance conservation and management of fisheries.

Customary Fishery Management Systems

Following Cinner and Aswani (2007), we define "customary management" as practices that are historically designed to regulate the use, access, and transfer of resources through sociocultural institutions. Customary management is informed by local and traditional ecological knowledge, developed through intergenerational accumulation of knowledge based on human interactions with the local environment, and embedded in customary land and sea tenure institutions. Although many customary management systems have ancient origins, it is critical to emphasize that these systems are dynamic and continually evolving through adaptive processes, including the introduction, loss, and combining of different knowledge types and understandings (Berkes et al. 1998).

In practice, customary fisheries management generally includes restrictions on one or more of the following: (1) spatial areas, (2) time, (3) gear or harvesting technology, (4) effort, (5) types of species that can be harvested, and (6) the amount of resource harvested (Table 7.1). It is important to note that customary management systems not only manage fisheries, but also have social objectives such as harmony and cultural continuity. Social features of these systems commonly emphasize local social order, identity, and place. They also promote historical continuity and link to territories and associated ancestors (Aswani and Ruddle 2013). Although our focus in this chapter is on the Asia-Pacific region, customary management systems are widespread in coastal cultures worldwide.

Conventional Fisheries Management

Conventional fisheries management, like customary systems, varies widely in relation to geography, environment, and the governance-system context. In the Asia-Pacific region, several countries (e.g., Indonesia, Philippines, and parts of Micronesia) have variable national and/or local legal and institutional systems that control the use of coastal and marine resources. Such institutions have evolved in response to the need to manage and conserve marine fisheries and, in particular, to protect habitats upon which important fisheries depend (e.g., coral reef, seagrass, mangrove, and estuarine systems). Typical interventions in conventional fisheries management include controls on fishing methods (e.g., the most destructive gear are banned), legal protection of threatened and vulnerable species (e.g., sea turtles, some finfish, sharks, and marine mammals), and restrictions on gear or species. One of the most common interventions is spatial management using marine protected areas (MPAs). These often include zoning to prescribe uses and to control fishing effort, gear, or access. Table 7.1 provides a comparison of customary and conventional fishery management and examples of integrated approaches.

HISTORICAL PERSPECTIVES ON CUSTOMARY FISHERIES MANAGEMENT

One of the difficulties associated with historical research on customary management systems is that researchers must rely on observational, descriptive, and reconstruction-based methods—often using poor and incomplete datasets—rather than the experimental approach that is a hallmark of scientific research (Diamond and Robinson 2010). To address this, researchers often rely on in-depth case studies that make use of a diversity of data types, including paleontological, archaeological, historical, and ecological data (Lotze et al. 2011; see Box 7.1 by Rafe Sagarin, who describes the importance of observation in ecology

TABLE 7.1 Comparisons of Customary and Conventional Fisheries Management and Application to Integrated Approaches

Customary Management	Description	Analog in Conventional Management	Examples of Integrated Approaches
Spatial restrictions	Temporary or permanent closure of areas to fishing	Marine protected areas (MPAs), temporary fisheries closures	Locally managed marine areas, with established no-take zones to replenish resources if needed; periodic harvests from community MPAs
Temporal restrictions	Restricting harvesting activities during specific days or periods; often short in duration, specific to certain species, and for a specific event (e.g., for religious ceremonies or to protect spawning aggregations)	Closed seasons	Community-based moon calendars that show which species are spawning and should be protected
Gear restrictions	Prohibitions or restrictions on certain harvesting methods or techniques; chiefly control of materials for fishing gear and boats, which have limited access to some fisheries resources	Gear prohibitions	Restrictions on certain gear (e.g., for gill or lay nets, or no spearfishing with SCUBA)
Effort restrictions	Limits on who can access certain areas (e.g., only residents of a community can access the adjacent reef); limiting who can harvest certain species, use certain gear, or fish certain areas	Permitting; territorial user-rights systems for fisheries; limited-entry fisheries	Community-based subsistence fishing areas with rules that are developed in an inclusive, placed-based manner; permitted access for local families or residents of a province or district
Species restrictions	Prohibitions on consumption of certain species, often related to class, gender, or lineage	Protection of vulnerable or endangered species	Bans on certain species until populations are regenerated; limits on harvest for culturally significant species or resources that contribute significantly to local food security
Catch restrictions	Restricting the quantity of a harvest; social norms discourage wasting and other harmful practices	Total allowable catch limits; individual transferable quotas for fisheries with catch limits	Communal harvest events to sustain connections to local resources; educational and outreach programs to connect community members and build social capital

| Aquaculture | Creation of fishponds, which sequester nutrients from uplands, reducing land-based pollution, and potentially take pressure off wild stocks | Modern aquaculture | Rebuild and revitalize fishponds so they can provide fisheries resources to communities; explore creation of Community Supported Fisheries models for local fishponds and generate revenue for restoration |
| Enforcement | Violations of customary restrictions result in sanctions or punishments that may be severe | Fines, penalties, license revocation | Develop and implement a penalty schedule of graduated sanctions that includes community service by violators in restoration activities |

Sources: Adapted from Cinner and Aswani (2007:203: table 1), McClenachan and Kittinger (2012: table 3), and Friedlander et al. (2013: table 1).

BOX 7.1   Viewpoint from a Practitioner: Marine Historical Ecology
and the Renewed Power of Observation
Rafe Sagarin

My first real scientific study in the early 1990s had fellow undergraduate Sarah Gilman and I poking through the tidepools by the Hopkins Marine Station in Monterey, California, in the dead of night counting invertebrates in the exact locations where they had been surveyed 60 years previously by graduate student Willis G. Hewatt. Not exactly earth-shattering science, but to our surprise, our counts revealed a dramatic pattern of change that fit the expectations—previously only shown in theory—of how species would respond to climate warming (Sagarin et al. 1999). Remarkably at that time, a handful of other equally simplistic studies, but these on birds and amphibians and butterflies and plants, were revealing similar results for the first time (Schneider and Root 2001). Just by counting animals in historical context, we and our colleagues—many of them also students—had discovered something timely and globally applicable.

Despite the success of the study, it defied all the rules of science that I had been taught: it wasn't experimental; we were trying to infer causation from correlation; and the whole thing didn't start with a nice set of hypotheses, but instead was a big old fishing expedition. It was my first clue that the way science "should" be is not the way that science actually is.

A decade later, I found myself on part of a voyage to retrace the storied 1940 expedition of two of my marine biology and literary heroes, Edward Ricketts and John Steinbeck, to the Gulf of California (Sea of Cortez). Their journey and the book they wrote about it, *Sea of Cortez* (Steinbeck and Ricketts 1941), was remarkable not just for the extensive surveys they did of intertidal life throughout the Gulf, but for how they weaved these observations within a philosophical narrative that espoused building observations, unbounded by predetermined hypotheses, into a holistic picture of an interconnected system—what we might now call a "social–ecological system." But despite their broad view and keen sense of both the little and big pictures, they lamented

at the end of the journey that they "could not yet relate the microcosm of the Gulf with the macrocosm of the sea." By contrast, owing to the passage of time and the deluge of troubling ecological observations from around the world's oceans, on our journey we were quite easily able to connect the microcosm of our observations—of climate-related distributional shifts, the destructive force of coastal development, and the loss of top predators and ascendance of others—with the macrocosm of what is happening to oceans around the world (Sagarin et al. 2008). Again I learned that simple observations are incredibly powerful, and their power is increased by some magnitude when they are repeated through time.

The continual surprises yielded by historical ecology are both a function of lowered expectations (how can such simple studies tell us anything important?) and the serendipitous byproducts of a recent adaptation of ecological science itself. We have been led to believe—by textbooks, by professors who contemptuously sneer that "correlation *does not* indicate causation," and by funders that continually reject discovery-driven science as "just a fishing expedition"—that the "best" science, the "hardest" science, is that which is strictly controlled, driven by well-circumscribed hypotheses, and infinitely replicable. But quietly, in the background, a paradigm shift has been occurring in our science (Sagarin and Pauchard 2012). The urgency of large-scale ecological change combined with our new abilities to observe this change—sometimes afforded by amazing new technologies, and sometimes just a gift of time and hindsight—is making us all question our notions of what science "should" be. In place of a strict prescription of the one right scientific method, all sorts of ways of understanding what actually "is" going on in our world—including citizen science, tracking animal-borne sensors, and digging up all sorts of historical records—are fast becoming the best tools we have to understand a complex and dynamic Earth system.

Rafe Sagarin is Program Manager for Oceans at the Institute of the Environment and Biosphere 2, University of Arizona.

and the value of these approaches both in historical research and in a current resurgence of natural-history approaches). Increasingly, marine historical ecology research has included a focus on understanding not only changes in species, habitats, and ecosystems, but also historical social dynamics and social–ecological relationships (Aswani and Allen 2009, Kittinger et al. 2011, Hunt and Lipo 2011). Below, we present two in-depth historical analyses that illustrate the importance of historical perspectives on customary management systems.

Customary Reef Fisheries Management Systems in Precontact Hawai'i

The Hawaiian Islands were among the last habitable places on Earth to be colonized by humans. Voyaging Polynesians arrived in Hawai'i at approximately AD 1250–1290 (Wilmshurst et al. 2011), and thereafter they established complex societies and resource production systems that supported a dense human population with complex sociopolitical systems (Vitousek et al. 2004, Kirch 1985). Polynesians introduced exotic species and utilized both terrestrial and marine ecosystems for basic subsistence, altering endemic populations of fauna and flora and transforming natural ecosystems into cultural landscapes and seascapes in the process (Kittinger et al. 2011, Athens 2009, Burney et al. 2001).

In the precontact period (before AD 1778), Hawaiian societies were able to use customary reef management systems to manage coral reefs sustainably and return high catch rates. Kittinger et al. (2011) reconstructed social–ecological relationships to test for sustainable levels of human use and impact in Hawaiian coral reefs. This approach used an intensive review and assessment of diverse information to assess changes in the ecological status of coral reef biota. The information used included archaeological deposits, ethnohistorical and anecdotal descriptions, and modern ecological and fishery data. The authors discovered previously undetected recovery periods in coral reefs in precontact Hawai'i, which they attributed to effective customary reef-management systems (Figure 7.1). These management systems were embedded in a larger-scale cultural adaptation to the ecological constraints of living in the world's most isolated archipelago. Precontact Hawaiian societies developed a broader subsistence base, relying increasingly on agriculture and animal husbandry to provide less risky sources of protein in their diets (Kirch 2002, Kirch and O'Day 2003).

Throughout the precontact period, reef fisheries remained a focus for foraging and a basis for food production systems. Incredibly, fisheries yields produced 15 mt year^{-1} over several centuries prior to Western contact (Figure 7.1), a production level much higher than has been proposed as the sustainable yield for island-based coral reef fisheries currently (Newton et al. 2007). These high yields were enabled by complex sociocultural institutions for reef fishery management. Customary management limited impacts by employing approaches such as time and area closures; regulation of gear, effort, and materials; social norms that discouraged overharvest and waste; restrictions on consumption of vulnerable species (including gender- and class-specific consumption restrictions); and place-based restrictions that allowed reef use only by community members living in adjacent watersheds (McClenachan and Kittinger 2012). Reef fisheries were multispecies and managed with

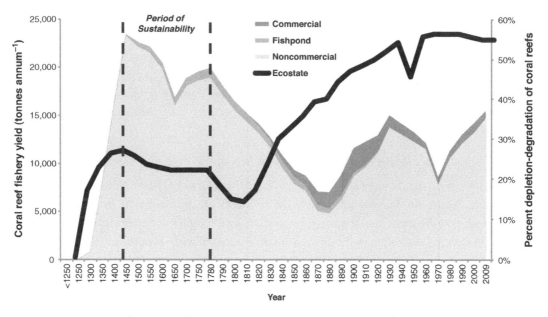

FIGURE 7.1 Historical analyses of coral reef ecosystem conditions and fishery yields in Hawai'i. Coral reef ecosystem reconstruction (black line) represents averages of changes in percent depletion–degradation in seven trophic guilds of coral reef biota (for detailed methods, see Kittinger et al. 2011). Coral reef fisheries yield reconstructions (shaded areas) were estimated using catch reconstruction methods. These combine per capita catch rates for noncommercial yield (subsistence, recreational, and cultural fishing) with reported fisheries yield data for commercial and fishpond aquaculture sectors (for detailed methods, see McClenachan and Kittinger 2012). Dashed lines bound a period of sustainability wherein customary reef management systems protected ecosystem integrity and provided high levels of ecosystem goods and services to precontact Hawaiian societies. Time is represented on the horizontal axis (AD 1250–2009); 50-year increments are used in the precontact period (AD 1250–1778), and decadal increments are used for the historical and modern period (AD 1778–2009).

adaptive strategies, many of which have modern analogs (Table 7.1). Rules were accompanied by robust sociocultural institutions that exhibited a high degree of compliance, given draconian punishments for rule violators and incentives for adherence (McClenachan and Kittinger 2012).

In Hawai'i, cultural adaptation to island environments resulted in active ecosystem engineering, with conserving mechanisms that included the development of customary management practices and a diversified subsistence strategy. These approaches likely evolved culturally to minimize risks of food insecurity in an island environment prone to resource patchiness and unpredictable disturbances such as droughts, floods, tsunamis, hurricanes, and large storms. Precontact Hawaiian societies were able to manage coral reefs sustainably by using customary management systems that protected ecosystem integrity and provided high levels of ecosystem goods and services (Figure 7.1). Such sustainability has often eluded current management, and the ability of customary systems to provide these linked social and environmental outcomes has become the focus of much scholarship and knowledge-to-action partnerships. It should also be noted, however, that the historical context is much different than the contemporary one. Hawaiian societies were most likely assisted by natural barriers that limited reef access (e.g., seasonal swell events and distance from human settlements), many of which are easily overcome today through technological advancements.

A Marquesan Coral Reef in Historical Context

At Anaho Bay, in the Marquesas Islands, understanding historical human and ecological interactions has been crucial for making present-day decisions around management of coastal fisheries and reef ecosystems. Aswani and Allen (2009) combined biological and anthropological data to understand the historical factors that have led to the current reef conditions in this bay. Marquesan coral reefs are of particular interest because they are so spatially limited, in sharp contrast to the coral-rich atolls of the nearby Tuamotu Islands and the reef-fringed volcanic Society Islands farther west. Archaeologically, the limited coral reefs of the Marquesas have been seen as a major challenge to human settlement, adaptation, and long-term survival in the islands. The materials recovered in archaeological deposits document the long-standing importance of reef resources to the local population, with evidence in the archaeological record for sustainable levels of extraction (Aswani and Allen 2009). These researchers found that for several centuries, a relatively stable fishery under the control of a customary system supported several hundred people.

The archaeological work of Allen (2006) also showed that during the seventeenth century, terrigenous sedimentation increased, possibly in response to increased El Niño–Southern Oscillation (ENSO) activity. Changing climatic conditions, in combination with greater cultural use of the valley interior, may have amplified erosion, leading to increased sedimentation of coral reefs. The nineteenth-century arrival of exotic European grazers (especially goats) appears to have further accelerated erosion and sediment mobilization at Anaho Bay. Increased siltation combined with ENSO events may also underlie more recent outbreaks of ciguatera poisoning, which have forced both residents and transient visitors to turn to other resources (Allen 2006). This interpretation is substantiated by biological results, which indicate that although large predatory reef fish are abundant, the reef is currently in a state of decline that was preceded by centuries of sustainable interactions. Such historical studies will be relevant for managing resources across the insular Pacific, as practitioners work to hybridize customary and conventional fisheries management approaches (Aswani and Ruddle 2013).

Evidence of Outcomes of Current Comanagement: A Comparative Approach

Although we cannot go back in time and directly evaluate historical customary management practices, research on contemporary implementation of these systems can shed light on both their historical efficacy and their use in modern contexts. Several studies have applied comparative analyses to examine the performance of modern comanagement, using data-driven analyses that compare managed and unmanaged areas and the factors that

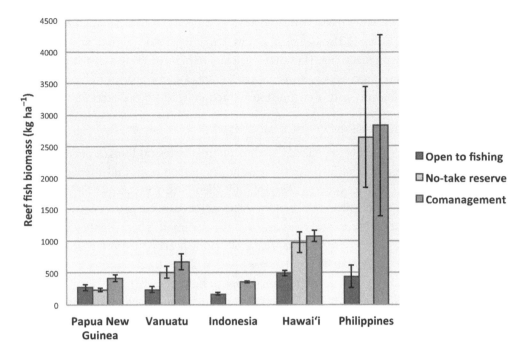

FIGURE 7.2 Comparison of coral reef fish biomass under open-access conditions, top-down government management areas, and comanagement arrangements incorporating aspects of local or customary management across sites in the Asia-Pacific region. Data are from Williams et al. (2008), Friedlander et al. (2007, 2013), McClanahan et al. (2006), and White et al. (2008). Estimates are from multiple sites at each location and do not consider differences in management area size, productivity, or other factors that may affect reef fish biomass and the efficacy of different management regimes.

affect success. Key factors include variables such as population size, wealth, market integration, and other socioeconomic characteristics of fisheries and social institutions (Cinner et al. 2012). These comparative analyses show that customary practices provide some conservation benefits, but effects were limited largely to targeted fish and invertebrates (rather than habitat builders, such as corals; Figure 7.2). For example, McClanahan et al. (1997) found that sacred sites in Kenya were able to increase fish catch in adjacent landing areas but were ineffective at protecting species diversity or ecological function (McClanahan et al. 1997). In Papua New Guinea and Indonesia, customary closures showed significantly higher biomass of target fish than open-access areas. These closures also were effective at conserving both long- and short-lived fish species (Cinner et al. 2005, Cinner and McClanahan 2006, Campbell et al. 2012). In a related study of 11 sites throughout Papua New Guinea and Indonesia, McClanahan et al. (2006) compared the ecological effectiveness of customary management, MPAs comanaged by communities and nongovernment organizations, and national marine parks. At all three sites practicing customary reef closures, there was higher fish biomass inside the reserves, whereas only one of four comanaged sites and none of the four national parks showed a difference in biomass for inside versus outside sites. However, none of the areas appeared to conserve other ecosystem indicators such as coral diversity, fish species richness, or fish abundance. Similar results have been found in other contexts, including Hawai'i (Friedlander et al. 2013) and Indonesia (Campbell et al. 2012).

In other cases, however, customary management practices were shown to benefit not only fish stocks, but also other aspects of fisheries ecosystems. For example, in the Aceh province of Indonesia, coral cover protected by customary management was almost four times higher than in nearby open-access areas subject to destructive fishing practices such as bombing and cyanide use (Baird et al. 2005). Similarly, clam species (*Polymesoda* spp. and *Anadara granosa*) were more abundant in fully protected and periodically protected areas under local management than in open sites in Roviana Lagoon, Solomon Islands, with rotating closure sites having more clams overall (S. Aswani, unpublished data). In this region, greater fish diversity, biomass of fish (particularly grazing species), and herbivory occurred inside the managed areas that integrated modern MPAs with customary management than in the adjacent areas that were open to local fishing (Aswani et al. 2007).

Although these customary management systems have been able to produce outcomes consistent with some conservation goals, the spatial scale at which they operate is often very small (i.e., on the order of tens of hectares; McClanahan et al. 2006). Whether these systems are large enough to sustain ecosystem processes and functions if conditions in the adjacent areas become highly degraded is a question of critical importance (Foale and Manele 2004). The small scale of these systems raises questions about the mechanisms driving differences observed inside customary management systems compared with other areas, and the types of ecosystem functions they can preserve (Jupiter et al. 2012). For example, are fish migrating from outside customary management areas or is growth and recruitment occurring inside these managed areas? Do customary management systems preserve broader ecosystem functions, such as predation and larval exchange (e.g., see Almany et al. 2013)? To answer these sorts of questions, more research is needed to (1) investigate ecological conditions before and after harvests following temporary closures, (2) monitor fish catch during regular fishing activities and from harvesting events, and (3) provide understanding of the spatial patterns of fishing effort to determine whether temporary closure areas become severely overfished after they are opened.

AN INSTITUTIONAL PERSPECTIVE ON INTEGRATING CUSTOMARY AND CONVENTIONAL MANAGEMENT SYSTEMS

Successfully integrating customary and conventional resource-management systems depends on how such actions can be implemented by human institutions. By "institutions," we refer to the diversity of conventions, norms, and formally sanctioned rules of society, implemented and structured through the formal and informal human organizations that mediate collective social actions (Ostrom 1990, Young 2002). Integrating indigenous and Western scientific knowledge and management can be difficult, because goals, social and ecological benefits, and spatial scales often differ (Bohensky and Maru 2011, Cinner and Aswani 2007). Comanagement is one productive strategy that managers and practitioners have explored for integrating these systems (Cinner et al. 2012, Berkes 2009). Comanagement can take many forms but generally involves sharing management authority between local resource users or community groups and central government authorities (Berkes 2010). Below, we review two key aspects of integrating customary and Western resource management in comanagement: (1) institutional design of integrated systems and (2) strategies to maintain the resilience and robustness of resource-management arrangements.

Given the diversity of human institutions involved in marine management, from community-based stewardship initiatives to international agreements for migratory stocks, it can be daunting to consider general ways in which customary practices can be implemented into conventional management efforts. A common characteristic of marine resources, however, is that they are primarily "common pool" resources. Such resources are those from which it is costly to exclude potential beneficiaries. Common-pool resources often generate collective-action problems, due to divergences between group and individual-level interest. In turn, these divergences complicate efforts to avoid the deterioration of a shared resource. Customary management systems for common-pool resources are often predicated on collective social action to regulate consumption, use, and maintenance (Basurto and Ostrom 2009, Ostrom 1990). By contrast, where such collective social actions are not present, resources can be overused and degraded to the point of collapse, a phenomenon referred to as the "tragedy of the commons" (Hardin 1968) and often couched in the language of Malthusian overfishing (Pauly 1990, McClanahan et al. 2008).

Practitioners seeking to design effective strategies to integrate customary and conventional management often draw on existing literature on common-pool governance regimes. For example, Elinor Ostrom proposed several key institutional design principles associated with long-enduring common-property resource systems (Cox et al. 2010, Ostrom 1990). Ostrom and others have used these principles to evaluate how well different governance arrangements preserve common-pool resources (Ostrom 1990, Cinner et al. 2012). These design principles provide a blueprint for developing effective management systems where coastal resource users and stakeholders can engage in collective social action for resource stewardship (Table 7.2).

The principles in Table 7.2 provide a framework for designing effective resource-comanagement arrangements, but how do such arrangements come to fruition in real-world contexts? The development of novel institutional arrangements is described as "emergence" in the scholarly literature (de Haan 2006), and the ability of such arrangements to persist through time is often referred to as "robustness" (Anderies et al. 2003). Below, we illustrate the emergence of comanagement arrangements using examples from the Southeast Asian countries of the Philippines and Indonesia. Both of these countries have a high dependence on coastal and marine resources, as well as the potential for historical marine resource-management practices to inform conventional fisheries management.

TABLE 7.2 Institutional Design Principles and Strategies for Incorporating Customary Practices and Modern Management into Comanagement Arrangements

Institutional Design Principle	Strategy for Integration into Comanagement Governance Arrangements
1. Clearly Defined Boundaries and Membership:	
A. The boundaries of the resource system must be well defined.	A. Boundaries are decided and agreed upon at the local level, with state-level recognition and codification in official rules and regulations; boundary revision processes should allow for local capacity to alter boundaries.
B. Individuals or households who have rights to withdraw resource units from the resource system must be clearly defined.	B. Membership rights should reflect customary practices, which are often based on traditional kinship groups, social networks, or place-based rights systems; in some situations, the state may maintain legal rights to the resource system but cede membership rules to local-level institutions.
2. Congruence between Appropriation and Provision Rules and Local Conditions:	
A. Appropriation rules restricting time, place, technology, and quantity of resource units are related to local conditions.	A. Rules are developed in a place-based manner at the local level in cooperation with state-level resource management authorities; flexibility is maintained for locally adaptive rules that comply with minimum requirements for ecological integrity and social inclusiveness set at the state level.
B. The benefits obtained by users, as determined by appropriation rules, are proportional to the amount of inputs required in the form of labor, material, or money, as determined by provision rules.	B. Local-level institutions establish processes that incentivize stakeholder participation levels by tying costs to benefits.
3. Collective-choice Arrangements: Most individuals affected by operational rules can participate in modifying operational rules.	Local decision-making processes are established that allow for multi-stakeholder participation; local planning and stakeholder engagement processes are developed in concert with state-level authorities and follow standards to ensure accessibility and transparency.

(continued)

TABLE 7.2 *(continued)*

Institutional Design Principle	Strategy for Integration into Comanagement Governance Arrangements
4. Monitoring: Monitors, who actively audit common-pool resource conditions and user behavior, are accountable to the users or are the users themselves.	State-level monitoring programs are reconfigured to address the geography of local-level resource systems; local resource users must be involved in the monitoring process in order to participate in local decision-making processes.
5. Graduated Sanctions: Users who violate operational rules are likely to receive graduated sanctions (depending on the seriousness and context of the offense) from other users, from officials accountable to these users, or from both.	State-level authorities and institutions enforce penalties, and rules for graduated sanctions are developed together by local and state-level authorities. Local-level and state-level enforcement programs are integrated.
6. Conflict-resolution Mechanisms: Users and their officials have rapid access to low-cost, local arenas to resolve conflict among users or between users and officials.	Local-level resource users develop a conflict resolution process that is accessible and fair for local resource users; process has a hierarchical design so that conflicts that cannot be resolved at the local level can be elevated upward through higher-level adjudication processes.
7. Minimal Recognition of Rights to Organize: The rights of users to devise their own institutions are not challenged by external governmental authorities.	State-level legislation establishes a clear process for local-level institutions to self-organize, with legal recognition of institutional rights to self-govern and low transaction costs for establishment and recognition; local planning and stakeholder engagement processes are developed in concert with state-level authorities and follow standards to ensure accessibility and transparency.
8. Nested Enterprises: Appropriation, provision, monitoring, enforcement, conflict resolution, and governance activities are organized in multiple layers of nested enterprises.	State- and local-level integration processes are established through comanagement legislation or associated policy, which define hierarchical structure for monitoring, enforcement, conflict resolution and decision-making processes; local-level institutions are linked together in learning networks that are supported by the state; communication and information flow through all levels is maintained through established processes and informal leadership networks.

Source: Institutional design principles modified from Cox et al. (2010).

Institutional Design and the Emergence of Integrated Systems

In the Philippines and Indonesia, a hybrid form of marine resource management has recently emerged. These management systems draw on both historical traditions and modern management and governance approaches. In both countries, responsibility for managing coastal resources has been decentralized to local government units. In the Philippines, the Local Government Code (1991) mandated coastal municipalities and cities to manage coastal areas to 15 km offshore (White et al. 2006). Each coastal municipality in the Philippines is composed of *barangays,* which are the equivalent of villages. Each *barangay* elects its leader, and *barangay* representatives sit on a municipal council. In coastal areas, the primary concerns of the *barangay* leader and council often include the regulation and management of their coastal reefs, local fisheries, shoreline access, beaches for tourism, and related resources (White et al. 2006). In a few areas of the Philippines, recognized indigenous groups can also apply for a Certificate of Ancestral Domain, which describes the extent of the groups' traditional lands and waters and allows these groups to develop management plans for terrestrial and coastal resources. In Indonesia, the customary rules under what is termed "*sasi* law," or use rights, are being considered within modern legislation at the local government level (Campbell et al. 2012).

While the local government units in the Philippines must follow national laws, they are free to prescribe local ordinances, develop local management plans, zone their coastal waters, license users, limit gear use, and institute other approaches. One result of this decentralization is that small, locally planned and implemented MPAs and resource-management schemes have sprung up throughout the archipelago. The planning and location of these small MPAs (10–100 ha) are normally determined by local use patterns, location of important fishing grounds, division of reef uses between fishers and tourists, local beliefs (e.g., not fishing in front of a cemetery), and other factors. The bottom line is that customary management factors into the planning and implementation of what is now a nationally recognized system of MPAs. In addition, many of the municipal governments working with their *barangay* units have formulated what they call "coastal resource management plans," which are integrated plans that cover their entire area of jurisdiction (White et al 2006). By definition, these plans must be community based and follow common principles of comanagement to be effective. Some of the key tenets of comanagement in the Philippines, and in other countries where it is being applied in an effective manner, are shown in Table 7.2.

Another example is the management of biocultural resources in the Papahānaumokuākea Marine National Monument, a large marine reserve in the northwestern Hawaiian Islands. As described in Box 7.2, management systems for this unique area were developed at the outset to bridge gaps among science, policy, and cultural knowledge and traditions. Such approaches, when undertaken in the early phases as they were in Papahānaumokuākea, can set the stage for better future management.

BOX 7.2   Viewpoint from Practitioners: Biocultural Resource Management in Papahānaumokuākea, Hawaiian Archipelago

'Aulani Wilhelm and Randall Kosaki

The 2010 inscription of Papahānaumokuākea as a World Heritage site by the United Nations Educational, Scientific and Cultural Organization (UNESCO) redefines the value of oceans to the collective heritage of humankind. The designation also recognizes the multifaceted relationship that indigenous people of Hawai'i still have with this ocean expanse. Like its prior national designations—first as a Coral Reef Ecosystem Reserve and then as a Marine National Monument—this distinction honoring both nature and culture was not bestowed by chance. It was the result of direct engagement by Native Hawaiians, in partnership with many others, in the design, advocacy, and implementation of the management regime put in place to protect more than three quarters of the Hawaiian Archipelago.

From a cultural perspective, this vast (362,074 km²) ocean domain is a place deeply rooted in the Native Hawaiian worldview. Crossing the Tropic of Cancer, Papahānaumokuākea is a physical manifestation of the intersection between *pō* (night, realm of the gods) and *ao* (daylight, realm of life). It is a seascape dominated by the coral polyp, the first Hawaiian ancestor born in the darkness of *pō*, as described in the Kumulipo genealogy of creation. As the westernmost place in our Hawaiian universe, many believe these islands and seas are the pathway upon which Native Hawaiians travel after death, returning to *pō*. This intimate kinship we have to Papahānaumokuākea has profound implications for contemporary management.

Because of both ancestral connections and ecological significance, our management regime emphasizes integration—the bridging of science, policy, and cultural knowledge, traditions, and practices—to forge successful management strategies and methodologies appropriate for both natural and cultural resource management. Our management is built on a foundation of Native Hawaiian values and practices that incorporate keen observation and understanding of the natural world; indigenous principles and philosophies; cultural norms and community relationships; and unique epistemologies deeply embedded in and formed by the relationship of people with place. While it is an ever evolving process to grow in these capacities, this approach is unusual in modern marine management by state and federal agencies.

A cornerstone of our work in managing Papahānaumokuākea has been the direct involvement of cultural practitioners in our policy, management, and research. This approach supports and respects a broad spectrum of methods for observation and knowledge acquisition. For example, cultural objectives are incorporated in the site's 15-year management plan, a milestone achievement for each of the seven administering agencies. Regulations governing the region were developed using Hawaiian terms and definitions to set permit criteria. Intertidal research is conducted in partnership with skilled fishermen who possess knowledge passed down through generations, and archaeology by native scholars has redefined theories of settlement and use of the area. Tradition-based ceremony is now conducted regularly as part of access protocols.

This biocultural approach has increased understanding of the physical, spiritual, and intellectual functions and role of places for people, shaping access requirements, best management practices, and required pre-access training. The success of our work has begun to demonstrate the broad application and relevance of traditions and practices to conservation science, education, policy, and law today.

'Aulani Wilhelm is Superintendent, and Randall Kosaki is Deputy Superintendent, of Papahānaumokuākea Marine National Monument.

Maintaining Resilience and Durability of the Comanagement Arrangement

As in the examples described above, customary management in many locations has been integrated with the goals, techniques, and institutions of conventional fisheries management. In the marine environment, these "hybrid" institutions may include customary governance structures (e.g., community or local councils) able to do one or more of the following: (1) allocate catch quotas in individual transferable quota systems (Adams 1998), (2) use traditional ecological knowledge to locate and temporarily restrict fishing in spawning aggregation sites of commercially valuable species (Graham and Idechong 1998, Drew 2005), (3) map vulnerable benthic habitats for integration into conservation plans (Aswani and Lauer 2006), (4) adaptively experiment with gear restrictions (Adams 1998, Cooke et al. 2000, McClanahan and Cinner 2008), (5) implement temporary closures to manage stocks with no previous commercial value (Ruttan 1998, Thorburn 2001, Hickey and Johannes 2002), and (6) establish community-owned and -managed MPAs (Cooke et al. 2000, Johannes 2002, Hickey and Johannes 2002, Aswani et al. 2007).

Hybrid community-based gear restrictions and MPAs based on customary governance structures have been developed in Palau, the Cook Islands, the Solomon Islands, Fiji, Samoa, Hawai'i, Vanuatu, and other nations and territories in the Asia-Pacific region (Graham and Idechong 1998, South and Vietayaki 1998, Johannes 2002, Cinner and Aswani 2007). Hybrid management regimes are often characterized by a higher level of institutional adaptability and flexibility. Such flexibility is less commonly found in the top-down governance programs implemented by state resource-management agencies, which tend to be constrained by managerial and statutory requirements (Aswani et al. 2007). Often this adaptability means that these hybrid management approaches are more robust—that is, these institutions are able to adapt successfully to changing social, political, and environmental conditions and ensure better ecological and socioeconomic outcomes as a result. However, in practice these systems face many challenges as societies undergo socioeconomic development. As described in Box 7.3, customary *qoliqoli* systems in Fiji can be undermined by economic market forces and other factors such as enforceability.

The institutional robustness of hybrid systems is associated with a number of factors, including the above-mentioned flexibility, locally relevant rules, and institutional arrangements that reflect design principles associated with long-enduring common-property management systems (Table 7.2). These hybrid systems can take advantage of both customary and conventional governance systems (such as village councils and village bylaws, respectively) to implement and manage marine resources. However, modern governance frameworks can hamper the implementation of hybrid institutions by legally restricting the capacity of local-level organizations to manage resources (Ruddle 1993, Vaughan 2012). In some countries, development of hybrid institutions will require changes in governance structures to move away from top-down, centralized systems to multiscale institutional arrangements that allow for flexible local decision making.

Although fundamental differences exist between customary and conventional management systems (Table 7.1), several key strategies have been developed to hybridize them.

In the Pacific Islands, increased modernization and globalization continue to advance from the main population centers outward to rural areas. In response, community-based fisheries resource management has been advanced by communities to preserve their own interests and complement the contemporary management systems that are in place. The dual system of coastal resource management in many of these islands today denotes the coexistence of the informal management system, devised and implemented by resource users and owners, with the formally instituted, government-led and legislation-based management system. In Fiji, for example, there are 410 registered customary fishing rights areas (*qoliqoli*), which are managed by subsistence fishers who have ownership rights over these areas (South and Veitayaki 1998). Resource management is now critical in these heavily exploited *qoliqoli,* because commercial fishing pressure from local users is no longer sustainable. Some local communities are committing to manage their fishing areas, but these are varied in their coverage. Although the specific approaches differ widely, from permanent no-take areas to the regulated use of marine resources in general, many Fijian communities have taken the hard decision to restrict the use of their marine resources, protect and manage their coastal resources, and search for alternative sources of livelihood. For example, management activities such as the ban on gill netting for reef fish or the protection of mangrove forests are restoring some species.

The major challenge is to support these self-organized initiatives, make them efficient in conserving resources and the natural environment, and convince the resource owners and users to commit to them. While the declaration of community-managed areas is fairly straightforward, the adherence and compliance of people to the management effort over the long term is more demanding. Coastal dwellers are part of contemporary economic circles that use their marine resources to support their development aspirations. For this reason, fishing activities normally intensify after the people have established their marine managed areas (MMAs), which become the targets of illegal

Cinner and Aswani (2007) drew on lessons from terrestrial systems, the literature on marine systems, and their experience studying customary and hybrid management to develop a heuristic for hybrid systems. Figure 7.3 provides an overview of some of the key differences between customary and contemporary fisheries management approaches, and a set of strategies that practitioners may employ to successfully hybridize these management approaches (Table 7.3).

Hybridizing customary management with conventional systems should ideally include extensive stakeholder engagement throughout design, implementation, and monitoring phases (Cooke et al. 2000, Mills et al. 2010, Kittinger 2013b). Inclusive approaches can help implement resource management and conservation in a culturally sensitive manner, increasing compliance, efficacy of conservation actions, and the robustness of fishing. In some areas of Fiji, the owners of traditional fishing grounds and customary fishing rights have taken little action to safeguard their *qoliqoli* and are still passive observers to the national government officials responsible for contemporary resource management decisions. Management and the conservation of fisheries are difficult to promote because they contradict the maximization of catch that is desired by most fishery operators. Management is also expensive because of enforcement costs and other requirements, which is unfair to pass on to the fishers, particularly if they are villagers. Fishers in Fiji, like their counterparts elsewhere in the world, leave home with the wish for a big and quick catch. This desire sometimes drives people to use fishing methods such as gill nets and dynamite, which are destructive to the coastal environment. This is why government support is required.

Recent experiences within the Pacific Islands show the popularity of community-based activities and the reliance of local communities on traditional resource-management practices. This integrated coastal management tradition is appropriate for inshore fisheries management because it covers issues on land as well as in the sea. However, customary systems that worked well in traditional societies face new challenges with increased socioeconomic development. Major challenges that can undermine traditional management in contemporary Fijian communities include the emphasis on production, an increased capacity and demand for resource exploitation, a lack of information on which to base management, and the socially destabilizing effects of the cash economy. The changes associated with modernization of society have given coastal communities the capability of overexploiting their resources, as the people are driven by the importance of the cash economy to maximize their productive capacity and, consequently, their impact on resources and the marine environment. These dual systems need to be successfully integrated and adapted to new socioeconomic contexts. The challenge is to integrate the two systems for the effective management of coastal resources and work toward a comanagement system that takes advantage of both systems' strengths.

Joeli Veitayaki is Associate Professor and Head of the School of Marine Studies, University of the South Pacific.

the comanagement arrangement to changing social and environmental conditions. It is important to note, however, that if improperly planned and implemented, efforts to hybridize these systems may do more harm than good. Poorly planned efforts may erode confidence not only in modern science and conservation organizations, but also in traditional authority (Gelcich et al. 2006). For example, attempts to develop customary management into comanagement arrangements have undermined and weakened traditional authorities and reduced the adaptive capacity of customary institutions in Chile and the Cook Islands (Tiraa 2006, Gelcich et al. 2006). Inadequate understanding of local power structures and the sociocultural aspects of customary institutions and local resource users can lead to increased balkanization and mistrust among stakeholder groups (Kittinger et al. 2012).

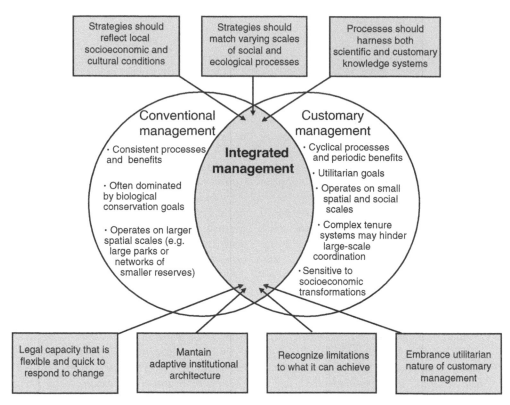

FIGURE 7.3 Properties of successful integrated, or "hybrid," management institutions. Customary and conventional management systems have contrasting goals, inferred social and ecological benefits, and spatial scales. The seven strategies (gray boxes) may help overcome these differences when managers, scientists, and communities hybridize the two systems (Table 7.3). (Adapted from Cinner and Aswani 2007.)

CONCLUSIONS

As we have shown, comanagement arrangements that integrate historically situated customary management practices show promise for sustainable management of small-scale fishery systems. Many examples are emerging of hybrid management composed of varying degrees of customary norms married with Western science and conventional governance regimes. This convergence is at least partly driven by the necessity of stemming degradation of coastal resources, with degradation causing communities and local officials to become more informed about the problems behind resource decline. Thus, customary practices are not being forgotten but are being integrated with modern planning and management to varying degrees.

Historical Ecology and Customary Management Systems

Historical research can shed light on the origins, development, and efficacy of customary management systems through time. It can also help elucidate the socioeconomic, governance,

TABLE 7.3 Key Strategies for Maintaining the Resilience of Integrated Comanagement of Marine Resources

Strategy	Description
1. Reflect local socioeconomic and cultural conditions.	Customary management strategies are diverse and heterogeneous, and specific hybridization strategies should reflect tenure, socioeconomic conditions, and local context.
2. Match varying social and ecological process scales.	Hybrid institutions should ideally attempt to match the varying spatial scales for resource use rights and ownership, as well as the scales of ecologically relevant processes such as herbivory, predation, and recruitment.
3. Harness both scientific and customary knowledge systems.	Harness western scientific data and analyses as well as traditional and local knowledge systems to assess and react to changes in social–ecological systems.
4. Incorporate legal capacity that is flexible and quick to respond to change.	Develop legal and policy capacity to enact and enforce decentralized management, ideally by strengthening the resilience of traditional authority structures upon which customary management practices depend.
5. Maintain adaptive institutional architecture.	Develop or maintain institutional arrangements and policies for comanagement that include aspects of key institutional design principles (Table 2) and that can be altered in response to changing social or environmental conditions.
6. Embrace utilitarian nature of customary management.	Preserving biodiversity and ecological integrity may be less important than utilitarian community goals such as ensuring traditional uses, resource security, and sociocultural connections to place.
7. Recognize limitations.	Hybridized management may be limited in the scope and scale of the threats it can address, and in its resilience to some socioeconomic processes, particularly those beyond the local scale.

political, and environmental conditions under which these systems evolved. As we have shown above, insights from historical ecological research can also reveal long-term trajectories of ecosystems and coastal societies, and the factors associated with sustainability or decline. Understanding these processes is critical for identifying how to integrate past lessons into modern contexts. There is evidence across the globe that customary management systems have provided for successful collective stewardship of fisheries resources in the past, and these approaches show much promise for managing fisheries more sustainably today. Comanagement approaches, as informed by historical ecological research, provide a mechanism to productively engage communities in stewardship and to increase participatory processes that link marine resource-management efforts across spatial scales.

Looking to the Future

This chapter illuminates the diversity of approaches used to integrate customary and contemporary management approaches and indicates an increasing trend in such approaches globally. Hybridized approaches will likely become stronger as more coastal communities recognize the limits of marine resources and explore new options for improving the status of their natural resource base. A simple story points to the future: an island community in the Philippines resisted for years the entrance of scuba-diving visitors to their reef because they believed that visitors would disturb fish and negatively affect catch. Then, one day, a group of fishers on the island suggested that an area would be designated for divers (away from the popular fishing area) and that they could pay a fee to the community for the right to dive. The ironic outcome of this story is that because fishing was ended in the diving area and more people were employed in tourism, fishing pressure declined; the community not only generated new income from tourism, but their fish catch increased. Finding bright spots such as this story—and helping spread and amplify the innovations and lessons learned—will help pave the way for sustainable small-scale fisheries into the future.

ACKNOWLEDGMENTS

The authors thank the editors of this book for their comments on drafts of this chapter. J.N.K. thanks Meg Caldwell and Larry Crowder from the Center for Ocean Solutions at Stanford University for support to engage this work, and J. Zachary Koehn for research support for this chapter. Additional data for Figure 7.2 from the Philippines were kindly provided by Aileen Maypa (University of Hawai'i) and Roxie Diaz and Dean Apistar (Coastal Conservation and Education Foundation, Inc., Cebu City, Philippines).

REFERENCES

Adams, T. (1998) The interface between traditional and modern methods of fishery management in the Pacific Islands. *Ocean and Coastal Management* 40, 127–142.

Agrawal, A. (2001) Common property institutions and sustainable governance of resources. *World Development* 29, 1649–1672.

Allen, M.S. (2006) New ideas about Late Holocene climate variability in the Central Pacific. *Current Anthropology* 47, 521–535.

Almany, G.R., Hamilton, R.J., Bode, M., et al. (2013) Dispersal of grouper larvae drives local resource sharing in a coral reef fishery. *Current Biology* 23, 626–630.

Anderies, J.M., Janssen, M.A., and Ostrom, E. (2003) Design principles for robustness of institutions in social–ecological systems. http://dlc.dlib.indiana.edu/dlc/bitstream/handle/10535/1777/Ostrom%2cElinor.pdf?sequence=1.

Aswani, S. (2011) Socioecological approaches for combining ecosystem-based and customary management in Oceania. *Journal of Marine Biology* 2011, article 845385.

Aswani, S., Albert, S., Sabetian, A., and Furusawa, T. (2007) Customary management as precautionary and adaptive principles for protecting coral reefs in Oceania. *Coral Reefs* 26, 1009–1021.

Aswani, S., and Allen, M.S. (2009) A Marquesan coral reef (French Polynesia) in historical context: an integrated socio-ecological approach. *Aquatic Conservation: Marine and Freshwater Ecosystems* 19, 614–625.

Aswani, S., and Lauer, M. (2006) Incorporating fishermen's local knowledge and behavior into geographical information systems (GIS) for designing marine protected areas in Oceania. *Human Organization* 65, 81–102.

Aswani, S., and Ruddle, K. (2013) Design of realistic hybrid marine resource management programs in Oceania. *Pacific Science* 67, 461–476.

Athens, J. S. (2009) *Rattus exulans* and the catastrophic disappearance of Hawai'i's native lowland forest. *Biological Invasions* 11, 1489–1501.

Baird, A., Campbell, S., Anggoro, A., et al. (2005) Acehnese reefs in the wake of the Asian tsunami. *Current Biology* 15, 1926–1930.

Basurto, X., and Ostrom, E. (2009) Beyond the tragedy of the commons. *Economia delle fonti di energia e dell'ambiente* 52, 35–60.

Bell, J. D., Kronen, M., Vunisea, A., et al. (2009) Planning the use of fish for food security in the Pacific. *Marine Policy* 33, 64–76.

Berkes, F. (2009) Evolution of co-management: role of knowledge generation, bridging organizations and social learning. *Journal of Environmental Management* 90, 1692–1702.

Berkes, F. (2010) Devolution of environment and resources governance: trends and future. *Environmental Conservation* 37, 489–500.

Berkes, F., Folke, C., and Colding, J. (1998) *Linking Social and Ecological Systems: Management Practices and Social Mechanisms for Building Resilience*. Cambridge University Press, Cambridge, UK.

Berkes, F., Mahon, R., McConney, P., et al. (2001) *Managing Small-scale Fisheries: Alternative Directions and Methods*. International Development Research Centre, Ottawa, ON.

Bohensky, E. L., and Maru, Y. (2011) Indigenous knowledge, science, and resilience: what have we learned from a decade of international literature on "integration"? *Ecology and Society* 16: article 6.

Burke, L., Reytar, K., Spalding, M., and Perry, A. (2011) *Reefs at Risk Revisited*. World Resources Institute, Washington, DC.

Burney, D. A., James, H. F., Burney, L. P., et al. (2001) Fossil evidence for a diverse biota from Kaua'i and its transformation since human arrival. *Ecological Monographs* 71, 615–641.

Campbell, S. J., Cinner, J. E., Ardiwijaya, R. L., et al. (2012) Avoiding conflicts and protecting coral reefs: customary management benefits marine habitats and fish biomass. *Oryx* 46, 486–494.

Chuenpagdee, R., Liguori, L., Palomares, M. L. D., and Pauly, D. (2006) Bottom-up, global estimates of small-scale marine fisheries catches. *Fisheries Centre Research Reports* 14(8).

Cinner, J. E., and Aswani, S. (2007) Integrating customary management into marine conservation. *Biological Conservation* 140, 201–216.

Cinner, J. E., Marnane, M. J., McClanahan, T. R., and Almany, G. R. (2005) Periodic closures as adaptive coral reef management in the Indo-Pacific. *Ecology and Society* 11, article 31.

Cinner, J. E., and McClanahan, T. R. (2006) Socioeconomic factors that lead to overfishing in small-scale coral reef fisheries of Papua New Guinea. *Environmental Conservation* 33, 73–80.

Cinner, J. E., McClanahan, T. R., MacNeil, M. A., et al. (2012) Comanagement of coral reef social–ecological systems. *Proceedings of the National Academy of Sciences USA* 109, 5219–5222.

Clarke, P., and Jupiter, S. D. (2010) Law, custom and community-based natural resource management in Kubulau District (Fiji). *Environmental Conservation* 37, 98–106.

Cooke, A. J., Polunin, N. V. C., and Moce, K. (2000) Comparative assessment of stakeholder management in traditional Fijian fishing-grounds. *Environmental Conservation* 27, 291–299.

Cox, M., Arnold, G., and Villamayor-Tomas, S. (2010) A review of design principles for community-based natural resource management. *Ecology and Society* 15, article 38.

de Haan, J. (2006) How emergence arises. *Ecological Complexity* 3, 293–301.

Delgado, C. L., Wada, N., Rosegrant, M. W., et al. (2003) Fish to 2020: Supply and demand in changing global markets. WorldFish Center Technical Report 62. International Food Policy Research Institute and WorldFish Center, Washington, DC.

Diamond, J. M., and Robinson, J. A. (2010) *Natural Experiments of History*. Harvard University Press, Cambridge, MA.

Drew, J. A. (2005) Use of traditional ecological knowledge in marine conservation. *Conservation Biology* 19, 1286–1293.

FAO (2012) *The State of World Fisheries and Aquaculture 2012*. Food and Agriculture Organization of the United Nations (FAO), Rome.

Foale, S., and Manele, B. (2004) Social and political barriers to the use of marine protected areas for conservation and fishery management in Melanesia. *Asia Pacific Viewpoint* 45, 373–386.

Friedlander, A. M., Brown, E. K., and Monaco, M. E. (2007) Coupling ecology and GIS to evaluate efficacy of marine protected areas in Hawaii. *Ecological Applications* 17, 715–730.

Friedlander, A. M., Shackeroff, J. M., and Kittinger, J. N. (2013) Customary marine resource knowledge and use in contemporary Hawai'i. *Pacific Science* 67, 441–460.

Gelcich, S., Edwards-Jones, G., Kaiser, M., and Castilla, J. (2006) Co-management policy can reduce resilience in traditionally managed marine ecosystems. *Ecosystems* 9, 951–966.

Graham, T., and Idechong, N. (1998) Reconciling customary and constitutional law: managing marine resources in Palau, Micronesia. *Ocean & Coastal Management* 40, 143–164.

Hardin, G. (1968) The tragedy of the commons. *Science* 162, 1243–1248.

Hickey, F. R., and Johannes, R. E. (2002) Recent evolution of village-based marine resource management in Vanuatu. *SPC Traditional Marine Resource Management and Knowledge Information Bulletin* 14, 8–21.

Hunt, T. L., and Lipo, C. (2011) *The Statues That Walked: Unraveling the Mystery of Easter Island*. Free Press, New York, NY.

Jacquet, J., and Pauly, D. (2008) Funding priorities: big barriers to small-scale fisheries. *Conservation Biology* 22, 832–835.

Johannes, R. E. (2002) The renaissance of community-based marine resource management in Oceania. *Annual Reviews in Ecology and Systematics* 33, 317–340.

Jupiter, S., Weeks, R., Jenkins, A., et al. (2012) Effects of a single intensive harvest event on fish populations inside a customary marine closure. *Coral Reefs* 31, 321–334.

Kirch, P. V. (1985) *Feathered Gods and Fishhooks: An Introduction to Hawaiian Archaeology and Prehistory*. University of Hawai'i Press, Honolulu.

Kirch, P. V. (2002) *On the Road of the Winds: An Archaeological History of the Pacific Islands before European Contact*. University of California Press, Berkeley, CA.

Kirch, P. V., and O'Day, S. J. (2003) New archaeological insights into food and status: a case study from pre-contact Hawaii. *World Archaeology* 34, 484–497.

Kittinger, J. N. (2013a) Human dimensions of small-scale and traditional fisheries in the Asia-Pacific region. *Pacific Science* 67, 315–325.

Kittinger, J. N. (2013b) Participatory fishing community assessments to support coral reef fisheries comanagement. *Pacific Science* 67, 361–381.

Kittinger, J. N., Bambico, T. M., Watson, T. K., and Glazier, E. W. (2012) Sociocultural significance of the endangered Hawaiian monk seal and the human dimensions of conservation planning. *Endangered Species Research* 17, 139–156.

Kittinger, J. N., Pandolfi, J. M., Blodgett, J. H., et al. (2011) Historical reconstruction reveals recovery in Hawaiian coral reefs. *PLoS ONE* 6, e25460.

Lotze, H. K., Erlandson, J. E., Hardt, M., et al. (2011) Uncovering the ocean's past. In *Shifting Baselines: The Past and Future of Ocean Fisheries* (J. B. C. Jackson, K. E. Alexander, and E. Sala, Eds.). Island Press, Washington, DC. pp. 137–162.

McClanahan, T., Castilla, J., White, A., and Defeo, O. (2009) Healing small-scale fisheries by facilitating complex socio-ecological systems. *Reviews in Fish Biology and Fisheries* 19, 33–47.

McClanahan, T. R., and Cinner, J. E. (2008) A framework for adaptive gear and ecosystem-based management in the artisanal coral reef fishery of Papua New Guinea. *Aquatic Conservation: Marine and Freshwater Ecosystems* 18, 493–507.

McClanahan, T. R., Glaesel, H., Rubens, J., and Kiambo, R. (1997) The effects of traditional fisheries management on fisheries yields and the coral-reef ecosystems of southern Kenya. *Environmental Conservation* 24, 105–120.

McClanahan, T. R., Hicks, C. C., and Darling, E. S. (2008) Malthusian overfishing and efforts to overcome it on Kenyan coral reefs. *Ecological Applications* 18, 1516–1529.

McClanahan, T. R., Marnane, M. J., Cinner, J. E., and Kiene, W. E. (2006) A comparison of marine protected areas and alternative approaches to coral reef management. *Current Biology* 16, 1408–1413.

McClenachan, L. E., and Kittinger, J. N. (2012) Multicentury trends and the sustainability of coral reef fisheries in Hawai'i and Florida. *Fish and Fisheries* 14, 239–255.

Mills, M., Pressey, R. L., Weeks, R., et al. (2010) A mismatch of scales: challenges in planning for implementation of marine protected areas in the Coral Triangle. *Conservation Letters* 3, 291–303.

Newton, K., Côté, I. M., Pilling, G. M., et al. (2007) Current and future sustainability of island coral reef fisheries. *Current Biology* 17, 655–658.

Ostrom, E. (1990) *Governing the Commons: The Evolution of Institutions for Collective Action.* Cambridge University Press, Cambridge, UK.

Pagdee, A., Kim, Y., and Daugherty, P. (2006) What makes community forest management successful: a meta-study from community forests throughout the world. *Society and Natural Resources* 19, 33–52.

Pauly, D. (1990) On Malthusian overfishing. *Naga, the ICLARM Quarterly* 13, 3–4.

Ruddle, K. (1993) External forces and change in traditional community-based fishery management systems in the Asia-Pacific region. *Maritime Anthropological Studies* 6, 1–37.

Ruttan, L. M. (1998) Closing the commons: cooperation for gain or restraint? *Human Ecology* 26, 43–66.

Sagarin, R. D., Barry, J. P., Gilman, S. E., and Baxter, C. H. (1999) Climate-related change in an intertidal community over short and long time scales. *Ecological Monographs* 69, 465–490.

Sagarin, R. D., Gilly, W. F., Baxter, C. H., et al. (2008) Remembering the Gulf: changes to the marine communities of the Sea of Cortez since the Steinbeck and Ricketts expedition of 1940. *Frontiers in Ecology and the Environment* 6, 374–381.

Sagarin, R., and Pauchard, A. (2012) *Observation and Ecology. Broadening the Scope of Science to Understand a Complex World.* Island Press, Washington, DC.

Schneider, S. H., and Root, T. L. (2001) *Wildlife Responses to Climate Change: North American Case Studies.* Island Press, Washington, DC.

South, G. R., and Veitayaki, J. (1998) The Constitution and indigenous fisheries management in Fiji. *Ocean Yearbook* 13, 452–466.

Steinbeck, J., and Ricketts, E. F. (1941) *Sea of Cortez; a leisurely journal of travel and research, with a scientific appendix comprising materials for a source book on the marine animals of the Panamic faunal province. By John Steinbeck and Edward F. Ricketts.* Viking Press, New York, NY.

Starr, R. M., Culver, C. S., and Pomeroy, C. (2010) Managing data-poor fisheries workshop: case studies, models and solutions. Final Report, Data-Poor Fisheries Workshop, 1–4 December 2008. California Sea Grant, University of California, San Diego, CA.

Teh, L. S. L., Teh, L. C. L., and Sumaila, U. R. (2013) A global estimate of the number of coral reef fishers. *PLoS ONE* 8, e65397.

Thorburn, C. (2001) The house that poison built: customary marine property rights and the live food fish trade in the Kei Islands, southeast Maluku. *Development and Change* 32, 151–180.

Tiraa, A. (2006) Ra'ui in the Cook Islands—today's context in Rarotonga. *SPC Traditional Marine Resource Management and Knowledge Information Bulletin* 19, 11–15.

Vaughan, M. B. (2012) Holoholo i ke kai o Hiala'a: collaborative community care and management of coastal resources: creating state law based on customary community rules to manage a near shore fishery in Hawai'i. PhD dissertation, Emmet Interdiscipliary Program in Environment and Resources, Stanford University.

Vitousek, P. M., Ladefoged, T. N., Kirch, P. V., et al. (2004) Soils, agriculture, and society in precontact Hawaii. *Science* 304, 1665–1669.

White, A., Deguit, E., Jatulan, W., and Eisma-Osorio, L. (2006) Integrated coastal management in Philippine local governance: evolution and benefits. *Coastal Management* 34, 287–302.

White, A. T. (2010) Status of coastal and marine resources: implications for fisheries management and poverty in Southeast Asia. In *Poverty Reduction through Sustainable Fisheries: Emerging Policy and Governance Issues in Southeast Asia* (R. M. Briones and A. G. Garcia, Eds.). ISEAS Publishing, Singapore. pp. 199–232.

White, A. T., Hale, L. Z., Renard, Y., and Cortesi, L. (1994) *Collaborative and Community-based Management of Coral Reefs: Lessons from Experience.* Kumarian Press, Sterling, VA.

White, A. T., Maypa, A., Tesch, S., et al. (2008) Summary field report: Coral Reef Monitoring Expedition to Tubbataha Reefs Natural Park, Sulu Sea, Philippines, March 26–April 1, 2008. Coastal Conservation and Education Foundation, Cebu City, Philippines.

Williams, I. D., Walsh, W. J., Schroeder, R. E., et al. (2008) Assessing the importance of fishing impacts on Hawaiian coral reef fish assemblages along regional-scale human population gradients. *Environmental Conservation* 35, 261–272.

Wilmshurst, J., Hunt, T. L., Lipo, C. P., and Anderson, A. J. (2011) High-precision radiocarbon dating shows recent and rapid initial human colonization of East Polynesia. *Proceedings of the National Academy of Sciences USA* 108, 1815–1820.

Young, O. R. (2002) *The Institutional Dimensions of Environmental Change: Fit, Interplay, and Scale.* MIT Press, Cambridge, MA.

PART THREE

RESTORING ECOSYSTEMS

Lead Section Editor: KERYN B. GEDAN

An interacting community of species—an ecosystem—is more than the sum of its parts. Complex patterns and processes emerge in ecosystems—including intricate food webs, niche differentiation (the specific roles that plants and animals adaptively create), and coupled nutrient cycles. When human activities disrupt these processes, through actions such as overfishing, habitat degradation, or eutrophication, the whole ecosystem can be knocked out of balance. Ecosystem-level changes that result from human activities—such as trophic cascades or shifts in ecological states—create complex conservation challenges.

Alongside fisheries, ecosystem changes were the impetus for many initial studies in marine historical ecology. In kelp forests, Dayton et al. (1998) documented numerous changes over the preceding 25+ years, including large oscillations in kelp density, shifts in kelp biomass, and the functional loss or extinction of many grazers and mesopredators that had far-reaching and indirect effects in kelp forest communities. Similarly, Jackson (1997) elucidated the major shift from coral to algal dominance in Caribbean coral reef ecosystems documented in the twentieth century. He famously showed that this shift had its roots in previous centuries of exploitation—including rampant harvesting of turtles and large marine vertebrates—leaving the reef fauna depauperate in grazers and predators. Aptly, these authors compared today's kelp and coral ecosystems to that of the Serengeti without large grazers or carnivores, and researching them to "studying the termites and the locusts while ignoring the elephants and the wildebeest" (Jackson 1997).

These studies underscore the tangled, interconnected nature of ecosystems, as well as the complex changes that ripple through these systems because of human actions. Today's conservation scientists acknowledge these connections and seek to recover and manage ecosystems holistically. As the authors in this section highlight, historical ecology has as much to offer conservation in practice.

First and foremost, historical information can provide a baseline against which to measure the effect of conservation actions and can provide information on alternative outcomes, broadening the portfolio of possibilities for restoration targets and expanding our vision of the possible. Knowing what existed before—how ecosystems functioned, how widespread they were, what factors affected their distribution, their carrying capacity and productivity—provides powerful guidance for helping determine the way forward in protecting and reviving marine ecosystems.

In this part Gedan, Breitburg, Grossinger, and Rick examine the multitude of ways that historical perspectives can inform ecosystem restoration. They also assess how the restoration applications of historical ecology are evolving, as restoration ecologists seek not only to recover ecosystems from human degradation but to build more resilient ecosystems that resist present human impacts and withstand others on the horizon, such as climate change.

In the following chapter, zu Ermgassen, Spalding, and Brumbaugh directly link ecosystems and people, examining how the capacity of marine ecosystems to support human communities has changed over time and how a quantification of historical ecosystem services can be used as an informative and relevant baseline for ecological restoration targets. They demonstrate this process with an example of oyster reefs in the United States, building estimates of early oyster-reef filtration capacity from historical data on habitat area and complexity and on estuary conditions such as salinity. Given that conservation activities are shaped by human priorities and the needs of people, setting the frame of reference in terms of changes in the production of ecosystem services (the benefits of ecosystems to people) is an idea with great scientific and public appeal.

In the final chapter of this part, Ban, Kittinger, Pandolfi, Pressey, Thurstan, Lybolt, and Hart give a step-by-step description of why, when, and how to incorporate historical data into the stages of systematic conservation planning for marine protected areas, one of the most prevalent tools in marine conservation today. They document numerous examples wherein historical data have been used successfully to set conservation goals and, importantly, how these efforts would have gone awry if history had been ignored.

All three chapters promote incorporating historical data into ecosystem conservation and restoration practice and stress that even the most modern approaches, including large-scale ecosystem restoration, systematic conservation planning, and ecosystem service assessments, can benefit from the inclusion of historical ecology. Even forward-looking marine conservation efforts must look back to learn from the past.

REFERENCES

Dayton, P. K., Tegner, M. J., Edwards, P. B., and Riser, K. L. (1998) Sliding baselines, ghosts, and reduced expectations in kelp forest communities. *Ecological Applications* 8, 309–322.
Jackson, J. B. (1997) Reefs since Columbus. *Coral Reefs* 16, S23–S32.

EIGHT

Historical Information for Ecological Restoration in Estuaries and Coastal Ecosystems

KERYN B. GEDAN, DENISE L. BREITBURG, ROBIN M. GROSSINGER, and TORBEN C. RICK

The appropriate role of historical information in ecosystem restoration is a topic of debate within restoration ecology, as the discipline and the practice of restoration adapt to keep up with the increasing demands and challenges of multiple human impacts and global climate change. Whereas historical data have traditionally been used to define restoration baselines and criteria for success, current practice emphasizes a less prescriptive and more process-based interpretation of historical data. In addition to these traditional applications, we discuss the broader use of historical information to describe landscape processes and linkages and to understand ecosystem trajectories and controls by examining historical responses of ecosystems to human impacts, disturbances, and climate changes. Historical ecology can also inform restoration sustainability and the identification of novel ecosystems. Despite the challenges of restoring ecosystems within larger, often highly modified, landscapes and the need to manage ecosystems to be resilient to global environmental change, ecosystem restorations make sense only in the light of history.

INTRODUCTION

Many assume that all restorations are, by definition, an attempt to recreate historical conditions. While a historical reference point is integral to some definitions of restoration—for example, "the return of an ecosystem to a close approximation of its condition prior to disturbance" (National Research Council 1992), a definition adopted by many—others emphasize the capacity of historical baselines to limit restoration potential and reject the idea in favor of more "forward-looking" practices (Figure 8.1). For example, Choi (2007) stated that "our restoration efforts for the future need not be constrained by 'historical-fidelity,'"

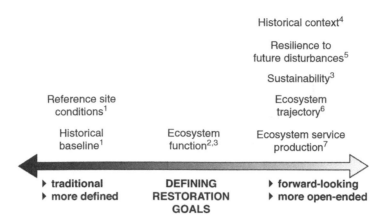

FIGURE 8.1 A conceptual model describes general types of information used to define restoration goals. Traditionally, historical baselines and reference site conditions were used to define restoration goals, but today a host of different approaches are used to plan and evaluate restoration activities. These approaches are more complex, but scientists hope that this approach will allow greater utility and sustainability of restored ecosystems. References: [1]White and Walker 1997, [2]Palmer et al. 1997, [3]Choi 2007, [4]Jackson and Hobbs 2009, [5]Choi et al. 2008, [6]Hughes et al. 2012, [7]Palmer and Filoso 2009.

a sentiment echoed by Hobbs et al. (2004). Choi (2007) continued, "A future-oriented restoration should aim to establish ecosystems that are able to persist in future environment[s]." There are those who describe the use of historical information in restoration activities as a desire for a "living museum," an impractical exhibition of species that does not evolve with a changing world (Buizer et al. 2012). As restoration ecology adapts to meet the needs of ecosystems and people in this era of rapid and large-scale environmental changes, some restoration ecologists have been dismissive of historical perspectives, finding them backward-looking and stagnant (Choi 2007, Buizer et al. 2012).

Historical records, however, actually reveal the dynamic nature of the environment, a revelation that is most useful for restoration (Jackson and Hobbs 2009). Used appropriately, historical information is a valuable tool that can inform planners and stakeholders about the options and potential for a given system. Moreover, historical data can provide important insights into ecological relationships and causes and effects of human actions and climate change in the past. As William Throop put it, "If we try to return a system to a previous trajectory, we are less likely to make new errors"(Hobbs et al. 2004). In fact, disregarding historical realities can lead to restoration interventions that will be inappropriate (e.g., single-channel in place of multichannel river restorations; Grossinger et al. 2007, Montgomery 2008) or unsuccessful. Finally, and perhaps most critically, by describing the altered status of a site or habitat to be restored, historical ecology provides the motivation and overarching context for restoration (Table 8.1).

TABLE 8.1 Advantages of Historical Data for Restoration

Advantage	Explanation
Context for restoration	Knowledge of the ecosystem prior to disturbance (i.e., its history) is the foundation for restoration activities.
Practical applications	As discussed in the chapter text, historical data can be used to set or guide restoration targets and in restoration design planning.
Including all appropriate stakeholders	At different times, different groups of people may have played a role in the history of the ecosystem, by using natural resources or by implementing conservation measures. Identifying these groups can expand the number of stakeholders in a restoration project and involve them up-front in the project planning to increase project awareness and buy-in.
Reuse in interpretive materials for public display	The communication of restoration impacts to sponsors and the public is critical for securing and maintaining restoration resources. "Past and present" comparisons can provide valuable graphics to communicate restoration effects (e.g., Figure 8.4).

In this chapter, we describe a number of constructive uses of historical data in restoration and give examples of restoration projects that have benefited from the inclusion of historical information. We use the term "historical" loosely to include information from paleobiology, archaeology, and history (Dietl and Flessa 2011, Szabo and Hédl 2011, Lyman 2012, Rick and Lockwood 2012, Wolverton and Lyman 2012) but generally limit these datasets to the Holocene (~10,000 years ago to present)—the interglacial period we are currently living through and a major period of human population growth and geographic expansion.

Historical data are not a prescription for restoration (see Szabo and Hédl 2011, Rick and Lockwood 2012). In many cases, it is impossible or infeasible to recreate previous ecosystem characteristics (Duarte et al. 2009); in other situations, restoration goals may conflict with historical realities. Historical data can be difficult or costly to acquire, and interpreting environmental changes that occurred centuries and millennia ago from often incomplete or fragmentary data sources can be challenging. Even in adverse restoration settings, however, historical data are an important foundation for restoration activities because they can inspire and guide restoration practitioners (see chapter 12, this volume), motivate conservation funding and action (Marsh et al. 2005, McClenachan et al. 2012), and provide a context for restoration targets and outcomes. Far from being backward-looking, applications of historical ecology and shifting baselines have always been oriented toward restoring and protecting natural systems and natural resources for the future (Pauly 1995, Jackson et al. 2011).

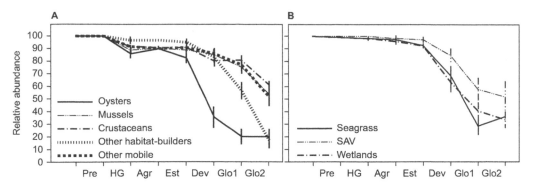

FIGURE 8.2 Ecological restoration is motivated by habitat degradation and loss that has occurred primarily during economic and industrial development (Dev) and globalization (Glo1 and Glo2) periods. The graphs show the average decline in abundance of (A) habitat-forming and other marine invertebrates and (B) vegetated marine habitats in 12 estuaries over historical cultural periods (for definitions of the periods and their abbreviations, see Table 8.2; SAV = submerged aquatic vegetation). (Reprinted from Lotze et al. 2006.)

Restore to When? Ecosystems Have a Long History of Human Interactions

Historically, human impacts may have accumulated within an ecosystem (e.g., Hughes 1994, Foster et al. 2003) like the layers of an onion. How many can or should be peeled away to reveal an ecosystem's "natural state"? What is the baseline for restoration?

While many treat European colonization as the commencement of human impacts in the United States and the industrial revolution as the intensification of impacts, human interactions with the land, the sea, and natural resources existed long before Columbus set sail (Redman 1999, Jackson et al. 2001, Kirch 2005, Rick and Erlandson 2008). Humans have lived along the coast and exploited marine ecosystems for 150,000 years in southern Africa, and likely since the initial colonization of Australia, the Americas, and elsewhere (Erlandson 2001, Marean et al. 2007, Erlandson and Rick 2010). The scale and extent of human impacts on marine ecosystems have increased through time with the growth of human populations and increasingly sophisticated technologies. Archaeological, paleobiological, and historical records often document the impacts and interactions of past human groups, and more recently these historical ecological datasets have been used to help set restoration goals.

Jackson et al. (2001) developed three general periods of human environmental interactions and impacts—aboriginal, colonial, and historical—each recognizing the differences in the scale, magnitude, and time frame of human impacts. Lotze et al. (2006) built on this scheme by devising a flexible anthropological framework for understanding the timeline of human impacts in estuaries that used periods of cultural development, rather than absolute time, to account for the variation in human activities at different places (Figure 8.2 and Table 8.2). In their study of 12 estuaries around the world, the period of economic development and globalization saw the largest changes in marine natural resources (Figure 8.2; Lotze et al. 2006).

TABLE 8.2 Historical Cultural Periods, with Descriptions of Social Attributes and Ecological Interactions with Ocean Resources during Each Period

Cultural Period	Human Presence, Technology, and Market Conditions
Prehuman (Pre)	No human presence; only natural disturbance of ecosystems
Hunter-gatherer (HG)	Premarket; low population numbers; seasonal settlements; individual resource use; subsistence exploitation
Agricultural (Agr)	Premarket; low population numbers; permanent settlements: individual or village-based resource use; subsistence and artisanal exploitation
Establishment (Est)	Establishment of local economy and market; European settlement in the New World; low population numbers; trade between colonies and empire home countries; mostly products needed to survive
Development (Dev)	Strong growth and expansion of economy, market, and trade; rapid population rise; commercialization of resource use; development of luxury and fashion markets, especially affecting birds (feather trade), mammals (whale baleen, seal and otter furs, walrus ivory), and reptiles (turtle shells, crocodile skins); onset of industrialization and technological progress; mass killing of mammals and birds with guns; fishing mostly inshore and seasonal, with selective and nondestructive gear (e.g., hook and line, light trawls towed by sail boats over soft bottoms)
Early Global (Glo 1)	Global economy and market develops; strong population increase; industrialization and technological progress toward more efficient, less selective, and more destructive gear (e.g., motor boats, steam trawlers) accelerating exploitation, bycatch, and habitat destruction; increasing fishing effort; fishing possible in any season but still mostly inshore and coastal
Late Global (Glo 2)	Global economy and market; increased industrial fishing (especially after World War II) extending offshore and toward the deep sea; multiple unselective and destructive gear (otter trawls, purse seines, long-lining, rock hopper gear, etc.); all habitats are being fished, and all fish are detectable; mass fishing; other human impacts increase (eutrophication, pollution, fish farming); conservation efforts increase

Source: Adapted from Lotze et al. (2006).
Note: Abbreviations (in parentheses) correspond to Figure 8.2.

The type, scale, and intensity of human impacts have shifted over time in ways that are important to consider for restoration purposes. Overexploitation of fish, shellfish, and marine mammals has (intermittently and unevenly) altered coastal ecosystems for millennia. At times, even hunter-gatherers regionally depleted resources, which sometimes had whole-ecosystem effects (Rick and Erlandson 2009, Erlandson and Rick 2010). For example, the possible depletion of sea otters (*Enhydra lutis*) by hunters in the eastern Pacific beginning ~9,000 years ago in California and ~3,000 years ago in the Aleutian Islands likely triggered trophic cascades—the expansion of abalone and urchin populations, which devoured kelp forests (Simenstad et al. 1978, Erlandson and Rick 2010). Millennia ago, these produced localized impacts that were not unlike the larger ecosystem effects of otter overexploitation by fur traders in the 1700s and 1800s (Estes and Duggins 1995).

The depletion and subsequent recovery of otters in the Pacific provides an example of the ability of historical information to identify ecosystems (in this case, kelp forest ecosystems) with high potential to benefit from restoration activities. On two occasions, separated by millennia, overexploitation triggered a similar ecosystem response, and the removal of the impact (harvest of sea otters) reverted the ecosystem to its prior state of kelp dominance (Simenstad et al. 1978). However, it is also telling that, in recent decades, the recovery of sea otters and kelp forests has been complicated by the co-occurrence of other large-scale stressors (e.g., killer whale population expansion; Estes et al. 1998) as well as protests from fishermen dependent on clams, abalones, urchins, and other species that are primary prey for otters.

Historical ecological datasets from kelp forests also offer glimpses of hope for marine ecosystems and organisms, including red abalones (*Haliotis rufescens*) that show resilience to millennia of Native American harvest and potential recovery in the future (Braje et al. 2009). In the Gulf of Maine, archaeological and historical data have also helped identify changes in the abundance of cod, lobsters, and other important kelp forest species through time, as well as past phase shifts and trophic cascades (Steneck et al. 2004, Bourque et al. 2008, Alexander et al. 2009). Similarly, historical data have provided key insight into ancient degradation and recovery of coral reefs (Knowlton and Jackson 2008, Kittinger et al. 2011). Long-term records of human interactions with, and impacts on, coastal marine ecosystems extend well beyond kelp forests and coral reefs, including mangroves, estuaries, and a variety of other ecosystems. Increasingly, researchers are demonstrating the importance of these data for informing contemporary restoration by providing historical baselines, by helping understand long-term ecosystem responses to climate change and anthropogenic activities (i.e., historical range of variability), by documenting novel or no-analog ecosystems, and by assisting in determining desired future conditions.

CASE STUDIES ILLUSTRATING THE RANGE OF APPLICATIONS OF HISTORICAL DATA

Using Historical Ecology to Define Ecological Restoration Targets and Strategies

The effects of ecological restoration are routinely evaluated in relation to two reference points: (1) historical conditions at the restoration site and (2) contemporary conditions at a less disturbed reference site with similar characteristics to the restoration site (Figure 8.1). White and Walker (1997) suggest that historical information about both restoration and reference site(s) provides important insights into the past state and range of natural ecosystem dynamics that restoration activities should be measured against.

Alternatively to using historical conditions as an actual target for restoration, historical reconstructions can be used to elucidate key attributes and likely settings for successful restoration efforts. With historical information on where the ecosystem thrived in the past and the abiotic characteristics of those settings, restoration practitioners can identify areas that are likely to support healthy habitat currently and into the future. These areas may be in the same place as they were historically, implying a literal site identification, or different locations if physical conditions have changed. In both cases, historical ecology helps define the characteristics and location of suitable habitat. Past habitat area and condition provide a relevant, if not rigid, baseline against which to evaluate restoration potential and to measure restoration success.

Oyster reefs are often a candidate for restoration in estuaries because of their degraded state (Figure 8.2), overexploitation to the point of ecological extinction in many places (Beck et al. 2011), and ecological value in terms of improving water quality and supporting a diversity of life, food webs, and adjacent marsh and seagrass habitats (Newell 1988, Grabowski and Peterson 2007). For survival of larval and adult life stages, oysters require appropriate physical conditions, including hard substrate, brackish to saltwater salinities, and retentive estuarine circulation (Mann and Evans 2004). Restoration ecologists and planners often use the historical distribution of oyster reefs as a tool for defining sites where remnant hard substrate may be able to support oyster reef communities once more.

In planning for large-scale (tributary-level) restoration efforts in the Chesapeake Bay, the historic Yates (1913) and Baylor (1896) maps of oyster grounds in Maryland and Virginia have served as the first layers for site selection for oyster reef restoration activities (Mann et al. 2009, Oyster Metrics Workgroup 2011). Contemporary information on substrate availability and navigation channels has helped further define suitable restoration sites (Mann et al. 2009). Historical information has also been used to guide restoration goals in the Chesapeake; accounts from the early period of heavy exploitation that described reefs as patchy, with areas devoid of oysters (Winslow 1882, Moore 1910), have led restoration planners to aim for 30% oyster cover within restored reefs (Oyster Metrics Workgroup 2011). While the 30% goal is somewhat arbitrary, the decision to aim for a restored habitat without full oyster cover, based on historical evidence, is logical. Given that reefs were not fully covered with oysters prior to impact, a restoration with a goal of complete oyster coverage would likely come up short. Although it may be infeasible to reach historical baselines in habitat area or condition, these baselines provide useful metrics for restoration progress, without which the expectations for the restoration may be set dangerously high, leading to perceptions of restoration failure, or too low, further contributing to shifting baselines.

As this example demonstrates, historical information does not stand alone to choose restoration sites or targets but is useful as a source of information and guidepost. Restoration must be jointly guided by ecological information and feasibility constraints. Perhaps most

FIGURE 8.3 (A) Wearing his customary overalls, former Maryland State Senator Bernie Fowler wades into the Patuxent River holding hands with U.S. Congressman Steny Hoyer and other Patuxent River wade-in participants, for his annual measurement of the "Sneaker Index," the depth at which Bernie can see his worn white sneakers. (B) Chesapeake Bay Program Director Nick DiPasquale (EPA) updates the Sneaker Index graph with 2013's reading of 34 inches (0.86 m). This depth is compared to a baseline from when Bernie was a child and could see his shoes under 63 inches (1.6 m) of water, an inventive 60-year data record that puts turbidity in a Chesapeake Bay tributary on a human scale, and one that has rallied public interest and attention.

importantly, being able to explain the relationship between historical baselines and restoration targets demonstrates an understanding of system controls that is necessary and can build public support (Figure 8.3; Hanley et al. 2009). Because historical data are inevitably subject to uncertainty and potential bias (Power et al. 2010, Lotze et al. 2011), the use of multiple sources and types of historical information and transparent assessment of uncertainty is critical to establishing confidence in our understanding of historical conditions (Table 8.3; Grossinger et al. 2005, 2007).

Historical Conditions Define a Range of Restoration Targets

While variables such as habitat extent and percent cover are captured in a single time point and may be obtained from documents like historical maps, the historical status of more dynamic ecosystem conditions is more difficult to quantify and often requires inference. Clever use of historical and paleoecological data can identify the natural range of dynamic ecosystem conditions to serve as baselines for restoration.

One case where historical information was used to establish the range of an important and dynamic ecosystem condition is that of dissolved oxygen concentration criteria in the Chesapeake Bay. Low dissolved oxygen is an important negative consequence of increased nutrient loads to estuaries and other aquatic systems (Diaz and Rosenberg 2008, Diaz and Breitburg 2009, Rabalais et al. 2010), and restoring dissolved oxygen to socially acceptable and scientifically justifiable levels is often a major focus of water-quality restoration. For

TABLE 8.3   Limitations of Historical Data for Ecological Restoration

Limitation	Explanation
Difficulty of finding data sources	Historical data are not always readily available and require additional resources to assemble. Moreover, a truncated or short-term record may misrepresent important dynamics (e.g., McClenachan et al. 2012: fig. 2).
Data patchiness	Data are not available for the site of interest, but are for other sites nearby. Can these data be extrapolated to the restoration site?
Difficult to interpret nontraditional data sources	One of historical ecology's greatest contributions has been to expand the type of records that may be considered to provide scientific evidence about the environment (e.g., Figure 8.3). However, data such as those collected from menus, mosaics, and other unstandardized sources can be difficult to interpret and integrate with other data sources.
Data certainty (e.g., on maps)	Some maps show it and others do not. Did the environmental feature exist? How accurate is the size of the feature?
Environmental filters of data	Proxy records such as cores of sediment or coral rubble, pollen, or fossil deposits can be contaminated by biotic and abiotic processes.
Cultural filters of data	In the case of man-made records, cultural biases may have influenced or exaggerated the data that were recorded. For example, the wealth of the New World was often inflated in journal writings sent back to Europe to voyage sponsors.
Distinguishing natural vs. cultural changes	Often, historical environmental dynamics do not have a clear driver, and changes could be attributed to natural or cultural factors. Spurious correlations will misguide restoration activities.

Notes: These problems are common to any application of historical environmental data and can be overcome by data validation: using multiple records (overlapping if possible), more than one data type (e.g., multiple proxies such as isotope data and diatom data from sediment cores), tying the record to a modern dataset, and/or using experiments and models to confirm findings. Then the quality of the data can be evaluated by the "preponderance of evidence." To assist restoration practitioners in deciding whether to act on the data, quantifying uncertainty and integrating it into the historical record is also helpful (Grossinger et al. 2007).

sessile fauna, the consequences of low dissolved oxygen range from reduced growth and reproduction to compromised immune functions to death (Diaz and Rosenberg 1995). Mobile species and life stages also shift distributions vertically and laterally in response to low dissolved-oxygen concentrations, and, as a result, trophic interactions can be markedly altered (Breitburg 1992, Breitburg et al. 2009, Diaz and Breitburg 2009). Because oxygen depletion in estuaries results from a combination of natural system characteristics (e.g., flushing rates, stratification, and depth) and anthropogenically derived nutrient loads that stimulate production and biomass, identification of historical conditions is critical for determining the role of human activities in water-quality degradation and for setting restoration targets.

Seasonally persistent and episodic low dissolved-oxygen concentrations in the mainstem estuary and deep tidal tributaries of the Chesapeake Bay were identified, during planning for regional water-quality restoration in the 1980s, as undesirable consequences of anthropogenic nutrient loading to the watershed. Water-quality criteria for dissolved oxygen now provide the basis for many of the nutrient-related restoration efforts (Batiuk et al. 2009). Several lines of historical evidence for worsening eutrophication were important in demonstrating that seasonal (late spring through early autumn) oxygen depletion worsened during the mid-twentieth century and that the oxygen depletion that occurred historically was less severe, extensive, and persistent. Biological proxies in long sediment cores indicated that the current salinity patterns in Chesapeake Bay were established during the late Holocene (2,300 years ago to present) and that this period could be used as a baseline against which modern conditions could be compared (Cronin and Ishman 2000, Cronin et al. 2000). Dissolved-oxygen proxy data suggest that oxygen concentrations substantially below 100% saturation levels were likely a natural feature of the deep channel of the mainstem Chesapeake Bay, but that prior to the seventeenth century, dissolved oxygen concentrations rarely, if ever, dropped below ~25% saturation (Karlsen et al. 2000). Shifts in the relative abundance of benthic and planktonic diatoms, as well as geochemical markers, indicated a reduction in water clarity and dissolved oxygen associated with mid-nineteenth-century land clearing, and a substantial further decline in these water-quality parameters with increasing population and use of high-nutrient fertilizers in the watershed during the second half of the twentieth century (Cooper and Brush 1991, Karlsen et al. 2000, Cronin and Vann 2003). Short-term increases in oyster growth in the Chesapeake Bay associated with increased nutrient loading have also been identified during the early to mid-nineteenth century using archaeological midden data (Kirby and Miller 2005), in general agreement with geological data. Instrumental and informal (Figure 8.3) datasets also concurred; research cruises conducted prior to the 1960s recorded higher dissolved-oxygen concentrations and less frequent observations of low dissolved-oxygen concentrations in the bottom layer of the water column than those measured in subsequent decades (Hagy et al. 2004).

Beyond establishing a better understanding of previous physical and chemical conditions of Chesapeake Bay, historical data on species with greatly reduced populations were used to determine historical ranges and habitat use. This was particularly important because habitat use for many bay species varies seasonally and with life stage, and because of a desire to protect important habitat for threatened and endangered species that are special targets of restoration and protection. Dissolved-oxygen requirements of Atlantic sturgeon (*Acipenser oxyrhynchus*) and short-nosed sturgeon (*A. brevirostrum*) became major determinants of minimum oxygen requirements set for shallow-water habitat in the mainstem bay and its subestuaries (EPA 2003b, Willard and Cronin 2007, Batiuk et al. 2009). Historical distributions were estimated from predictions of physiological and behavioral responses to low oxygen, historical records of sightings and landings, and predictions of historical distributions of dissolved oxygen, temperature, and salinity in the Chesapeake Bay system (Secor and Gunderson 1998, EPA 2003a).

Results of these historical reconstructions, along with contemporary data on effects of low dissolved oxygen on other native species, and considerations of financial, technical, and political feasibility, were then used to set specific targets for various habitats and water-column depths within the Chesapeake Bay system (Batiuk et al. 2009). Shallow-water habitat was predicted to have been oxygen rich historically, and the requirements of species such as sturgeon that are very sensitive to low oxygen exposure were used to set dissolved-oxygen criteria for shallow and nearshore bay waters. By contrast, research predicting historical hypoxia in the deep channel of Chesapeake Bay provided a basis for setting different criteria for this habitat than for surface waters, and setting those criteria at levels that are more possible to attain. A range of scenarios were then run using models predicting effects of nutrient loads on dissolved oxygen in order to set requirements for reducing nutrient discharges (EPA 2003b). Historical ecological data describing a wide range of different aspects of the estuarine ecosystem were an essential (although not sufficient) component to a robust process for setting regulatory standards and restoration targets.

Using Historical Data to Understand Landscape Connections

Marine ecosystems are linked to one another by tides, currents, resources, and environmental gradients, and restored ecosystems should be as well. Replicating these linkages as well as possible can improve restoration outcomes. When ecosystem linkages are severed by modern land uses, historical data can reveal the identity and features of cross-ecosystem linkages prior to their disruption. For example, the restoration of hydrology is one of the most critical components of tidal wetland restoration (Zedler and Callaway 2000, Neckles et al. 2002). Historical maps of tidal wetlands have provided lost information about pre-impact channel width, creek sinuosity, and pond abundance (Grossinger 1995, Hood 2004, Van Dyke and Wasson 2005), and these features can be recreated in restored marshes, through channel and pond dredging and reestablishing the outlets of tidal creeks, or used to evaluate the return to pre-impact conditions of restored marsh waterways (Van Dyke and Wasson 2005, Grossinger 2012).

The conversion of San Francisco Bay's tidal marshlands is one of the more remarkable cases of estuarine habitat loss and restoration, and a case in which historical data played an invaluable role in understanding landscape pattern and process and in guiding restoration targets and approaches. From 1860 to 1970, 80% of the historical tidal marsh in the San Francisco Bay was diked and filled for conversion into urban and agricultural land, managed marsh for hunting clubs, and salt production ponds (Atwater et al. 1979, Goals Project 1999). A goal-setting process during the 1990s for the regional integration of tidal wetland restoration activities included deliberate attention to the region's cultural and ecological history (Goals Project 1999, Grossinger et al. 2005).

Historical reconstruction of the intertidal ecosystem enabled historical data to be used in a number of ways. Quantifying the types and amounts of different major intertidal habitats (e.g., tidal marsh, tidal flats, and shallow waters) and their spatial relationships provided a context for setting restoration targets, resulting in widely viewed bar charts showing decline and proposed recovery by subregion (Goals Project 1999). In addition to providing technical information for restoration, maps of past and present wetland extent became a

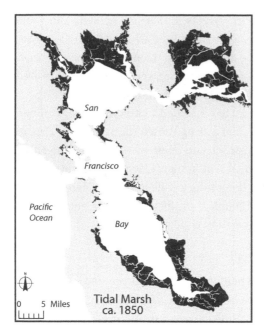
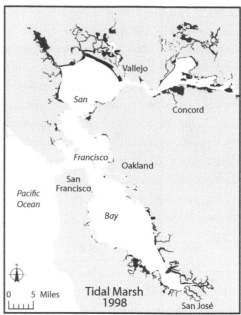

FIGURE 8.4 Maps like these (from the San Francisco Estuary Institute, comparing the distribution of tidal marsh in the San Francisco region in 1850 and 1998) motivate public support for restoration activities. (Graphic credit: Micha Salomon, SFEI.)

FIGURE 8.5 Restoration of tidal exchange to a former tidal marsh in South San Francisco Bay in 2006. The salt pond levee was breached at the location of a former channel mouth, identified in historical maps (and still evident topographically). (Photo credit: Paul Amato, EPA.)

central part of the public campaign for restoration (Figure 8.4). At a finer scale, a number of largely forgotten wetland features were recognized, such as marsh pannes and sandy beaches; these became targets for certain settings and ecological objectives (Williams and Faber 2001, Grossinger et al. 2005). Highly accurate and detailed historical mapping converted into GIS layers has been used to identify "ghost" channels and breach locations for the reintroduction of tidal flows to diked lands (Figure 8.5; Grossinger 2012). More recently, historical interpretation, drawing particularly on herbarium specimens, early flora, and map analyses (Grossinger 1995, Baye et al. 2000) and on investigation of contemporary remnants (Vasey et al. 2012), has led to increased attention to the conservation significance of landward ecotone environments, important landscape linkages that represent areas of lost and potential future plant biodiversity (Vasey et al. 2012). Historical analysis of the estuary has since been extended upstream into the Sacramento–San Joaquin Delta (Whipple et al. 2012).

The Baylands Goals Project led to the 2003 acquisition of a 6,111-ha tract of commercial salt ponds in the South Bay for restoration, the largest tidal-marsh-restoration project on the Pacific coast, as well as large projects in other parts of the bay. As of 2013, several large salt ponds had been breached and 40% of a slated 3,035 ha of tidal marsh had been restored (South Bay Salt Pond Restoration Project 2013).

WHAT WILL BE THE ROLE OF HISTORICAL ECOLOGY IN RESTORATIONS IN THE FUTURE?

The response of ecosystems to previous environmental changes is especially relevant given the rapid changes in environmental conditions predicted to occur in this century. Restored habitats must be resilient to species invasions, global warming, sea-level rise, and changing precipitation regimes, circulation patterns, and ocean chemistry, or else the ecological benefits will be short lived. Anticipated changes are likely to dramatically affect coastal ecosystems (Scavia et al. 2002, Harley et al. 2006, Molnar et al. 2008). As restoration ecology adapts to meet the needs of the conservation challenges of the twenty-first century (Hobbs et al. 2011), what will be the new and continued roles of historical ecology in coastal marine restoration?

Historical Data Can Inform Restoration Sustainability

Recognition that restoration needs to be more forward-oriented to be resilient to environmental change has moved restoration ecologists away from using historical baselines as explicit targets (Harris et al. 2006). Indeed, it is futile to restore an ecosystem to a historical state when environmental conditions have changed or will change to make the restoration unsustainable. Pitcher et al. (2005) developed a framework for restoring fisheries that straddles these ostensibly conflicting ideals (Box 8.1). In another example, in the Hudson River in New York, the Haverstraw Bay–Tappan Zee area has been targeted for oyster reef restoration because the area supported an eastern oyster fishery in the eighteenth and nineteenth centuries. A historical perspective seems to support this action, as paleoecological and

The compelling contributions in this book raise two key questions: "What was the past like?" and "How can this information shape the future?" The second question, a topic that many who would happily support historical reconstruction tend to balk at, implies that the discipline termed "historical ecology" should be adjusted to "applied historical ecology." And the applied aspect is decidedly the most controversial, as reflected, for example, in reactions to my "Back to the Future" program for historical, ecosystem-based restoration ecology in the oceans (see arguments in Pitcher and Lam 2010, Campbell et al. 2009, Pitcher 2005). This brief note aims to substantiate the use of theory in both ecological reconstruction and the restoration of marine ecosystems in the belief that the more explicit inclusion of theory will clarify these issues.

The first question—"What was the past like?"—is not without challenges. Even the core discipline of past reconstruction, history, can be equivocal about the exact nature of the past. Although fabrication of past events is regarded as a serious transgression, political,

social, and cultural interpretations, inevitably influenced by contemporary culture, are a legitimate part of the discipline, however dated and far-fetched they may seem a few years after publication. On the other hand, historians who analyze the past through the lens of environment and ecology are often marginalized (see Bolster 2008). Some historians, like the late Barbara Tuchman (1912–1989) and today's Simon Schama, adopt a fully social–ecological perspective using theoretical insights from both psychological and environmental disciplines. For example, Tuchman's simile of "a distant mirror" captures the difficulty in knowing exactly what happened in medieval France 700 years ago; her work shows that historical veracity is better assured when analysis is imbued with the appropriate theory covering political, military, and social aspects.

For the historical reconstruction of past ecosystems, the relevant theoretical issues are (1) the constraint put on relative biomass by the fundamental thermodynamic requirement of mass balance; (2) changes in the baseline amounts of primary production; (3) the scope, role, and

archaeological studies indicate that oysters were once plentiful in the region and, indeed, that multicentennial warming since the Little Ice Age (450–250 years ago) is bringing the estuary's temperature closer to optimal for the eastern oyster (Carbotte et al. 2004). However, other predicted environmental changes may decrease the likelihood of restoration success. Low-salinity events capable of killing oysters are predicted to increase in frequency in the Hudson River as the northeastern United States receives greater precipitation; therefore, Levinton et al. (2011) have predicted that oyster restorations in this region may be in jeopardy. Restoration practitioners in the Hudson River estuary must navigate the waters that span historical precedent, predicted climate conditions, feasibility, and the desires of restoration stakeholders.

relative abundance of forage fish; (4) the ecosystem consequences of the loss or restoration of keystone species whose influence, often through habitat structure rather than food web dynamics, is disproportionate to their biomass; (5) the ecosystem consequences of local extinctions and invasions; and (6) the effects of human activities on local habitats and ecosystem structure. These issues have been discussed before (Ainsworth et al. 2008, Pitcher et al. 2005) and will be explored synoptically elsewhere.

The second question—"How can this information shape the future?"—has deeper issues to address. A fishery manager in Puget Sound once commented, "I enjoy seeing an account of what has happened here, and appreciate knowing more clearly what we have lost, but I do not for the life of me see how we managers can use this detailed information from historical reconstructions when the climate and landscape has changed so much." My thesis here is that such cognitive confusion, and subsequent dysfunctional management goals and stasis, may be alleviated if management protocols have stronger and more ecosystem-based theoretical underpinnings. Restorations based on, but not necessarily exactly recapitulating, the relatively recent past of both habitats and organisms have the advantage that they encapsulate viable ecosystems and could provide known benefits to humans. Given the obvious uncertainties in forecasting the future, scenarios of possible futures have to be compared. Moreover, for management of restoration ecology based on past ecosystems, the relevant theoretical issues are (1) the reversibility of past depletions in the face of potentially ratchet-like processes (Pitcher 2001); (2) ecosystem-based stock assessment; (3) ecosystem-based habitat evaluations; (4) clear ecosystem-based reference points that can define progress in restoration, such as optimal restorable biomass and "lost valleys" (Ainsworth and Pitcher 2008, 2010); and (5) fishery goals that encapsulate policy based on sustainability theory, such as the no-discards "clean fleet" (Pitcher and Ainsworth 2010). Restoration ecology of the oceans is in its infancy and needs more theory: the contributions here provide encouragement that we can move in that direction.

Tony J. Pitcher is Professor and Director of Policy and Ecosystem Restoration in Fisheries Research Unit, Fisheries Centre, University of British Columbia.

Long-term climatic cycles, with frequencies of decades to millennia, can also affect restoration outcomes and may be a proxy for contemporary climate changes. Importantly, historical and paleoecological perspectives can shed light on the relationships between ecosystem conditions and climate cycles that might be missed by the shorter span of instrumental records. For example, Willard and Cronin (2007) discussed how climate in the Chesapeake Bay region is linked to hemispheric climate patterns such as the North Atlantic Oscillation and the important role of this climate cycle in determining the outcome of environmental management activities, such as nutrient reduction efforts, that are also shaped by natural precipitation regimes.

Historical data can help establish a long-term (prehistoric and historical) range of ecological variability, or the responses of ecosystems across centuries, millennia, or to climate change and anthropogenic activities (Szabo and Hédl 2011, Rick and Lockwood 2012). Although historical data do not provide a one-to-one correlation for restoration, they can be part of the larger dialogue that helps perform the best possible restoration in light of increasingly challenging and uncertain future conditions.

Historical information can also help us understand what makes ecosystems sustainable and resilient. In doing so, historical data continue to be a vital component of ecological restorations in the face of global environmental change (Harris et al. 2006, Jackson and Hobbs 2009). By helping us understand how systems work—rather than just how they were (Safford et al. 2012)—historical ecology can help us identify specific characteristics and processes that will enable systems to maintain desired ecological functions through environmental perturbations (Grossinger 2012). In San Francisco Bay–Delta wetland restorations, for example, a historical analysis of changes in bathymetry and sediment supply using historical U.S. Coast (and Geodetic) Survey maps has been an integral part of understanding short- and long-term marsh restoration potential (Jaffe et al. 2007). Sediment released by hydraulic mining during the nineteenth-century California Gold Rush, which caused rapid bay filling and marsh expansion in the late nineteenth century (Jaffe et al. 2007), has largely moved through the estuary, while dams have greatly reduced the sediment input from rivers. The effects of these historical impacts on sediment supply will dictate both the pace of development of restored salt marshes, which will vary by sub-estuary (Goals Project 1999), and their ability to keep up with sea-level rise (Callaway et al. 2007).

As the Definition of "Restoration" Expands, Historical Data Ground Restoration Activities

To accommodate changing environmental conditions and societal needs, the definition of "restoration" has expanded to include activities that are not founded on a history of impact, like the restoration of ecosystem services, habitat construction, and the creation of artificial habitats (Hobbs and Cramer 2008). The production of ecosystem services, or the goods and services provided by nature that are necessary for human existence and welfare, is increasingly the target of restoration activities (McCay et al. 2003, Peterson et al. 2003, Grabowski and Peterson 2007, Palmer and Filoso 2009, Hobbs et al. 2011). In cases where in situ restorations are infeasible, habitats are being created de novo to recover ecosystem services of lost or degraded ecosystems. Creating habitat where it never existed has expanded the definition of "restoration," but proponents of this expansion of restoration emphasize that such decisions about what is possible and what is desirable must be "rooted in historical understanding" (Hobbs et al. 2009). Accurately and convincingly determining whether altered ecosystems have the potential to recover historical characteristics, or rather should be managed as hybrid or novel systems, requires a strong understanding of historical ecological conditions, their physical drivers, and causes of change (e.g., Hobbs et al. 2009, Jackson and Hobbs 2009). For example, estimates of the historical provision of ecosystem services may serve as a useful restoration target to be met by a range of project designs (see chapter 9, this volume).

Novel Ecosystems

Novel or no-analog ecosystems that contain species and abiotic conditions that have not previously coexisted are forming more frequently as a result of climate change, species introductions, and other impacts pushing species and places together in new combinations (Hobbs et al. 2006). Novel ecosystems may still provide desirable ecosystem services to society, and may even, in the future, be incorporated into restoration frameworks (Hobbs and Cramer 2008). What role, if any, should historical information play in the evaluation and management of these new systems?

For one, the identification of a novel ecosystem requires historical norms. Acceptance of a novel ecosystem is usually done in recognition that some historical conditions have changed irreversibly (e.g., as a result of physical transformation, irrevocable species invasion, extinction of keystone species, or global climate change), with significant effects on the ecosystem that cannot be overcome by restoration interventions (Choi 2007). For example, Ruesink et al. (2005) and the National Research Council (2004) have explored the acceptability of restoring lost oyster reefs with an introduced oyster species, given the degree of impact on the native species and its chance of recovery, as well as the likelihood of establishment of the introduced species and the odds that its introduction will be accompanied by unintended consequences. In the Gulf of Maine, overfishing of cod and high biomass of lobster were considered a no-analog condition (Steneck et al. 2004), but a recent analysis of shell midden records in Maine discovered evidence of a prior trophic cascade between 4,350 and 1,200 years ago due to harvest of apex predators by Native Americans (Bourque et al. 2008). However, the notable absence of lobsters and crabs from shell midden records leaves open the possibility that the current state is indeed novel or, alternatively, that lobsters and crabs are missing from middens because of lack of harvest or poor preservation (Bourque et al. 2008). In another case, Kowalewski et al. (2000) used paleoecological data from AD 950–1950 to document dramatic recent changes in the Colorado River estuary after the river was dammed, including the disappearance of trillions of marine bivalves, along with fishes and marine mammals, to its present condition of desert.

Moreover, as Jackson and Hobbs (2009) pointed out, no-analog communities without modern counterparts are a persistent feature of the past 2.5 million years, as communities have assembled and disassembled with each glaciation. Given this perspective, paleoecological records may offer unique insight into the dynamics of novel ecosystems. As historical ecological datasets continue to grow, they will likely provide further examples of ecological novelty in the past and help us better contextualize the formation of novel ecosystems in the present and future.

CONCLUSIONS

While many factors must be taken into consideration in siting and designing an ecological restoration, a historical context provides a critical foundation. Although historical baselines are now seen as less prescriptive for ecological restorations (Figure 8.1; Choi 2007, Choi et al. 2008, Hobbs and Cramer 2008, Hobbs et al. 2011), they remain important guideposts, telling us what has occurred and, at least in some cases, what could occur again. They help us understand how systems work, both in the past and in the present, which is essential to accurately define potential trajectories, whether they veer toward historical conditions, novel habitats, or somewhere in between. Particularly in this time of "a new ecological world order" (Hobbs et al. 2006), historical information provides important grounding to restoration ecology, which otherwise has the capacity to produce environments that are untethered to an ecological reality and difficult to evaluate (e.g., "open-ended restoration"; Hughes et al. 2012). Although historical reconstructions or baselines are sometimes presumed to be in opposition to the consideration of novel ecosystems, these concepts are integral parts of the complex cultural and scientific conversation about what is possible and what is desirable (Hobbs et al. 2004).

Above all, a history of an ecosystem provides the context for restoration, by depicting what types of impacts an ecosystem has experienced and how those impacts have changed its structure and function. Historical information can also inform restoration activities in other ways: as a reference for restoration targets; to infer the range of conditions that have occurred through time (i.e., historical range of variability) and could be acceptable restoration outcomes; for understanding landscape connections between habitats and their implications for the target ecosystem; to help determine desired future conditions; and to understand and maximize the sustainability of restoration activities. As restoration ecology stretches to meet the needs of our society in a time of new environmental pressures and conservation challenges, historical ecology will continue to provide vital information that shapes and guides restoration activities.

REFERENCES

Ainsworth, C., and Pitcher, T. J. (2008) Back to the future in Northern British Columbia: evaluating historic marine ecosystems and optimal restorable biomass as restoration goals for the future. In *Reconciling Fisheries with Conservation: Proceedings of the Fourth World Fisheries Congress* (J. L. Nielsen, J. J. Dodson, K. Friedland, T. R. Hamon, J. Musick, and E. Verspoor, Eds.). Symposium 49. American Fisheries Society, Bethesda, MD. pp. 317–329.

Ainsworth, C. H., and Pitcher, T. J. (2010) A bioeconomic optimization approach for rebuilding marine communities: British Columbia case study. *Environmental Conservation* 36, 1–11.

Ainsworth, C. H., Pitcher, T. J., Heymans, J. J., and Vasconcellos, M. (2008) Reconstructing historical marine ecosystems using food web models: northern British Columbia from pre-European contact to present. *Ecological Modeling* 216, 354–368.

Alexander, K. E., Leavenworth, W. B., Cournane, J., et al. (2009) Gulf of Maine cod in 1861: historical analysis of fishery logbooks, with ecosystem implications. *Fish and Fisheries* 10, 428–449.

Atwater, B. F., Conard, S. G., Dowden, J. N., et al. (1979) History, landforms, and vegetation of the estuary's tidal marshes. In *San Francisco Bay: The Urbanized Estuary* (T. J. Conomos, Ed.). American Association for the Advancement of Science, San Francisco, CA. pp. 347–386.

Batiuk, R. A., Breitburg, D. L., Diaz, R. J., et al. (2009) Derivation of habitat-specific dissolved oxygen criteria for Chesapeake Bay and its tidal tributaries. *Journal of Experimental Marine Biology and Ecology* 381, S204–S215.

Baye, P. R., Faber, P. M., and Grewell, B. (2000) Tidal marsh plants of the San Francisco Estuary. In *Baylands Ecosystem Species and Community Profiles: Life Histories and Environmental Requirements of Key Plants, Fish and Wildlife* (P. R. Olofson, Ed.). San Francisco Bay Regional Water Quality Control Board, Oakland, CA. pp. 9–32.

Baylor, J. B. (1896) Method of defining and locating natural oyster beds, rocks and shoals. *Oyster Records* [pamphlets, one for each Tidewater, VA, county]. Board of Fisheries of Virginia.

Beck, M. W., Brumbaugh, R. D., Airoldi, L., et al. (2011) Oyster reefs at risk and recommendations for conservation, restoration, and management. *BioScience* 61, 107–116.

Bolster, W. J. (2008) Putting the ocean in Atlantic history: maritime communities and marine ecology in the northwest Atlantic, 1500–1800. *American Historical Review* 113, 19–47.

Bourque, B. J., Johnson, B. J., and Steneck, R. S. (2008) Possible prehistoric fishing effects on coastal marine food webs in the Gulf of Maine. *Human Impacts on Ancient Marine Ecosystems*, 165–185.

Braje, T. J., Erlandson, J. M., Rick, T. C., et al. (2009) Fishing from past to present: continuity and resilience of red abalone fisheries on the Channel Islands, California. *Ecological Applications* 19, 906–919.

Breitburg, D. L. (1992) Episodic hypoxia in Chesapeake Bay: interacting effects of recruitment, behavior, and physical disturbance. *Ecological Monographs*, 525–546.

Breitburg, D. L., Hondorp, D. W. Davias, and R. J. Diaz (2009) Hypoxia, nitrogen and fisheries: Integrating effects across local and global landscapes. *Annual Reviews in Marine Science* 1, 329–350.

Buizer, M., Kurz, T., and Ruthrof, K. (2012) Understanding restoration volunteering in a context of environmental change: in pursuit of novel ecosystems or historical analogues? *Human Ecology* 40, 153–160.

Callaway, J. C., Parker, V. T., Vasey, M. C., and Schile, L. M. (2007) Emerging issues for the restoration of tidal marsh ecosystems in the context of predicted climate change. *Madroño* 54, 234–248.

Campbell, L. M., Gray, N. J., Hazen, E. L., and Shackeroff, J. M. (2009) Beyond baselines: rethinking priorities for ocean conservation. *Ecology and Society* 14, article 14.

Carbotte, S. M., Bell, R. E., Ryan, W. B. F., et al. (2004) Environmental change and oyster colonization within the Hudson River estuary linked to Holocene climate. *Geo-Marine Letters* 24, 212–224.

Choi, Y. D. (2007) Restoration ecology to the future: a call for new paradigm. *Restoration Ecology* 15, 351–353.

Choi, Y. D., Temperton, V. M., Allen, E. B., et al. (2008) Ecological restoration for future sustainability in a changing environment. *Ecoscience* 15, 53–64.

Cooper, S. R., and Brush, G. S. (1991) Long-term history of Chesapeake Bay anoxia. *Science* 254, 992–996.

Cronin, T. M., and Ishman, S. E. (2000) Holocene paleoclimate from Chesapeake Bay based on ostracodes and benthic foraminifera from Marion-Dufresne core MD99–2209. In *Initial Report on IMAGES V Cruise of the Marion-Dufresne to the Chesapeake Bay* (T. M. Cronin, Ed.). Open-file Report 00–306. U.S. Geological Survey, Reston, VA.

Cronin, T. M., and Vann, C. D. (2003) The sedimentary record of climatic and anthropogenic influence on the Patuxent Estuary and Chesapeake Bay ecosystems. *Estuaries* 26, 196–209.

Cronin, T. [M.], Willard, D., Karlsen, A., et al. (2000) Climatic variability in the eastern United States over the past millennium from Chesapeake Bay sediments. *Geology* 28, 3–6.

Diaz, R. J., and Breitburg, D. L. (2009) The hypoxic environment. *Fish Physiology* 27, 1–23.

Diaz, R. J., and Rosenberg, R. (1995) Marine benthic hypoxia: a review of its ecological effects and the behavioural responses of benthic macrofauna. *Oceanography and Marine Biology: An Annual Review* 33, 245–303.

Diaz, R. J., and Rosenberg, R. (2008) Spreading dead zones and consequences for marine ecosystems. *Science* 321, 926–929.

Dietl, G. P., and Flessa, K. W. (2011) Conservation paleobiology: putting the dead to work. *Trends in Ecology & Evolution* 26, 30–37.

Duarte, C. M., Conley, D. J., Carstensen, J., and Sánchez-Camacho, M. (2009) Return to Neverland: shifting baselines affect eutrophication restoration targets. *Estuaries and Coasts* 32, 29–36.

EPA (2003a) Ambient water quality criteria for dissolved oxygen, water clarity and chlorophyll-a for Chesapeake Bay and its tidal tributaries. U.S. Environmental Protection Agency Region III Chesapeake Bay Program Office, Annapolis, MD, Region III Water Protection Division, Philadelphia, PA, and Office of Water, Office of Science and Technology, Washington, DC.

EPA (2003b) Biological Evaluation for the Issuance of Ambient Water Quality Criteria for Dissolved Oxygen, Water Clarity and Chlorophyll a for the Chesapeake Bay and Its Tidal Tributaries. U.S. Environmental Protection Agency Region III Water Protection Division, Philadelphia, PA.

Erlandson, J. M. (2001) The archaeology of aquatic adaptations: paradigms for a new millennium. *Journal of Archaeological Research* 9, 287–350.

Erlandson, J. M., and Rick, T. C. (2010) Archaeology meets marine ecology: the antiquity of maritime cultures and human impacts on marine fisheries and ecosystems. *Annual Review of Marine Science* 2, 231–251.

Estes, J. A., and Duggins, D. O. (1995) Sea otters and kelp forests in Alaska: generality and variation in a community ecological paradigm. *Ecological Monographs* 65, 75–100.

Estes, J. A., Tinker, M. T., Williams, T. M., and Doak, D. F. (1998) Killer whale predation on sea otters linking oceanic and nearshore ecosystems. *Science* 282, 473–476.

Foster, D., Swanson, F., Aber, J., et al. (2003) The importance of land-use legacies to ecology and conservation. *BioScience* 53, 77–88.

Goals Project (1999) *Baylands Ecosystem Habitat Goals: A Report of Habitat Recommendations Prepared by the San Francisco Bay Area Wetlands Ecosystem Goals Project.* U.S. Environmental Protection Agency, San Francisco, Calif./S.F. Bay Regional Water Quality Control Board, Oakland, CA.

Grabowski, J. H., and Peterson, C. H. (2007) Restoring oyster reefs to recover ecosystem services. *Theoretical Ecology Series* 4, 281–298.

Grossinger, R. (1995) Historical evidence of freshwater effects on the plan form of tidal marshlands in the Golden Gate estuary. M.S. thesis, University of California, Santa Barbara, CA.

Grossinger, R. (2012) *Napa Valley Historical Ecology Atlas: Exploring a Hidden Landscape of Transformation and Resilience.* University of California Press, Berkeley, CA.

Grossinger, R., Askevold, R., and Collins, J. N. (2005) T-sheet user guide: application of the historical U.S. Coast Survey maps to environmental management in the San Francisco Bay area. SFEI Report No. 427. San Francisco Estuary Institute, Oakland, CA.

Grossinger, R., Striplen, C. J., Brewster, E., et al. (2007) Historical landscape ecology of an urbanized California valley: wetlands and woodlands in the Santa Clara Valley. *Landscape Ecology* 22, 103–120.

Hagy, J. D., Boynton, W. R., Keefe, C. W., and Wood, K. V. (2004) Hypoxia in Chesapeake Bay, 1950–2001: long-term change in relation to nutrient loading and river flow. *Estuaries and Coasts* 27, 634–658.

Hanley, N., Ready, R., Colombo, S., et al. (2009) The impacts of knowledge of the past on preferences for future landscape change. *Journal of Environmental Management* 90, 1404–1412.

Harley, C. D., Hughes, A. R., Hultgren, K. M., et al. (2006) The impacts of climate change in coastal marine systems. *Ecology Letters* 9, 228–241.

Harris, J. A., Hobbs, R. J., Higgs, E., and Aronson, J. (2006) Ecological restoration and global climate change. *Restoration Ecology* 14, 170–176.

Hobbs, R. J., Arico, S., Aronson, J., et al. (2006) Novel ecosystems: theoretical and management aspects of the new ecological world order. *Global Ecology and Biogeography* 15, 1–7.

Hobbs, R. J., and Cramer, V. A. (2008) Restoration ecology: interventionist approaches for restoring and maintaining ecosystem function in the face of rapid environmental change. *Annual Review of Environment and Resources* 33, 39–61.

Hobbs, R. J., Davis, M. A., Slobodkin, L. B., et al. (2004) Restoration ecology: the challenge of social values and expectations. *Frontiers in Ecology and the Environment* 2, 43–48.

Hobbs, R. J., Hallett, L. M., Ehrlich, P. R., and Mooney, H. A. (2011) Intervention ecology: applying ecological science in the twenty-first century. *BioScience* 61, 442–450.

Hobbs, R. J., Higgs, E., and Harris, J. A. (2009) Novel ecosystems: implications for conservation and restoration. *Trends in Ecology & Evolution* 24, 599–605.

Hood, G. W. (2004) Indirect environmental effects of dikes on estuarine tidal channels: thinking outside of the dike for habitat restoration and monitoring. *Estuaries and Coasts* 27, 273–282.

Hughes, F. M., Adams, W. M., and Stroh, P. A. (2012) When is open-endedness desirable in restoration projects? *Restoration Ecology* 20, 291–295.

Hughes, T. P. (1994) Catastrophes, phase shifts, and large-scale degradation of a Caribbean coral reef. *Science* 265, 1547–1551.

Jackson, J. B., Alexander, K. E., and Sala, E. (Eds.) (2011) *Shifting Baselines: The Past and the Future of Ocean Fisheries.* Island Press, Washington, DC.

Jackson, J. B., Kirby, M. X., Berger, W. H., et al. (2001) Historical overfishing and the recent collapse of coastal ecosystems. *Science* 293, 629–637.

Jackson, S. T., and Hobbs, R. J. (2009) Ecological restoration in the light of ecological history. *Science* 325, 567–569.

Jaffe, B. E., Smith, R. E., and Foxgrover, A. C. (2007) Anthropogenic influence on sedimentation and intertidal mudflat change in San Pablo Bay, California: 1856–1983. *Estuarine, Coastal and Shelf Science* 73, 175–187.

Karlsen, A. W., Cronin, T. M., Ishman, S. E., et al. (2000) Historical trends in Chesapeake Bay dissolved oxygen based on benthic foraminifera from sediment cores. *Estuaries and Coasts* 23, 488–508.

Kirby, M. X., and Miller, H. M. (2005) Response of a benthic suspension feeder (*Crassostrea virginica* Gmelin) to three centuries of anthropogenic eutrophication in Chesapeake Bay. *Estuarine, Coastal and Shelf Science* 62, 679–689.

Kirch, P. V. (2005) Archaeology and global change: the Holocene record. *Annual Review Environment and Resources* 30, 409–440.

Kittinger, J. N., Pandolfi, J. M., Blodgett, J. H., et al. (2011) Historical reconstruction reveals recovery in Hawaiian coral reefs. *PLoS ONE* 6, e25460.

Knowlton, N., and Jackson, J. B. (2008) Shifting baselines, local impacts, and global change on coral reefs. *PLoS Biology* 6, e54.

Kowalewski, M., Serrano, G. E. A., Flessa, K. W., and Goodfriend, G. A. (2000) Dead delta's former productivity: two trillion shells at the mouth of the Colorado River. *Geology* 28, 1059–1062.

Levinton, J., Doall, M., Ralston, D., et al. (2011) Climate change, precipitation and impacts on an estuarine refuge from disease. *PLoS ONE* 6, e18849.

Lotze, H. K., Erlandson, J. M., Hardt, M. J., et al. (2011) Uncovering the ocean's past. In *Shifting Baselines: The Past and the Future of Ocean Fisheries* (J. B. Jackson, K. E. Alexander, and E. Sala, Eds.). Island Press, Washington, DC. pp. 137–161.

Lotze, H. K., Lenihan, H. S., Bourque, B. J., et al. (2006) Depletion, degradation, and recovery potential of estuaries and coastal seas. *Science* 312, 1806–1809.

Lyman, R. L. (2012) A warrant for applied palaeozoology. *Biological Reviews* 87, 513–525.

Mann, R., and Evans, D. A. (2004) Site selection for oyster habitat rehabilitation in the Virginia portion of the Chesapeake Bay: a commentary. *Journal of Shellfish Research* 23, 41–49.

Mann, R., Harding, J. M., and Southworth, M. J. (2009) Reconstructing pre-colonial oyster demographics in the Chesapeake Bay, USA. *Estuarine, Coastal and Shelf Science* 85, 217–222.

Marean, C. W., Bar-Matthews, M., Bernatchez, J., et al. (2007) Early human use of marine resources and pigment in South Africa during the Middle Pleistocene. *Nature* 449, 905–908.

Marsh, H., De'Ath, G., Gribble, N., and Lane, B. (2005) Historical marine population estimates: triggers or targets for conservation? The dugong case study. *Ecological Applications* 15, 481–492.

McCay, F., Peterson, C. H., DeAlteris, J. T., and Catena, J. (2003) Restoration that targets function as opposed to structure: replacing lost bivalve production and filtration. *Marine Ecology Progress Series* 264, 197–212.

McClenachan, L., Ferretti, F., and Baum, J. K. (2012) From archives to conservation: why historical data are needed to set baselines for marine animals and ecosystems. *Conservation Letters* 5, 349–359.

Molnar, J. L., Gamboa, R. L., Revenga, C., and Spalding, M. D. (2008) Assessing the global threat of invasive species to marine biodiversity. *Frontiers in Ecology and the Environment* 6, 485–492.

Montgomery, D. R. (2008) Dreams of natural streams. *Science* 319, 291–292.

Moore, H. F. (1910) Condition and extent of oyster beds in the James River, Virginia. U.S. Bureau of Fisheries Document No. 729.

National Research Council (1992) *Restoration of Aquatic Ecosystems: Science, Technology, and Public Policy*. National Academies Press, Washington, DC.

National Research Council (2004) Non-native oysters in the Chesapeake Bay. Committee on Non-native Oysters in the Chesapeake Bay, Ocean Studies Board, Division on Earth and Life Studies, Washington, DC.

Neckles, H. A., Dionne, M., Burdick, D. M., et al. (2002) A monitoring protocol to assess tidal restoration of salt marshes on local and regional scales. *Restoration Ecology* 10, 556–563.

Newell, R. I. E. (1988) Ecological changes in Chesapeake Bay: are they the result of overharvesting the American Oyster, *Crassostrea virginica*? In *Understanding the Estuary: Advances in Chesapeake Bay Research* (pp. 536–546). Publication 129, Chesapeake Research Consortium, Solomons, MD.

Oyster Metrics Workgroup (2011) Restoration goals, quantitative metrics and assessment protocols for evaluating success on restored oyster reef sanctuaries. Report to the Sustainable Fisheries Goal Implementation Team of the Chesapeake Bay Program.

Palmer, M. A., Ambrose, R. F., and Poff, N. L. (1997) Ecological theory and community restoration ecology. *Restoration Ecology* 5, 291–300.

Palmer, M. A., and Filoso, S. (2009) Restoration of ecosystem services for environmental markets. *Science* 325, 575–576.

Pauly, D. (1995) Anecdotes and the shifting baseline syndrome of fisheries. *Trends in Ecology & Evolution* 10, 430.

Peterson, C. H., Grabowski, J. H., and Powers, S. P. (2003) Estimated enhancement of fish production resulting from restoring oyster reef habitat: quantitative valuation. *Marine Ecology Progress Series* 264, 249–264.

Pitcher, T. J. (2001) Fisheries managed to rebuild ecosystems: reconstructing the past to salvage the future. *Ecological Applications* 11, 601–617.

Pitcher, T. J. (2005) 'Back to the Future': A fresh policy initiative for fisheries and a restoration ecology for ocean ecosystems. *Philosophical Transactions of the Royal Society of London Series B* 360, 107–121.

Pitcher, T. J., and Ainsworth, C. H. (2010) Resilience to change in two coastal communities: using the maximum dexterity fleet. *Marine Policy* 34, 810–814.

Pitcher, T. J., Ainsworth, C. H., Buchary, E. A., et al. (2005) Strategic management of marine ecosystems using whole-ecosystem simulation modelling: the 'back-to-the-future' policy approach. In *Strategic Management of Marine Ecosystems* (E. Levner, I. Linkov, and J.-M. Proth, Eds.). NATO Science Series IV, vol. 50. Springer, Dordrecht, The Netherlands. pp. 199–258.

Pitcher, T. J., and Lam, M. (2010) Fishful thinking: rhetoric, reality and the sea before us. *Ecology and Society* 15, 12–29.

Power, A., Corley, B., Atkinson, D., et al. (2010) A caution against interpreting and quantifying oyster habitat loss from historical surveys. *Journal of Shellfish Research* 29, 927–936.

Rabalais, N. N., Diaz, R. J., Levin, L. A., et al. (2010) Dynamics and distribution of natural and human-caused hypoxia. *Biogeosciences* 7, 585–619.

Redman, C. L. (1999) *Human Impact on Ancient Environments.* University of Arizona Press, Tucson, AZ.

Rick, T. C., and Erlandson, J. M. (2008) *Human Impacts on Ancient Marine Ecosystems: A Global Perspective.* University of California Press, Berkeley, CA.

Rick, T. C., and Erlandson, J. M. (2009) Coastal exploitation. *Science* 325, 952–953.

Rick, T. C., and Lockwood, R. (2012) Integrating paleobiology, archeology, and history to inform biological conservation. *Conservation Biology* 27, 45–54.

Ruesink, J. L., Lenihan, H. S., Trimble, A. C., et al. (2005) Introduction of non-native oysters: ecosystem effects and restoration implications. *Annual Review of Ecology, Evolution, and Systematics* 36, 643–689.

Safford, H. D., North, M., and Meyer, M. D. (2012) Climate change and the relevance of historical forest conditions. In *Managing Sierra Nevada Forests* (M. North, Ed.). USDA Forest Service General Technical Report PSW-GTR-237. pp. 23–45.

Scavia, D., Field, J. C., Boesch, D. F., et al. (2002) Climate change impacts on US coastal and marine ecosystems. *Estuaries and Coasts* 25, 149–164.

Secor, D. H., and Gunderson, T. E. (1998) Effects of hypoxia and temperature on survival, growth, and respiration of juvenile Atlantic sturgeon, *Acipenser oxyrinchus. Fishery Bulletin* 96, 603–613.

Simenstad, C. A., Estes, J. A., and Kenyon, K. W. (1978) Aleuts, sea otters, and alternate stable-state communities. *Science* 200, 403–411.

South Bay Salt Pond Restoration Project (2013) Track our progress. www.southbayrestoration.org /track-our-progress/.

Steneck, R. S., Vavrinec, J., and Leland, A. V. (2004) Accelerating trophic-level dysfunction in kelp forest ecosystems of the western North Atlantic. *Ecosystems* 7, 323–332.

Szabo, P., and Hédl, R. (2011) Advancing the integration of history and ecology for conservation. *Conservation Biology* 25, 680–687.

Van Dyke, E., and Wasson, K. (2005) Historical ecology of a central California estuary: 150 years of habitat change. *Estuaries and Coasts* 28, 173–189.

Vasey, M. C., Parker, V. T., Callaway, J. C., et al. (2012) Tidal wetland vegetation in the San Francisco Bay–Delta Estuary. *San Francisco Estuary and Watershed Science* 10.

Whipple, A., Grossinger, R., Rankin, D., et al. (2012) Sacramento–San Joaquin Delta Historical Ecology investigation: exploring pattern and process. SFEI-ASC Historical Ecology Program Publication No. 672. San Francisco Estuary Institute, Richmond, CA. www.sfei.org /DeltaHEStudy.

White, P. S., and Walker, J. L. (1997) Approximating nature's variation: selecting and using reference information in restoration ecology. *Restoration Ecology* 5, 338–349.

Willard, D. A., and Cronin, T. M. (2007) Paleoecology and ecosystem restoration: case studies from Chesapeake Bay and the Florida Everglades. *Frontiers in Ecology and the Environment* 5, 491–498.

Williams, P., and Faber, P. (2001) Salt marsh restoration experience in San Francisco Bay. *Journal of Coastal Research* 27, 203–211.

Winslow, F. (1882) Methods and results. Report of the oyster beds of the James River, Virginia and of Tangier and Pocomoke Sounds, Maryland and Virginia. Government Printing Office, Washington, DC.

Wolverton, S., and Lyman, R. L. (Eds.) (2012) *Conservation Biology and Applied Zooarchaeology.* University of Arizona Press, Tucson, AZ.

Yates, C. C. (1913) *Summary of survey of oyster bars of Maryland, 1906–1912.* Government Printing Office, Washington, DC.

Zedler, J. B., and Callaway, J. C. (2000) Evaluating the progress of engineered tidal wetlands. *Ecological Engineering* 15, 211–225.

NINE

Estimates of Historical Ecosystem Service Provision Can Guide Restoration Efforts

PHILINE S. E. ZU ERMGASSEN, MARK D. SPALDING, and ROBERT D. BRUMBAUGH

Restoration is undertaken not only to reverse habitat losses but also to recover the many valuable ecosystem services associated with coastal habitats. While ecosystem services are increasingly being used to define restoration objectives for a number of marine and terrestrial habitats, estimates of historical ecosystem service delivery are rare, in part because of the difficulty of making such estimates. However, by combining historical data with an understanding of the habitat characteristics (e.g., density and habitat complexity) and environmental conditions that influence service provision, (e.g., salinity and location relative to other habitats) estimates of historical ecosystem services can be used to target restoration efforts and management practices toward the desired outcomes. Oyster reefs have suffered an estimated 85% decline globally over the past 150 years, and there are growing efforts to restore oyster reefs at a large scale to recover oyster fishery, fish production, water quality, and other ecosystem services. In this chapter, we explore the estimation of historical provision of ecosystem services in oyster reefs as a case study to understand the ecological and socially relevant reference points that these estimates provide for restoration goals.

INTRODUCTION

The degradation and destruction of habitat has been prevalent throughout human history (Miller et al. 2005) but has increased in pace in recent decades (Roberts 2002, Waycott et al. 2009, Fearnside 2005). The resulting loss of habitat has led to substantial declines in biodiversity and ecosystem service provision (Brooks et al. 2002, Balvanera et al. 2006). Recognition of these declines has motivated an increase in restoration efforts in recent decades (Moreno-Mateos et al. 2012, Brumbaugh et al. 2010), with a growing emphasis on ecological restoration for ecosystem service provision over the past few years (Trabucchi et al. 2012). Ecosystem services are benefits that humans derive from ecosystems (Millennium Ecosystem Assessment 2005). These include both goods—natural products such as wood or food, and services—and less tangible commodities such as water purification (Costanza et al. 1997).

Although ecosystem service provision has only recently come to be recognized as a key element of our interaction with the natural world, there has been rapid progress in developing the science to quantify services (Rozas et al. 2005, Duffy and Kahara 2011, Brozozowska et al. 2012, Costanza et al. 2008, Barbier et al. 2011). The clarification of the link between ecology and the economy has resulted in ecosystem services being given increased emphasis in decision-making processes at many levels (Molnar and Kubiszewski 2012, UK National Ecosystem Assessment 2011). This includes a call for conservation initiatives to incorporate ecosystem services into their conservation planning and assessment (Egoh et al. 2007), a call that many restoration efforts have answered. Recovery of ecosystem services is now a central aim of many restoration projects, and these stated aims should therefore be reflected in the quantitative goals set.

Using ecosystem services as a restoration goal is not a new concept; restoration of coastal wetlands has led the way in incorporating ecosystem services in restoration objectives (Ehrenfeld 2000), with extensive valuation of coastal protection functions of natural habitats in particular (Costanza et al. 2008, Barbier et al. 2011). This increased understanding of ecosystem service provision has also led to an increase in ecosystem services being taken into account in decision making, often through the use of decision support tools such as InVEST (Tallis and Polasky 2009). The incorporation of ecosystem services into conservation planning has played a critical role in increasing support for conservation efforts from a broad range of stakeholders (Goldman et al. 2008). Whether or not ecosystem services can be quantified economically, their quantification and inclusion in the decision-making progress can add significant value to restoration projects (White et al. 2012). Furthermore, research into other conservation initiatives have shown that such indirect measures (i.e., measuring area as opposed to the quantity of service or biodiversity one seeks to achieve) are less effective at ensuring that goals are met than using direct measures of the services themselves (Ferraro and Simpson 2002). With this in mind, the field of restoration ecology should aim to develop the science that would allow estimates of the quantity or quality of ecosystem service to be determined, as well as tools for monitoring progress toward those goals.

Despite recognition that incorporating ecosystem services into restoration goal setting and planning may dramatically improve the outcome and value of restoration (Chan et al. 2006, Rondinini and Chiozza 2010), and that historical data are critical in understanding and potentially responding to shifting baselines (Swetnam et al. 1999; and see chapter 8, this volume), the historical function of an ecosystem is rarely quantitatively considered when setting restoration goals. Gaining an insight into the historical provision of services allows the full implications of the loss of habitat to be expressed in terms of broad appeal to the public and to groups for whom traditional biodiversity arguments may be weak. This is of particular importance given the human propensity for adapting baselines over very short periods (Turvey et al. 2010), which may compromise the perceived value of species for which there is no market in the present. Assessing the historical value of ecosystem services requires a good understanding of the historical ecology of the habitat, because a single hectare of a habitat 200 years ago may have been different in terms of species, productivity, and biomass from an apparently similar hectare today. Those differences, in turn, would influence levels of ecosystem service provision. Thus, an understanding of the functions provided by a habitat historically can provide a more useful baseline for considering goals based on ecosystem service delivery.

Here, we explore how historical data can be used to estimate historical ecosystem service delivery and how such data may be applied to assist in the setting of restoration targets. We present results from a case study habitat, oyster reefs, to illustrate the application of this approach, and we discuss the implications of the approach in terms of a broader range of habitat types, species, and ecosystems. We review which data may be necessary to build an understanding of historical ecosystem service delivery and discuss the nature of services that may be appropriate targets of goal setting. Finally, we introduce various challenges and opportunities presented by this approach to restoration goal setting.

Which Data Are Required to Estimate the Historical Contribution of Habitats?

Universally, one would expect all services to be affected by the spatial extent of the relevant habitat, so this is a key data need in determining the historical level of ecosystem service provision. However, spatial extent is often not the sole attribute needed for such analyses, because most ecosystem services are driven by processes that are also influenced by other attributes of the habitat, such as biomass, density, habitat structure, and species richness (Barbier et al. 2011, Hooper et al. 2005, zu Ermgassen et al. 2013). For example, carbon storage by forests is a function of both habitat extent and tree size and density (biomass), rugosity or habitat structure may influence the diversity or abundance of associated species (McClanahan 1994), and biodiversity likely underpins the resilience of ecosystem service provision in the event of environmental disturbance (Worm et al. 2006).

Applying Habitat Data to Ecosystem Service Estimates

Once the underlying drivers of the ecosystem service have been identified (frequently these are related to the size distribution and density of the foundation species, which may be expressed as biomass), it is necessary to determine the nature of the relationship between the underlying ecosystem attribute and the service. Although services that involve the direct extraction of goods may scale linearly in response to changes in species density or biomass, most services are likely to be subject to threshold densities below which ecosystem service provision is impaired nonlinearly (Koch et al. 2009). A range of possible relationships between density and ecosystem service provision are outlined in Figure 9.1. The point along the density gradient at which the threshold or tipping point in service provision lies is rarely known. This represents a major challenge in the assessment of most ecosystem services.

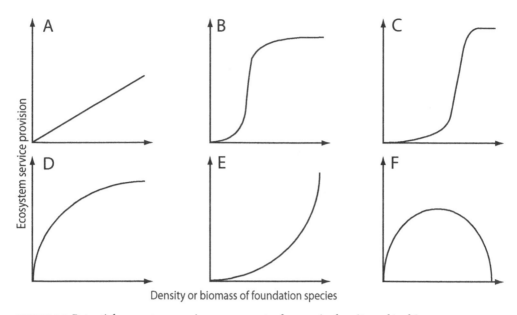

FIGURE 9.1 Potential ecosystem service responses to changes in density and/or biomass over a fixed area. (A) Directly linear response, such as that found in forest carbon stocks with tree biomass (Smith et al. 2004). (B) Threshold response where functionality is retained over a range of densities, such as the relationship between fish abundance and oyster density (Humphries et al. 2011, Soniat et al. 2004). (C) Threshold response where functionality is lost quickly following initial degradation. This response may be anticipated for the coastal defense function of seagrasses, where functionality is lost below 1,000 shoots m^{-2} (Koch et al. 2009). (D) Logarithmic function, where the ecosystem service is lost more slowly than anticipated with density. This response has been documented in bee pollination, where, following the loss of bees, remaining bees carry more pollen per individual (Harder and Thomson 1989). (E) Exponential function, expected where facilitation is the dominant interaction at increasing densities. This has been observed in the freshwater phytoplankton grazer, *Daphnia*, under conditions of phosphorus limitation (Sommer 1992). (F) Monotonic relationship, as expected where competition is the dominant interaction at high densities. This trend has been observed in rates of detrital breakdown with increasing densities of caddis flies (Klemmer et al. 2012).

A great deal of ecosystem service assessment has focused on obtaining monetary values, as the very concept of ecosystem services relies on the presence of human beneficiaries (Millennium Ecosystem Assessment 2005). From the perspective of establishing historical baselines, monetary values could be somewhat misleading, because the cultural setting, market structure, and other factors may have changed. The purpose of understanding historical extent and condition is to provide a metric against which to assess the potential present benefits, were the historical habitat still intact. To this end, while we are considering historical processes, their translation into potential human benefit is best stated as potential modern-day value of a former condition. For example, it is more informative to use the current market price than the historical market price of a target species to frame the debate surrounding how much we have lost and how much we should aim to restore.

DATA CHALLENGES

Quantifying the historical extent and subsequent change of a habitat can be challenging, especially in the marine environment. While historical data are available for many terrestrial habitats over longer time scales (Scales 2011, Hall et al. 2002), most subtidal habitats remained largely unexplored and unmapped until the latter half of the twentieth century (Giri et al. 2007, Waycott et al. 2009, Palandro et al. 2003). Furthermore, although models based on climatic variables have been developed to estimate the historical distribution of habitats in terrestrial systems (e.g., Goldewijk 2001), a lack of understanding of the driving variables and a lack of data have thus far precluded a similar approach being taken in the marine environment (Stevens and Connolly 2004). Even for habitats that have been mapped, differing technologies, habitat definitions, or survey aims can result in maps that are difficult or impossible to compare over time (Lausch and Herzog 2002). In order to assess changes in extent, differences between mapping definitions and mapping resolutions need to be resolved, and the assumptions clearly stated in subsequent analyses.

Similarly, while density or biomass data are now recorded more frequently, there are few examples where such data were documented historically (Hall et al. 2002, Lambin 1999), although abundant qualitative data indicate significant changes over time (Roberts 2007). Presence–absence data are more widely available, but they may poorly reflect changes in the community over time (Sagarin et al. 1999) and do not allow for estimation of biomass-related ecosystem services. As a result, either it may not be possible to quantify the contribution of historical habitat to ecosystem services, or it may be necessary to rely on proxies to provide estimates of historical values. The most likely proxy would be an undisturbed modern system, should one exist.

There are a number of further challenges to estimating the level of ecosystem services provided by a historical habitat, many of which have yet to be addressed even in modern settings. Our understanding of ecosystem service delivery and the drivers of those services is still incomplete (Kremen 2005), and the impacts of habitat fragmentation, fragment size, and landscape-scale effects on local services remain largely unknown (Boström et al. 2011, Harwell et al. 2011). Once such relationships are more clearly understood, it will be possible to appropriately value the more intact habitats that were present historically.

Abiotic factors such as temperature, nutrient inputs, and heavy-metal concentrations may all influence the degree to which ecosystem services are provided. For example, the level of denitrification that can be achieved in aquatic sediments is driven, in part, by the input of organic nitrogen (King and Nedwell 1985); therefore, if inputs were lower historically, the dentrification provided by the same area of sediment would also have been lower. Data on the historical levels of such abiotic factors are, however, often not available. In these cases, it may not be possible to determine the historical level of services, but it may be possible to estimate the services that would have been provided, had the historical extent still been in place under current abiotic conditions. Such a measure would, in fact, be more appropriate for assessing the value of the historical habitat in the modern landscape, while still providing a useful baseline for consideration when setting restoration goals. Measurements based on current conditions provide a more tangible and honest unit for considering the costs and benefits of conservation actions taken today.

USING HISTORICAL ECOLOGY TO INFORM RESTORATION GOALS FOR THE EASTERN OYSTER: A CASE STUDY

The eastern oyster, *Crassostrea virginica* (Gmelin, 1791), is a biogenic reef-building bivalve (Stenzel 1971). This species has a wide distribution along the Atlantic and Gulf coasts of North America and into the Caribbean, where it is found in brackish estuaries and inlets along the coast. Oyster reefs are built up over decades to centuries by the successive settlement of juvenile oysters on the oyster-shell substrate of older generations. The resulting habitat is a raised, complex, three-dimensional structure (Figure 9.2) that attracts a large number of associated species and provides a great diversity of microhabitats (Nestlerode et al. 1998, Luckenbach et al. 2005). An oyster reef's height and width are determined by the flow and tidal regime of the estuary, but they frequently exceed 1 m and may exceed 10 m in height (Ritter 1895, McCormick-Ray 2005).

The eastern oyster was once abundant on the Atlantic and Gulf coasts of the United States, where it formed reefs large enough to impede navigation and provided an essential form of sustenance to early explorers (Lafon 1806, Wharton 1957). Following European colonization, widespread overexploitation passed in a wave down the eastern coast of North America, which resulted in the virtual collapse of this ecosystem in most estuaries (Figure 9.3; Kirby 2004). The introduction of new diseases, pollution, and changes to coastal hydrology through engineering have all contributed to a sustained decline in oyster abundance over the past 200 years (Seavey et al. 2011, Wilberg et al. 2011, MacKenzie 2007). In the United States, native oyster ecosystems have suffered losses of 64% in area and 88% in biomass over the past 120 years (zu Ermgassen et al. 2012).

Oysters are filter-feeding, habitat-forming bivalves. When present in sufficient abundance, they can exert a strong influence over estuarine ecology, biodiversity, and water chemistry and are widely considered a foundation species (Coen et al. 2007). Oyster reef ecosystems provide a suite of ecosystem services, including filtration of the water column and enhancement of denitrification in the surrounding sediments (Piehler and Smyth 2011, Grizzle et al. 2008), both of which may lead to improved water quality. Oyster reefs may also perform coastal protection functions in shallow subtidal and intertidal locations by reducing wave energy (Scyphers et al. 2011, Piazza et al. 2005) and are recognized as essential fish habitat (Coen and Grizzle 2007). Oyster reefs may enhance fish and large crustacean biomass by 2.6 t ha^{-1} yr^{-1} (Peterson et al. 2003). A recent economic analysis of the ecosystem services provided by oyster reefs estimated that the value of the full suite of benefits may be

FIGURE 9.2 Recently restored reef in the Virginia Coast Reserve (photo courtesy of Mark Spalding, The Nature Conservancy).

as high as \$99,000 ha^{-1} yr^{-1} (Grabowski et al. 2012). The case for oyster restoration to recover ecosystem services is therefore strong (Coen et al. 2007).

Oyster Restoration

Historically, oyster restoration was focused on oyster fisheries enhancement (MacKenzie 1996), but the growing recognition of the dramatic loss of this ecosystem and the potential to regain ecosystem services has shifted the emphasis to restoration of the habitat as opposed to fishable individuals in many regions (Coen and Luckenbach 2000). The urgent need to address losses has attracted substantial restoration investment over the past couple of decades

FIGURE 9.3 (A) An oyster-shucking house in Baltimore Maryland, date unknown
(ca. 1900). Note the barrows of oysters being transported to the right-hand side of the
image and that the ground is built up by previously shucked oysters. (B) Pile of oyster
shell outside a shucking house where shell is being loaded onto boats to be returned to
exploited oyster beds, South Carolina, 1938. (Images courtesy of the NOAA Historic
Fisheries Collection.)

(NOAA 2012, Coen et al. 2007). Restoration funding and the enthusiasm of stakeholders are often the result of interest in the ecosystem services, as opposed to the habitat alone (Rice 2000, Brumbaugh et al. 2006, Coen et al. 2007). Nevertheless, despite the provision of many ecosystem services being heavily dependent on other measures of habitat quality such as density (zu Ermgassen et al. 2013, Luckenbach et al. 2005), areal extent remains the primary restoration goal and reported metric for oyster restoration projects (NOAA 2012).

This reliance on areal extent as an indicator of progress is, to some degree, misplaced. It is widely recognized that habitat degradation is a leading threat to ecosystems (Foley et al. 2007), and historical ecology has illustrated that areal extent may be a poor indicator of habitat status and of the services provided by oyster reefs (zu Ermgassen et al. 2012, 2013).

Although a number of challenges to using ecosystem services as an alternative to more traditional spatial extent or abundance goals and metrics have been identified (Peterson and Lipcius 2003), it is now feasible to quantify the contribution of some ecosystem components to ecosystem service delivery (e.g., zu Ermgassen et al. 2013, Peterson et al. 2003, Piehler and Smyth 2011). Such quantitative estimates of services, and appropriate methodologies for monitoring, can be used to set restoration targets where ecosystem service recovery is the aim.

Setting Restoration Targets

On a small scale, it may be appropriate to use increases in the desired ecosystem service or services as a restoration target. For example, restoration of a single site may aim to achieve oyster populations capable of filtering a volume equivalent to the water that passes over the reef at high or low tide, or to remove a given quantity of organic nitrogen from the system per unit time. At larger scales such as whole estuaries, knowledge of the historical level of ecosystem service provision may be a useful guide for setting restoration goals. Historical information, as with more traditional restoration goals, could play a critical role in preventing a shifting baseline in service provision by providing an insight into the impact of the loss of services on the estuary. The historical baseline is also a useful tool for expressing the costs or benefits of change in service provision over time. This novel unit for expressing the loss can provide strong motivation to restore, with appeal beyond biodiversity values alone (Goldman and Tallis 2009).

Filtration is a key ecosystem service provided by shellfish reefs and beds (Coen et al. 2007). Oysters remove particles larger than 5 µm from the water column with high efficiency (Riisgaard 1988), ingesting edible particles and binding inedible particles in mucus before depositing them onto the sediment. Particles are thus removed from the water column and organically enrich the sediments, except where they are resuspended by wave action. This drawdown can result in greater water clarity and enhanced growth of submerged aquatic vegetation such as seagrasses (Wall et al. 2011). The process of enriching the sediments may also enhance rates of denitrification in the sediments, permanently removing organic nitrogen from the system in the form of N_2 (Newell et al. 2002). Historically, oyster filtration was a dominant ecological function in many estuaries that would have underpinned many ecological processes and the community structure within the estuary (Newell 1988, zu Ermgassen et al. 2013).

Although bivalve filtration has been well studied in the laboratory (e.g., Walne 1972, Shumway et al. 1985), remarkably few field studies have quantified observed levels of seston drawdown over shellfish beds (Grizzle et al. 2006). Ex situ measurements of filtration may not capture the full range of factors that affect filtration by an in situ oyster population and are therefore challenging to extrapolate into the field at large scales, where adequate abiotic data may be lacking. To overcome this challenge, a model of filtration by oyster reefs, based on data collected in situ and modified by water temperature and oyster population metrics, was recently proposed by zu Ermgassen et al. (2013). The total volume of water filtered scales with oyster biomass, which is a function of oyster density and size distribution, as well as the area of oyster habitat. The volume filtered is not synonymous with the quantity of seston drawdown, because such measures also depend on the seston concentration and composition. Total filtration capacity is nevertheless a useful measure for considering the potential ecological function provided by oyster populations within an estuary and can be used to determine whether the oyster population is capable of full estuary filtration. Full estuary filtration was defined as the situation in which the oyster population filters more than the entire volume of the estuary within the residence time of water in that estuary (Figure 9.4). Although full estuary filtration is not necessarily indicative of the whole estuary being filtered (due to refiltration, resuspension, and imperfect mixing), it provides some indication of whether the oyster population has the potential to affect suspended sediments and phytoplankton at a large scale within the estuary (Smaal and Prins 1993, Dame 2011).

By applying their filtration model to historical data on the oyster population in 13 estuaries around the United States, zu Ermgassen et al. (2013) found that oysters had historically achieved full estuary filtration in six of the estuaries. In the case of Matagorda Bay, Texas, the historical population filtered the volume of the estuary in less than 3 hours, whereas the population remaining today requires nearly 180 days to filter the same volume. The baseline used by the authors was circa 1900, which represented a nonpristine baseline throughout much of the range, but in particular in the northeastern Atlantic, where the baseline postdates the collapse of the oyster fishery and population in the region (zu Ermgassen et al. 2012, Kirby 2004). For example, the mean oyster density recorded in Tangier and Pocomoke Sound circa 1900 was just 2 m^{-2}, and it was noted at the time that this represented a twofold to sevenfold decline from just 30 years earlier, a period that was also postdecline (zu Ermgassen et al. 2012, 2013). Therefore, although only six of the estuaries studied contained oyster populations capable of full estuary filtration, it is likely that this number would have been higher under pristine conditions (zu Ermgassen et al. 2013, Newell 1988).

Where the aim of restoration is to restore the lost filtration by oysters within an estuary, evidence of the historical function and level of services can provide a useful guideline for long-term, large-scale restoration goals. Here, we suggest a potential approach to setting such restoration goals. First, we selected nine estuaries for which there is strong evidence of full estuary filtration historically: Tangier and Pocomoke Sound (Virginia and Maryland),

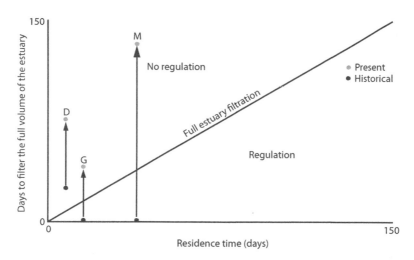

FIGURE 9.4 Graph illustrating estuary filtration (adapted from Dame 2011). Seston may be regulated by the oyster population in estuaries in which the latter filters the full volume of the estuary faster than the bay volume is turned over (within its residence time). This is equivalent to the area below the line, which represents full estuary filtration. Estuaries that fall above the line are unlikely to have seston regulated by the oyster population at a large scale. Example bays from zu Ermgassen et al. (2013) are illustrated. Black circles represent the historical population, and gray circles the present population in Delaware Bay (D), Galveston Bay (G), and Matagorda Bay (M). Arrows illustrate the direction of change from historical to present. Note that the oyster populations in Galveston and Matagorda Bays achieved full estuary filtration historically but not presently, whereas those in Delaware Bay did not achieve full estuary filtration in either period. The baseline (ca. 1900) represents a shifted baseline in Delaware Bay, which may contribute to this.

York River (Virginia), James River (Virginia), West Mississippi Sound (Mississippi and Louisiana), Galveston Bay (Texas), Matagorda Bay (Texas), San Antonio Bay (Texas), Aransas Bay (Texas), and Corpus Christi Bay (Texas) (zu Ermgassen et al. 2013, Newell 1988). These included the six estuaries identified in zu Ermgassen et al. (2013) and three tributaries of Chesapeake Bay for which there is strong evidence of full estuary filtration by oysters at a pristine baseline (Newell 1988, zu Ermgassen et al. 2013). We propose a restoration goal of returning the oyster population to the point where it may have a large-scale impact on the estuary through water filtration. We have chosen to approximate this by determining the population level required to achieve full estuary filtration. Here, we calculate the number of hectares of restoration required to achieve full estuary filtration in summer months. A similar approach could be taken to address filtration in spring, when phytoplankton blooms may be problematic in some estuaries (Paerl 1998). We estimated the area of additional restoration required to achieve our goal at the current and historical oyster densities and at 15 oysters m^{-2} at 1 g dry tissue mass each. The latter is the density agreed by the Oyster Metric

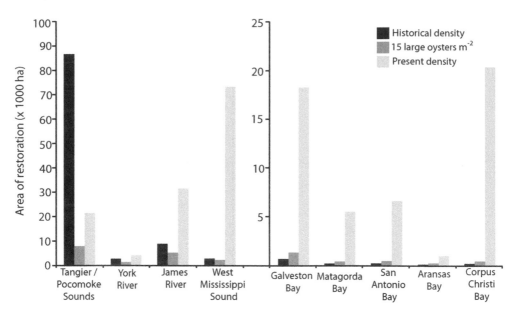

FIGURE 9.5 Bar chart showing area of reef required to achieve full estuary filtration at historical density (black), at 15 oysters (of 1 g dry tissue mass each) m^{-2} (dark gray), and at present density (light gray) for each estuary. Note the different y-axis scales for clarity.

Working Group to represent restored reefs in the Chesapeake (Oyster Metrics Working Group 2011). The resulting goals (Figure 9.5) illustrate the importance of density as well as areal extent in goal setting for ecosystem services such as filtration.

Such estimates for goal setting are not absolute, and the ecosystem service model used here is not spatially explicit within the estuary. While filtration may have whole-estuary impacts (Dame 2011, Fulford et al. 2010), the benefits of filtration are more often localized (Smith et al. 2009, Newell and Koch 2004). Thus, whole-estuary estimates of services based on historical states provide a useful framework to inform initial goal-setting efforts, but such large-scale targets proposed here should be further informed by local conditions and stakeholder interests, as the delivery of many ecosystem services is spatially variable (O'Higgins et al. 2010, Scyphers et al. 2011), and ecosystem services may have higher value in some sites within an estuary than in others (Barbier et al. 2011). Clearly, in some settings, restoration to historical levels is far beyond current capacity and unrealistic (Figure 9.5). Even in these locations, however, consideration of filtration capacity or other ecosystem services may be a useful tool for goal setting on a sub-estuary scale.

Other Ecosystem Services

Although we have considered just one ecosystem service in this example, most ecosystems provide an array of services. The inclusion of multiple ecosystem services in setting restoration or conservation goals often adds considerable value to conservation efforts (Rondinini and Chiozza 2010, Wainger et al. 2010, Bennett et al. 2009). This level of planning involves inevitable tradeoffs between these services, with the provision of one (e.g., timber extraction) affecting other services (e.g., carbon stocks) (Chan et al. 2011). Similarly, tradeoffs between services must be considered when planning where to undertake restoration or other conservation actions, because the provision of different services varies spatially; for example, oyster reefs may promote seagrass growth and recovery in appropriate locations (Wall et al. 2011), but oyster reefs are likely to benefit non-oyster-fishery species to a lesser degree near seagrass beds (Grabowski et al. 2005). Although efforts to set goals based on multiple services are challenging to undertake in practice, progress is being made (Rondinini and Chiozza 2010, Wainger et al. 2010, Bennett et al. 2009).

The approach considered above, in which historical levels of service are used as an ecological guide for setting large-scale restoration goals, may not be appropriate for all services. Ecosystem services were not measured historically, and we therefore rely on modern measurements to relate to historical data. For some ecosystem services, such as denitrification and fisheries, many factors other than oyster densities have changed over the past centuries, such that it is not presently possible to make estimates of the historical value of services. Modern measurements combined with knowledge of the historical extent and densities on reefs in the past can, however, be used to determine the level of services that would be delivered if the historical extent of oyster reefs still remained. Such data can provide a compelling story for avoiding shifting baselines.

Mechanisms for Application

Ecosystem services are critically important in engaging stakeholders in restoration (Coen et al. 2007). For ecosystems that have been in decline for centuries, like the U.S. Gulf Coast (Box 9.1), data regarding the historical level of services provided can be useful for conceptualizing the degree and impact of loss. Information related to absolute changes in the extent of habitats alone is rarely sufficient to compel societies to set meaningful conservation or restoration targets. More compelling is the knowledge that certain services or benefits are reduced or no longer available in relation to historical baselines. Using ecosystem services allows the value of these lost services to be estimated, thereby putting the losses into a context that is more compatible with our current decision-making structure, as well as winning support for conservation from new and diverse stakeholders and funders (Goldman and Tallis 2009, Goldman et al. 2008). This is particularly valuable where baselines may be shifted to the degree that many of the social and cultural values associated with the habitat have been lost.

Setting goals on the basis of ecosystem service provision is also a viable alternative where traditional approaches are not possible (Hobbs et al. 2009). Restoration goals are traditionally set either in relation to less disturbed but ecologically similar reference sites or in relation to a historical baseline (see chapter 8, this volume). These traditional restoration goals are, however, not always easily defined. For example, where the larger-scale physical processes that determine the status of a site have become disrupted, restoration to "historical" may not be possible (Hobbs et al. 2009, Moreno-Mateos et al. 2012). Similarly, some habitats

Katie Arkema

On July 26, 2012, The RESTORE (Resources and Ecosystems Sustainability, Tourist Opportunity, and Revived Economies of the Gulf States) Act was signed into law by President Obama. The act dedicates 80% of all Clean Water Act penalties paid by those responsible for the 2010 U.S. Gulf oil disaster to Gulf Coast restoration. The challenge now is to determine which restoration projects should be funded and where new coastal habitat is needed to provide which functions.

As zu Ermgassen and colleagues discuss in this chapter, state agencies, environmental NGOs, and academic institutions increasingly account for benefits that humans receive from nature (or "ecosystem services") to prioritize where and how best to spend funding for restoration. One of the most important benefits provided to Gulf Coast residents by coastal habitats is protection from ocean flooding and shoreline erosion. From the Florida Keys to Padre Island, Texas, large stretches of mangrove forests, wetlands, seagrass meadows, and coral and oyster reefs have the potential to reduce the numbers of vulnerable people and property most at risk from coastal hazards (Arkema et al. 2013). Sadly, these ecosystems have suffered extensive losses and degradation. Historical data suggest that the extent of coastal wetlands in the past would have better protected the shoreline from the devastation of

storms like Hurricane Katrina (Day et al. 2007). Increasing the coastal resilience of the Gulf is a central objective of the restoration activities funded by the RESTORE Act.

At the Natural Capital Project, we are working with The Nature Conservancy (TNC) to provide information about where coastal habitats are most critical for providing protection from coastal hazards to inform investments in restoration. We are using spatial models that integrate data on exposure to storms and sea-level rise, habitat distribution and abundance, and location of vulnerable people and property values. Historical habitat data are key to helping identify suitable sites for restoration. In Florida, TNC is combining the approaches they have developed for coral (TNC 2013) and oyster reef (Coastal Resilience 2013) restoration with our maps that indicate where protection provided by habitats is greatest. Working with the 23 Gulf counties and state and federal stakeholders, they are using this information to help target restoration projects funded through RESTORE legislation. The hope is that, armed with historical and contemporary data about where restoration of coastal habitats will be most important for people and property, we can increase the resilience of coastal communities to hazards while maintaining the full suite of benefits that people have long relied on from their natural environment.

Katie Arkema is Senior Scientist at the Natural Capital Project, Stanford Woods Institute for the Environment, Stanford University.

may have no appropriate reference sites remaining nor sufficient information to determine even fairly basic descriptors of a historical habitat (Roberts 2007).

As society's ability to estimate the production of services from intact habitats improves, it seems likely that policy and economic mechanisms will emerge and influence management decisions, including future conservation or restoration investments. Already, a number of ecosystem services associated with bivalve reefs and beds are being advanced as candidates for inclusion in market-based management frameworks, such as nutrient trading schemes (Higgins et al. 2011). These markets are, however, based on larger ecosystem management objectives (e.g., total maximum daily loads), and it is unlikely that the historical baselines will be a primary driver in such mechanisms. Nevertheless, historical values can raise awareness of the potential for increased ecosystem service provision through restoration and, hence, the potentially important role that habitat restoration may play in provision of marketable services.

Many restoration programs are undertaken for ecosystem services, yet most are still monitored and assessed by their spatial extent and, perhaps, biodiversity (Goldman et al. 2008). Using ecosystem service provision to set restoration goals and monitoring for ecosystem services directly are important steps to take toward delivering accountable restoration projects. Stakeholders benefit from this approach, as there is greater transparency regarding whether the promised benefits are being delivered. It should be noted that an emphasis on ecosystem services can be compatible with biodiversity aims. While there are inevitable tradeoffs between the two under some circumstances (e.g., forest biodiversity vs. timber extraction; Chan et al. 2011), a focus on ecosystem services compatible with biodiversity in conservation planning can result in win–win scenarios (Chan et al. 2006).

Historical data can play an important role in underpinning the use of ecosystem services in setting restoration goals. Without the historical context, one can quickly lose sight of what is possible. Traditional habitat-loss arguments provide us with the sense of responsibility that motivates many people to engage in restoration, but the potential to recover ecosystem services can engage a whole new group of stakeholders who are eager to see the benefits of their labors.

ACKNOWLEDGMENTS

This work was supported by a grant from the National Fish and Wildlife Foundation (award no. 2009–0078–000) and by the National Partnership between The Nature Conservancy and the National Oceanic and Atmospheric Administration Community-based Restoration Program (award nos. NA07NMF4630136 and NA10NMF463008). Additional funding support for the project was provided by the The Nature Conservancy–Shell Partnership and The Turner Foundation, Inc. The authors thank Dr. R. Grizzle for discussions that have underpinned much of this work.

REFERENCES

Arkema, K. K., Guannel, G., Verutes, G., et al. (2013) Coastal habitats shield people and property from sea level rise and storms. *Nature Climate Change* 3, 913–918.

Balvanera, P., Pfisterer, A. B., Buchmann, N., et al. (2006) Quantifying the evidence for biodiversity effects on ecosystem functioning and services. *Ecology Letters* 9, 1146–1156.

Barbier, E. B., Hacker, S. D., Kennedy, C., et al. (2011) The value of estuarine and coastal ecosystem services. *Ecological Monographs* 81, 169–193.

Bennett, E. M., Peterson, G. D., and Gordon, L. J. (2009) Understanding relationships along multiple ecosystem services. *Ecology Letters* 12, 1394–1404.

Boström, C., Pittman, S. J., Simenstad, C., and Kneib, R. T. (2011) Seascape ecology of coastal biogenic habitats: advances, gaps, and challenges. *Marine Ecology Progress Series* 427, 191–217.

Brooks, T. M., Mittermeier, R. A., Mittermeier, C. G., et al. (2002) Habitat loss and extinction in the hotspots of biodiversity. *Conservation Biology* 16, 909–923.

Brozozowska, R., Sui, Z., and Ho Kang, K. (2012) Testing the usability of sea mussel (*Mytilus* sp.) for the improvement of seawater quality—an experimental study. *Ecological Engineering* 39, 133–137.

Brumbaugh, R. D., Beck, M. W., Coen, L. D., et al. (2006) A practitioners guide to the design & monitoring of shellfish restoration projects. The Nature Conservancy, Arlington, VA.

Brumbaugh, R. D., Beck, M. W., Hancock, B., et al. (2010) Changing a management paradigm and rescuing a globally imperiled habitat. *National Wetlands Newsletter* (November–December), 16–20.

Chan, K. M. A., Hoshizaki, L., and Klinkenberg, B. (2011) Ecosystem services in conservation planning: targeted benefits vs. co-benefits or costs? *PLoS ONE* 6, e24378.

Chan, K. M. A., Shaw, M. R., Cameron, D. R., et al. (2006) Conservation planning for ecosystem services. *PLoS Biology* 4, e379.

Coen, L. D., Brumbaugh, R. D., Bushek, D., et al. (2007) Ecosystem services related to oyster restoration. *Marine Ecology Progress Series* 341, 303–307.

Coen, L. D., and Grizzle, R. E. (2007) The importance of habitat created by molluscan shellfish to managed species along the Atlantic Coast of the United States. Atlantic States Marine Fisheries Commission Habitat Management Series No. 8.

Coen, L. D., and Luckenbach, M. W. (2000) Developing success criteria and goals for evaluating oyster reef restoration: ecological function or resource exploitation? *Ecological Engineering* 15, 323–343.

Costanza, R., d'Arge, R., de Groot, R., et al. (1997) The value of the world's ecosystem services and natural capital. *Nature* 387, 253–260.

Costanza, R., Perez-Maqueo, O., Martinez, M. L., et al. (2008) The value of coastal wetlands for hurricane protection. *Ambio* 37, 241–248.

Dame, R. F. (2011) *Ecology of Marine Bivalves: An Ecosystem Approach.* CRC Press, Boca Raton, FL.

Day, J. W., Boesch, D. F., Clairain, E. J., et al. (2007) Restoration of the Mississippi Delta: lessons from Hurricanes Katrina and Rita. *Science* 315, 1679–1684.

Duffy, W. G., and Kahara, S. N. (2011) Wetland ecosystem services in California's Central Valley and implications for the Wetland Reserve Program. *Ecological Applications* 21, S18–S30.

Egoh, B., Rouget, M., Reyers, B., et al. (2007) Integrating ecosystem services into conservation assessments: a review. *Ecological Economics* 63, 714–721.

Ehrenfeld, J. G. (2000) Defining the limits of restoration: the need for realistic goals. *Restoration Ecology* 8, 2–9.

Fearnside, P. M. (2005) Deforestation in Brazilian Amazonia: history, rates, and consequences. *Conservation Biology* 19, 680–688.

Ferraro, P. J., and Simpson, R. D. (2002) The cost-effectiveness of conservation payments. *Land Economics* 78, 339–353.

Foley, J. A., Asner, G. P., Costa, M. H., et al. (2007) Amazonia revealed: forest degradation and loss of ecosystem goods and services in the Amazon Basin. *Frontiers in Ecology and the Environment* 5, 25–32.

Fulford, R. S., Breitburg, D. L., Luckenbach, M., and Newell, R. I. E. (2010) Evaluating ecosystem response to oyster restoration and nutrient load reduction with a multispecies bioenergetics model. *Ecological Applications* 20, 915–934.

Giri, C., Zhu, Z., Tieszen, L. L., et al. (2007) Mangrove forest distributions and dynamics (1975–2005) of the tsunami-affected region of Asia. *Journal of Biogeography* 35, 519–528.

Goldewijk, K. K. (2001) Estimating global land use change over the past 300 years: the HYDE Database. *Global Biogeochemical Cycles* 15, 417–433.

Goldman, R. L., and Tallis, H. (2009) A critical analysis of ecosystem services as a tool in conservation projects. *Annals of the New York Academy of Sciences* 1162, 63–78.

Goldman, R. L., Tallis, H., Kareiva, P., and Daily, G. C. (2008) Field evidence that ecosystem service projects support biodiversity and diversify options. *Proceedings of the National Academy of Sciences USA* 105, 9445–9448.

Grabowski, J. H., Brumbaugh, R. D., Conrad, R. F., et al. (2012) Economic valuation of ecosystem services provided by oyster reefs. *BioScience* 62, 900–909.

Grabowski, J. H., Hughes, A. R., Kimbro, D. L., and Dolan, M. A. (2005) How habitat setting influences restored oyster reef communities. *Ecology* 86, 1926–1935.

Grizzle, R. E., Greene, J. K., and Coen, L. D. (2008) Seston removal by natural and constructed intertidal eastern oyster (*Crassostrea virginica*) reefs: a comparison with previous laboratory studies, and the value of in situ methods. *Estuaries and Coasts* 31, 1208–1220.

Grizzle, R. E., Greene, J. K., Luckenbach, M. W., and Coen, L. D. (2006) A new in situ method for measuring seston uptake by suspension-feeding bivalve molluscs. *Journal of Shellfish Research* 25, 643–649.

Hall, B., Motzkin, G., Foster, D. R., et al. (2002) Three hundred years of forest and land-use change in Massachusetts, USA. *Journal of Biogeography* 29, 1319–1335.

Harder, L. D., and Thomson, J. D. (1989) Evolutionary options for maximizing pollen dispersal of animal-pollinated plants. *American Naturalist* 133, 323–344.

Harwell, H. D., Posey, M. H., and Alphin, T. D. (2011) Landscape aspects of oyster reefs: effects of fragmentation on habitat utilization. *Journal of Experimental Marine Biology and Ecology* 409, 30–41.

Higgins, C. B., Stephenson, K., and Brown, B. L. (2011) Nutrient bioassimilation capacity of aquacultured oysters: quantification of an ecosystem service. *Journal of Environmental Quality* 40, 271–277.

Hobbs, R. J., Higgs, E., and Harris, J. A. (2009) Novel ecosystems: implications for conservation and restoration. *Trends in Ecology & Evolution* 24, 599–605.

Hooper, D. U., Chapin, F. S., Ewel, J. J., et al. (2005) Effects of biodiversity on ecosystem functioning: a consensus of current knowledge. *Ecological Monographs* 75, 3–35.

Humphries, A. T., La Peyre, M. K., Kimball, M. E., and Rozas, L. P. (2011) Testing the effect of habitat structure and complexity on nekton assemblages using experimental oyster reefs. *Journal of Experimental Marine Biology and Ecology* 409, 172–179.

King, D., and Nedwell, D. B. (1985) The influence of nitrate concentration upon the end-products of nitrate dissimilation by bacteria in anaerobic salt marsh sediment. *FEMS Microbiology Letters* 31, 23–28.

Kirby, M. (2004) Fishing down the coast: historical expansion and collapse of oyster fisheries along continental margins. *Proceedings of the National Academy of Sciences USA* 101, 13096–13099.

Klemmer, A. J., Wissinger, S. A., Greig, H. S., and Ostrofsky, M. L. (2012) Nonlinear effects of consumer density on multiple ecosystem processes. *Journal of Animal Ecology* 81, 770–780.

Koch, E. W., Barbier, E. B., Silliman, B. R., et al. (2009) Non-linearity in ecosystem services: temporal and spatial variability in coastal protection. *Frontiers in Ecology and the Environment* 7, 29–37.

Kremen, C. (2005) Managing ecosystem services: what do we need to know about their ecology? *Ecology Letters* 8, 468–479.

Lafon, B. (1806) General chart of the territory of Orleans also including western Florida and a portion of the territory of the Mississippi according to the most recent observations by Barthélémy Lafon, engineer/geographer in New Orleans. Puiquet, Paris.

Lambin, E. F. (1999) Monitoring forest degradation in tropical regions by remote sensing: some methodological issues. *Global Ecology and Biogeography* 8, 191–198.

Lausch, A., and Herzog, F. (2002) Applicability of landscape metrics for the monitoring of landscape change: issues of scale, resolution and interpretability. *Ecological Indicators* 2, 3–15.

Luckenbach, M. W., Coen, L. D., Ross, P. G., and Stephen, J. A. (2005) Oyster reef habitat restoration: relationships between oyster abundance and community development based on two studies in Virginia and South Carolina. *Journal of Coastal Research* 40, 64–78.

MacKenzie, C. L., Jr. (1996) History of oystering in the United States and Canada, featuring the eight greatest oyster estuaries. *Marine Fisheries Review* 58, 1–78.

MacKenzie, C. L., Jr. (2007) Causes underlying the historical decline in eastern oyster (*Crassostrea virginica* Gmelin, 1791) landings. *Journal of Shellfish Research* 26, 927–938.

McClanahan, T. R. (1994) Kenyan coral reef lagoon fish: effects of fishing, substrate complexity, and sea urchins. *Coral Reefs* 13, 231–241.

McCormick-Ray, J. (2005) Historical oyster reef connections to Chesapeake Bay—a framework for consideration. *Estuarine, Coastal and Shelf Science* 64, 119–134.

Millennium Ecosystem Assessment (2005) *Ecosystems and Human Well-being: Synthesis*. Island Press, Washington, DC.

Miller, G. H., Fogel, M. L., Magee, J. W., et al. (2005) Ecosystem collapse in pleistocene Australia and a human role in megafaunal extinction. *Science* 309, 287–290.

Molnar, J. L., and Kubiszewski, I. (2012) Managing natural wealth: research and implementation of ecosystem services in the United States and Canada. *Ecosystem Services* 2, 45–55.

Moreno-Mateos, D., Power, M. E., Comin, F. A., and Yockteng, R. (2012) Structural and functional loss in restored wetland ecosystems. *PLoS Biology* 10, e1001247.

Nature Conservancy (n.d.) Florida: protecting oceans and coasts. www.nature.org/ourinitiatives /regions/northamerica/unitedstates/florida/howwework/florida-oceans-coasts.xml.

Nestlerode, J. A., Luckenbach, M. W., and O'Beirn, F. X. (1998) Use of underwater video to monitor and quantify use of constructed oyster reef habitats by mobile commercially and ecologically important species. *Journal of Shellfish Research* 17, 1309.

Newell, R. I. E. (1988) Ecological changes in Chesapeake Bay: are they the result of overharvesting the American oyster, *Crassostrea virginica*? In *Understanding the Estuary: Advances in Chesapeake Bay Research* (M. P. Lynch and E. C. Krome, Eds.). CRC Publication No. 129. Chesapeake Research Consortium, Solomons, MD. pp. 536–546.

Newell, R. I. E., Cornwell, J. C., and Owens, M. S. (2002) Influence of simulated bivalve biodeposition and microphytobenthos on sediment nitrogen dynamics: a laboratory study. *Limnology and Oceanography* 47, 1367–1379.

Newell, R. I. E., and Koch, E. W. (2004) Modeling seagrass density and distribution in response to changes in turbidity stemming from bivalve filtration and seagrass sediment stabilization. *Estuaries* 27, 793–806.

NOAA (2012) National Oceanic and Atmospheric Administration Restoration Atlas. www.habitat .noaa.gov/restoration/restorationatlas/.

O'Higgins, T. G., Ferraro, S. P., Dantin, D. D., et al. (2010) Habitat scale mapping of fisheries ecosystem service values in estuaries. *Ecology and Society* 15, article 7.

Oyster Metrics Working Group (2011) Restoration goals, quantitative metrics and assessment protocols for evaluating success on restored oyster reef sanctuaries. www.chesapeakebay.net /channel_files/17932/oyster_restoration_success_metrics_final.pdf.

Paerl, H. W. (1998) Coastal eutrophication and harmful algal blooms: importance of atmospheric deposition and groundwater as "new" nitrogen and other nutrient sources. *Limnology and Oceanography* 42, 1154–1165.

Palandro, D., Andrefouet, S., Dustan, P., and Muller-Karger, F. E. (2003) Change detection in coral reef communities using Ikonos satellite sensor imagery and historic aerial photographs. *International Journal of Remote Sensing* 24, 873–878.

Peterson, C. H., Grabowski, J. H., and Powers, S. P. (2003) Estimated enhancement of fish production resulting from restoring oyster reef habitat: quantitative valuation. *Marine Ecology Progress Series* 264, 249–264.

Peterson, C. H., and Lipcius, R. N. (2003) Conceptual progress towards predicting quantitative ecosystem benefits of ecological restorations. *Marine Ecology Progress Series* 264, 297–307.

Piazza, B. P., Banks, P. D., and La Peyre, M. K. (2005) The potential for created oyster shell reefs as a sustainable shoreline protection strategy in Louisiana. *Restoration Ecology* 13, 499–506.

Piehler, M. F., and Smyth, A. R. (2011) Habitat-specific distinctions in estuarine denitrification affect both ecosystem function and services. *Ecosphere* 2, article 12.

Rice, M. A. (2000) A review of shellfish restoration as a tool for coastal water quality management. *Environment Cape Cod* 3, 1–8.

Riisgaard, H. U. (1988) Efficiency of particle retention and filtration rate in 6 species of northeast American bivalves. *Marine Ecology Progress Series* 45, 217–223.

Ritter, H. P. (1895) Report on a reconnaissance of the oyster beds of Mobile Bay and Mississippi Sound, Alabama. *Bulletin of the U.S. Fisheries Commission* 15, 325–340.

Roberts, C. M. (2002) Deep impact: the rising toll of fishing in the deep sea. *Trends in Ecology & Evolution* 17, 242–245.

Roberts, C. (2007) *The Unnatural History of the Sea: The Past and Future of Humanity and Fishing.* Island Press, Washington, DC.

Rondinini, C., and Chiozza, F. (2010) Quantitative methods for defining percentage area targets for habitat types in conservation planning. *Biological Conservation* 143, 1646–1653.

Rozas, L. P., Caldwell, P., and Minello, T. J. (2005) The fishery value of salt marsh restoration projects. *Journal of Coastal Research* SI40, 37–50.

Sagarin, R. D., Barry, J. P., Gilman, S. E., and Baxter, C. H (1999) Climate-related change in an intertidal community over short and long time scales. *Ecological Monographs* 69, 465–490.

Scales, I. R. (2011) Farming at the forest frontier: land use and landscape change in western Madagascar, 1896–2005. *Environment and History* 17, 499–524.

Scyphers, S. B., Powers, S. P., Heck, K. L., and Byron, D. (2011) Oyster reefs as natural breakwaters mitigate shoreline loss and facilitate fisheries. *PLoS ONE* 6, e22396.

Seavey, J. R., Pine, W. E., Frederick, P., et al. (2011) Decadal changes in oyster reefs in the Big Bend of Florida's Gulf Coast. *Ecosphere* 2, article 114.

Shumway, S. E., Cucci, T. L., Newell, R. C., and Yentsch, C. M (1985) Particle selection, ingestion, and absorption in filter-feeding bivalves. *Journal of Experimental Marine Biology and Ecology* 91, 77–92.

Smaal, A. C., and Prins, T. C. (1993) The uptake of organic matter and the release of inorganic nutrients by bivalve suspension feeder beds. In *Bivalve Filter Feeders in Estuarine and Coastal Ecosystem Processes* (R. F. Dame, Ed.). Springer-Verlag, Heidelberg, Germany. pp. 273–298.

Smith, J. E., Heath, L. S., and Woodbury, P. B. (2004) How to estimate forest carbon for large areas from inventory data. *Journal of Forestry* 102, 25–31.

Smith, K. A., North, E. W., Shi, F. Y., et al. (2009) Modeling the effects of oyster reefs and breakwaters on seagrass growth. *Estuaries and Coasts* 32, 748–757.

Sommer, U. (1992) Phosphorus-limited *Daphnia*: intraspecific facilitation instead of competition. *Limnology and Oceanography* 37, 966–973.

Soniat, T. M., Finelli, C. M., and Ruiz, J. T. (2004) Vertical structure and predator refuge mediate oyster reef development and community dynamics. *Journal of Experimental Marine Biology and Ecology* 310, 163–182.

Stenzel, H. B. (1971) Oysters. In *Treatise on Invertebrate Paleontology, Part N* (R. C. Moore, Ed.). University of Kansas Press, Lawrence, KS. pp. 953–1224.

Stevens, T., and Connolly, R. M. (2004) Testing the utility of abiotic surrogates for marine habitat mapping at scales relevant to management. *Biological Conservation* 119, 351–362.

Swetnam, T. W., Allen, C. D., and Betancourt, J. L. (1999) Applied historical ecology: using the past to manage for the future. *Ecological Applications* 9, 1189–1206.

Tallis, H., and Polasky, S. (2009) Mapping and valuing ecosystem services as an approach for conservation and natural-resource management. *Annals of the New York Academy of Sciences* 1162, 265–283.

Trabucchi, M., Ntshotsho, P., O'Farrell, P., and Comin, F. A. (2012) Ecosystem service trends in basin-scale restoration initiatives: a review. *Journal of Environmental Management* 111, 18–23.

Turvey, S. T., Barrett, L. A., Yujiang, H. A. O., et al. (2010) Rapidly shifting baselines in Yangtze fishing communities and local memory of extinct species. *Conservation Biology* 24, 778–787.

UK National Ecosystem Assessment (2011) Synthesis of key findings. http://uknea.unep-wcmc .org/Resources/tabid/82/Default.aspx.

Wainger, L. A., King, D. M., Mack, R. N., et al. (2010) Can the concept of ecosystem services be practically applied to improve natural resource management decisions? *Ecological Economics* 69, 978–987.

Wall, C., Peterson, B., and Gobler, C. (2011) The growth of estuarine resources (*Zostera marina, Mercenaria mercenaria, Crassostrea virginica, Argopecten irradians, Cyprinodon variegatus*) in response to nutrient loading and enhanced suspension feeding by adult shellfish. *Estuaries and Coasts* 34, 1262–1277.

Walne, P. R. (1972) The influence of current speed, body size and water temperature on the filtration rate of five species of bivalves. *Journal of the Marine Biological Association of the United Kingdom* 52, 345–374.

Waycott, M., Duarte, C. M., Carruthers, T. J. B., et al. (2009) Accelerating loss of seagrasses across the globe threatens coastal ecosystems. *Proceedings of the National Academy of Sciences USA* 106, 12377–12381.

Wharton, J. (1957) *The Bounty of the Chesapeake: Fishing in Colonial Virginia.* University Press of Virginia, Charlottesville, VA.

White, C., Halpern, B. S., and Kappel, C. V. (2012) Ecosystem service tradeoff analysis reveals the value of marine spatial planning for multiple ocean uses. *Proceedings of the National Academy of Sciences USA* 109, 4696–4701.

Wilberg, M. J., Livings, M. E., Barkman, J. S., et al. (2011) Overfishing, disease, habitat loss, and potential extirpation of oysters in upper Chesapeake Bay. *Marine Ecology Progress Series* 436, 131–144.

Worm, B., Barbier, E. B., Beaumont, N., et al. (2006) Impacts of biodiversity loss on ocean ecosystem services. *Science* 314, 787–790.

zu Ermgassen, P. S. E., Spalding, M. D., Blake, B., et al. (2012) Historical ecology with real numbers: past and present extent and biomass of an imperilled estuarine ecosystem. *Proceedings of the Royal Society of London Series B* 279, 3393–3400.

zu Ermgassen, P. S. E., Spalding, M. D., Grizzle, R. E., and Brumbaugh, R. D. (2013) Quantifying the loss of a marine ecosystem service: filtration by the eastern oyster in U.S. estuaries. *Estuaries and Coasts* 36, 36–43.

TEN

Incorporating Historical Perspectives into Systematic Marine Conservation Planning

NATALIE C. BAN, JOHN N. KITTINGER, JOHN M. PANDOLFI, ROBERT L. PRESSEY, RUTH H. THURSTAN, MATT J. LYBOLT, and SIMON HART

Historical perspectives are highly relevant to marine conservation, yet rarely integrated into ocean planning efforts. By its nature, marine conservation planning is forward looking—concerned with measures that should be taken in the future. It usually focuses on mitigating anticipated adverse changes caused by current and future human activities, with the implicit assumption that present or recent conditions should be maintained. In this chapter, we show that without incorporating historical data and analysis, such approaches will, in the best case, cause us to aim too low; and in the worst case, they can result in inappropriate targets for planning and management. We review the role that historical perspectives can provide in marine conservation planning, highlight planning exercises in which this has occurred or has been discussed, and provide recommendations for researchers and planning practitioners. Using the systematic conservation planning framework, we show that each planning stage can greatly benefit from a historical perspective and illustrate that failure to consider historical information reduces the effectiveness of marine conservation planning. We posit that historical perspectives may shift the conservation focus from restoring previous ecosystem states to recovering critical ecosystem functions and processes that maintain resilience. Historical perspectives can fundamentally change the conservation vision for a region, providing a window into possibilities for the future.

INTRODUCTION

The charge to restore commits us to a state of permanent irony. We will never decide to what point in the past we should restore the land—and can never, in any event, actually get back there.

FIRE ECOLOGIST STEPHEN J. PYNE, 1999

Marine historical ecology has revealed striking declines in abundance and biodiversity (e.g., Jackson et al. 2001, Pandolfi et al. 2003, Sala and Knowlton 2006, Willis et al. 2010, Cardinale et al. 2011), prompting global concern and efforts to implement conservation measures (Convention on Biological Diversity 2010). Many conservation measures focus on designating places in the ocean where human activities are restricted or prohibited—for example, marine managed areas and marine protected areas (MPAs). Other conservation approaches include gear restrictions, changes in fisheries management (e.g., individual transferable quotas), and alternative livelihood strategies. In this chapter, we focus primarily on MPAs because conservation planning has focused on such spatial tools and, hence, historical marine ecology has the potential to contribute greatly. MPAs encompass a range of spatial measures, from limited restrictions of human uses to fully protecting areas from all extractive human uses (also known as no-take zones and marine reserves; Kelleher and Kenchington 1992). Although MPAs, especially no-take areas, have been shown to be effective for increasing the size and biomass of exploited species (Halpern and Warner 2002, Stewart et al. 2009) and supplement fished areas (Harrison et al. 2012), only ~1% of the ocean is currently protected (Wood et al. 2008, Mora and Sale 2011). There is a clear need to expand MPAs and explore other conservation strategies to curb further biodiversity declines and preserve critical ocean ecosystem services (Pauly et al. 2002, Worm et al. 2006).

The favored approach for protecting marine biodiversity is through creating networks of MPAs, which, unlike individual MPAs, can be managed in a broader spatial context as a system (Roberts et al. 2003, Fernandes et al. 2005, University of Queensland 2009). The network approach is preferred because it considers emergent properties of systems, including complementarity, redundancy (Margules and Pressey 2000), and connectivity (Almany et al. 2009). Planning networks can ensure that known aspects of biodiversity are represented and that species and ecosystems can persist.

A framework for implementing a network approach in MPA design has emerged out of terrestrial and marine conservation planning, called "systematic conservation planning" (Margules and Pressey 2000). Systematic conservation planning comprises a planning model in stages for practitioners to implement conservation actions in a target region. Systematic conservation planning allows practitioners to develop quantitative conservation objectives and then facilitates the design of priority conservation areas and actions to achieve those objectives. Additional advantages include efficient use of limited resources to achieve explicit conservation objectives (e.g., related to biodiversity, ecosystem services, and livelihoods), defensibility and accountability in the face of competition for natural resources, and flexibility in accommodating opportunities and constraints (Margules and Pressey 2000). This approach also allows for incorporating a portfolio of management strategies and spatial approaches into planning, rather than focusing only on no-take areas.

The systematic conservation planning approach is increasingly being used by conservation practitioners—for example, by governments and nongovernmental organizations (NGOs). In a review of conservation planning by conservation NGOs, Pressey and Bottrill (2009) found that many of the stages in the systematic conservation planning framework were implemented by practitioners. Perhaps the best example of the use of the systematic conservation planning framework is the rezoning process of the Great Barrier Reef (Fernandes et al. 2005). Other marine examples of the framework's application include MPA design in the Channel Islands, California (Airamé et al. 2003), and in Kimbe Bay, Papua New Guinea (Green et al. 2009).

Conservation planning is, by nature, forward looking, yet it needs to be grounded to have a chance of success. Thus, it must also engage with the past, for example by understanding the trajectory the planning region is on, and the historical factors that have influenced the current state in that planning region. Without a long-term historical perspective, planning will include only a basic understanding of the systems they seek to protect and enhance. Including long-term data can help planners and managers better identify the direct and underlying causes of decline in natural features and the real rates of ecological change.

The purpose of this chapter is to illustrate the important role that historical perspectives and information can play in marine conservation planning and highlight how such perspectives might benefit planning. Our focus here is conservation-oriented planning, although the insights presented may also be applied to other planning initiatives. Marine conservation planning usually focuses on mitigating impacts of human activities and, thus, assumes that present or recent ecological conditions should be maintained. However, this assumption will often cause us to aim too low, limited by shifting baselines that lack historical context (shifting baselines syndrome; Pauly 1995). Many of the classic examples are from fisheries, where today's "good day of fishing" produces catches that are many times smaller and less abundant than those a few generations ago (Rosenberg et al. 2005, McClenachan 2009). We provide an overview of the stages of systematic conservation planning and highlight how each can benefit from historical perspectives and data, drawing on examples from planning practice. By "historical perspectives," we refer to a diversity of ways in which history is considered and cognitively incorporated by people, considering qualitative, quantitative, formal, and informal data and information (e.g., ranging from people's internalized perceptions of history to quantitative reconstructions of marine biomass). We use the terms "historical information" and "historical data" interchangeably to refer more specifically to quantitative characterizations of change over time as well as the diverse information sources used to produce them.

HISTORICAL PERSPECTIVES IN MARINE CONSERVATION PLANNING

The stages of systematic conservation planning serve as a useful framework for examining the utility of historical perspectives. We group the framework's eleven stages (Pressey and Bottrill 2009) into five categories: setting the stage (scoping, involving stakeholders, identifying context), vision (defining goals), data (collecting social and biodiversity data and determining quantitative objectives), actions (gap analysis, selecting actions, and applying actions), and review (monitoring) (for a description of each stage, see Figure 10.1 and Table 10.1). The stages are linked and feed back to one another, and planning is meant to occur iteratively, through 5- or 10-year revisions of a conservation plan. The framework is

TABLE 10.1 Eleven Stages in the Process of Systematic Conservation Planning

Stage	Description
1. Scoping and costing the planning process	Decisions are necessary on the boundaries of the planning region, the composition and required skills of the planning team, the available budget, and how each step in the process will be addressed, if at all.
2. Identifying and involving stakeholders	Stakeholders (those who will influence or be affected by conservation actions arising from the planning process) need to be identified and involved in appropriate ways throughout the planning process.
3. Describing the context for conservation areas	The planning team describes the social, economic, and political setting for conservation planning, identifying the types of threats to natural features and the broad constraints on, and opportunities for, conservation actions.
4. Identifying conservation goals	A broad vision statement for the region needs to be drafted and progressively refined into qualitative goals about biodiversity (e.g., representation and persistence), ecosystem services, livelihoods, and other concerns.
5. Collecting data on socioeconomic variables and threats	Relevant spatially explicit data will include variables such as tenure, extractive uses (i.e., threats), costs of conservation, and constraints and opportunities to which planners can respond.
6. Collecting data on biodiversity and other natural features	The planning team will collect spatially explicit data on biodiversity that include habitat types, focal species, and ecological processes.
7. Setting conservation objectives	Goals need to be interpreted as quantitative conservation objectives for each spatial feature, and, where necessary, qualitative objectives need to be related to configuration, past disturbance, and other criteria.
8. Reviewing current achievement of objectives	Remote data, and perhaps also field surveys, are used to estimate the extent to which objectives have already been achieved in areas considered to be adequately managed for conservation.
9. Selecting additional conservation areas	With stakeholders, this stage requires decisions about the location and configuration of additional conservation areas that complement the existing ones in achieving objectives.
10. Applying conservation actions to selected areas	Application of conservation actions requires a variety of technical analyses and institutional arrangements to ensure that areas are given the most feasible and appropriate conservation management.
11. Maintaining and monitoring conservation areas	Activities ensure that individual areas are managed to promote the long-term persistence of the values for which they were established, including monitoring the effectiveness of management actions.

Source: Adapted from Pressey and Bottrill (2009).

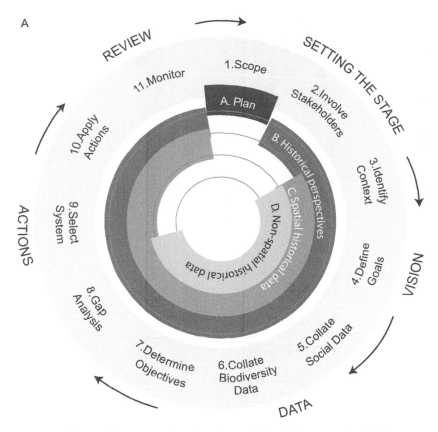

A

FIGURE 10.1 Systematic conservation planning and historical perspectives. Different ways to consider historical perspectives in conservation planning are organized by planning stage. (A) The scoping stage needs to plan for inclusion and collection of historical information. (B) Historical perspectives (i.e., all the ways in which history is considered and cognitively incorporated by people, qualitative and quantitative) influence all the ways in which history can be considered in the planning process. (C) Spatially explicit historical data are useful throughout the planning process, starting with the context. (D) Nonspatial historical data (e.g., historical catch records for a region that cannot be traced back to specific places and, thus, cannot be mapped) can be incorporated into the vision and data phases but do not directly influence the spatial selection of new marine protected areas.

adaptable and constantly evolving to improve planning practice, and, hence, opportunities exist for historical perspectives to inform future versions. The stages serve as a guide for conservation planners; in practice, planning efforts can use only some of the stages and change the order in which some or all are addressed. Furthermore, conservation planning must weave together ideas and techniques from different disciplines and areas of expertise and, ideally, would also use historical perspectives in each of its stages, as we discuss here. There is also increasing recognition of the importance of dynamic aspects of ecosystems and

TABLE 10.2   Examples of How Different Sources of Marine Historical Ecology Data Could Be Used to Set Goals for Conservation Planning

Case Study Location (Reference)	Description of Study	What Conservation Goals/Objectives Could Come from the Study?	What Might Be Missing if Historical Data Were Ignored?
Palaeontology			
Moreton Bay, Australia (Lybolt et al. 2011)	Dead corals were used to record historical ranges of variation in the extent, water depth, and species composition of coral reefs in Moreton Bay during the Holocene. The study found that reef development was episodic during the Holocene, fluctuating with environmental variations, and that Moreton Bay was inhospitable to coral reefs for about half of the Holocene. A significant change in coral species composition, from branching *Acropora* to massive corals, occurred after European transformation of the bay's catchments. There is limited potential for Moreton Bay to serve as a refugium for coral species pushed south by climatic warming, because of ongoing anthropogenic impacts and the natural historical instability of reef formation in Moreton Bay.	Goals: (1) Include existing coral reefs and suitable substrata in no-take or no-use zones to create potential for them to persist or recolonize. (2) Use increasing abundance of *Acropora* as a guide to the success of catchment management.	Goal without this historical information would be to restore coral communities as massive, slow-growing communities instead of the fast-growing acroporid-dominated communities that characterized the past, and to protect living coral reefs and past reef substrata as a refugium for species shifted south by climate change, the feasibility of which was questioned by this study.

Genetics

Global whale study (Roman and Palumbi 2003)	Mitochondrial DNA sequence-variation models were used to reconstruct the genetic diversity of North Atlantic whales (humpback, fin, and minke). Genetic diversity was found to be higher than expected, indicating that past population sizes of North Atlantic whales were much greater than previous historical estimates suggested. These results indicate that past fin and humpback whale populations could have been 6 and 20 times larger, respectively, than populations present today.	Goals: (1) Prevent increases in hunting quotas and/or maintain current protection of whale populations to enable further recovery. (2) Monitor additional sources of mortality (e.g., boating strikes and pollution) and determine whether these are inhibiting recovery rates or are likely to reduce population persistence in the future. (3) Implement management zones that protect breeding or nursery grounds with management of fishing, boating, and tourist activities.	Without these data, it would be assumed that populations of North Atlantic whales were closer to their prewhaling baselines; hence, decisions regarding exploitation or levels of protection might not be accurate, or goals might not be ambitious enough to ensure long-term persistence of populations.

Historical photographs

San Francisco Bay (www.sfei.org/he)	Historical source materials (i.e., coastal survey maps, past descriptions of habitats, past activities and land use) were collated to describe the extent of transformation of habitats and ecosystem linkages and to provide tools for restoration and protection of San Francisco Bay wetlands.	Goals: (1) Help scientists and managers develop strategies for more integrated and functional landscape management. (2) Establish coastal protected areas that represent habitat linkages and processes between land and sea. (3) Protect or restore fragmented habitats that were once more extensive, to enable recovery to former conditions.	Without this information, conservation planning might not recognize past landscapes or lost processes and their potential for restoration.
Florida (McClenachan 2009)	Historical photographs were used to measure changes, from 1956 to 2007, in size and composition of recreationally fished species landed by fishers around Key West, Florida. Results show that catches during the 1950s were dominated by large predators such as groupers and sharks, but that by 2007 this had shifted to smaller species such as snapper, while the average length of sharks had declined by 50%.	Goals: With this historical information, goals may become more ambitious as stakeholders and managers realize the potential for recovery of predatory species. (1) Establish large protected areas that encompass spawning or nursery grounds for groupers and coastal shark species. (2) Complement spatial management by managing recreational and commercial fishing activities for recovery of groupers and sharks.	Without this information, goals might set inappropriate or unambitious targets for restoration that do not allow recovery of higher-level predators.

(continued)

TABLE 10.2 (*continued*)

Case Study Location (Reference)	Description of Study	What Conservation Goals/Objectives Could Come from the Study?	What Might Be Missing if Historical Data Were Ignored?
Historical descriptions			
Caribbean (Jackson 2001)	Historical accounts indicate clearly that green sea turtles have been massively reduced in numbers in the Caribbean. The decline of turtles could have predisposed seagrasses to extensive mortality from a wasting disease in the 1980s, with losses of associated species. Grazing by turtles reduces factors associated with the disease, including sulfide toxicity and hypoxia in sediments, self-shading, and infection by slime mold.	Goals: Establish extensive management areas that would have dual purposes: (1) no-take zones for the recovery of ecological functional groups associated with seagrass and other ecosystems; and (2) large-scale experiments, with outside control areas, to establish the role of grazing by turtles in reducing the incidence and extent of wasting disease of seagrass.	Goals without historical information: (1) lower abundance and distribution of both herbivore and seagrass communities; and (2) management of disease as predominantly a climate change issue—recovery through physiology instead of trophodynamics.
Fisheries catch records			
Global (Zeller et al. 2007)	Small-scale fisheries catches were estimated for island areas in the western Pacific between 1950 and 2002, using a combination of small-scale reports and catch series for specific areas, interpolation of reported catches between periods, and expansion of local catches to island- or country-wide estimates where data were missing. Results showed that catches for all islands combined declined by 77%, in contrast to reported data, which showed increasing trends over time but covered only a limited number of commercial fisheries.	Goals: (1) Set management goals that aim to recover fish abundances to earlier states. (2) Integrate local restoration actions with objectives for the broader region. (3) Provide alternative income strategies to reduce dependency on coastal resources in order to enable fish population recovery and, hence, improve the food-security outlook for coastal communities.	Without this information, managers have only commercial-catch trends data to implement conservation measures. Without a historical baseline, major declines would be missed entirely and the marine environment managed for status quo, thus increasing the likelihood of future fish population collapses.

Multiple methods combined (including archaeology and oral history)

Global reef assessment (Pandolfi et al. 2003)	Records were compiled, extending back thousands of years, of the status and trends of seven major guilds of carnivores, herbivores, and architectural species from 14 coral reef regions from around the world. Large animals declined before small animals and architectural species, and Atlantic reefs declined before reefs in the Red Sea and Australia. However, the trajectories of decline were markedly similar worldwide. All reefs were substantially degraded long before outbreaks of coral disease and bleaching and global climate change.	Goal: Utilize trajectories of change in managing local reefs, because different drivers may be occurring in different places. Managing for local stressors is important because global climate change is just one of many factors influencing the degradation of these reefs. Management goals can be informed by historical data at specific sites, and historical scales of change provide measures of success for management actions (Pandolfi et al. 2005).	Goals without historical information: Recently bleached or diseased reefs would be managed for the impending climate change only, which could be seen to override any local stressors. Management is conducted in the absence of historical ecosystem states, so success of management actions is provided only by a much shifted baseline.
Hawai'i (Kittinger et al. 2011)	Ecological changes to coral reefs since the arrival of Polynesians were reconstructed, using archaeology, ethnohistory, anecdotal descriptions, and modern ecological and fishery data. The results indicated both reduction and recovery of ecological functional groups, with reefs around the main Hawaiian islands mostly degraded and those around the northwestern Hawaiian islands mostly in good condition or recovering currently.	Goals: (1) Establish an integrated system of marine management zones, including extensive no-take areas, to represent the variety of ecosystems related to coral reefs, with the aim of promoting recovery of all seven ecological functional groups, toward prehuman levels. (2) Complement marine management zones with management of catchments to limit the adverse impacts of terrestrial runoff on coral reefs.	Goals without this historical information might have focused on regions, functional groups, or species that might not need management as much as others.

human uses in planning (Pressey et al. 2007, Lybolt et al. 2011, Ban et al. 2012, Levy and Ban 2013). Here, we discuss how different kinds of historical perspectives and data are relevant throughout the planning process (Figure 10.1 and Table 10.1) and show how a range of methods exist to gather historical information (e.g., Table 10.2).

Setting the Stage for Planning

Conservation planning begins when the need for planning for biodiversity conservation in a specific region is recognized. Often the motivation for a planning process is provided by a sense that biodiversity generally, and harvested natural resources specifically, have been lost or reduced in comparison to some past date. Sometimes a catalytic event of some kind, such as a natural disaster or ecological catastrophe, causes initiation. Other times it is the gradual and cumulative degradation of a site or habitat type that motivates the change. This first stage is perhaps where historical perspectives are most commonly considered in the current practice of conservation planning, because historical processes often motivate the need for action.

Three stages characterize the initial phase of the conservation planning process. First, scoping and costing the process involves deciding on the boundaries of the planning region, building a planning team, determining the budget, and deciding how each subsequent stage will be addressed. Second, stakeholders need to be identified and a strategy developed for their involvement throughout the planning process. Third, the context for MPAs needs to be described, including the social, economic, and political setting as well as the constraints on, and opportunities for, conservation actions (Pressey and Bottrill 2009).

Scoping and Costing the Planning Process

Historical perspectives should be considered at the beginning of a conservation planning process so that subsequent stages can be designed to collect and consider relevant historical information (Figure 10.1 and Table 10.1). In particular, members of the planning team might need to be dedicated to collecting or synthesizing historical data, and a budget and timeline will be needed for data collection so that the information is available at relevant stages. Considering historical perspectives early on is important, because often the planning process is initiated as a result of some historically based issue (i.e., long-term degradation). Understanding this is critical if the goal is to recover the system, or at least to protect the remaining processes so that the system does not collapse. For example, historical data on the declining state of the Great Barrier Reef (Pandolfi et al. 2003) were used to raise awareness of adverse changes and help clarify the need for greater levels of protection (Fernandes et al. 2009). Information about past states and trajectories, and the spatial extent of ecosystems, can help define the boundaries of the planning region. If an effort is made to incorporate historical perspectives, some members of the planning team should spearhead this effort and set aside a suitable budget to enable integration of historical perspectives and data. This will also help outline how historical perspectives can be built into the subsequent planning stages.

Identifying and Engaging Stakeholders

Conservation planning processes must also grapple with strategies to effectively identify, engage, and build productive relationships with stakeholders. A historical perspective can help shape who should be engaged and how. Stakeholders include a diverse set of actors and organizations who will affect or be influenced by planning processes. There are many approaches to stakeholder identification (Mitchell et al. 1997, Ravnborg and Westermann 2002), analysis (Pomeroy and Douvere 2008, Reed et al. 2009), and engagement (Lynam et al. 2007, Reed 2008). Although systematic conservation planning outlines a sequential process, it recognizes that stakeholder engagement is an ongoing and iterative process that permeates the planning process, and that different people will be involved in different ways throughout (Pressey and Bottrill 2009).

A historical perspective not only shapes stakeholder engagement but can help make it more comprehensive and effective. Understanding past human uses of a region might, for example, help identify stakeholder groups. For example, long-term residents often hold a rich reservoir of information on social and ecological changes and could help influence planning goals and aspirations for recovery based on long-term baselines (for an example describing how the historical perspectives of commercial fishers and other stakeholders were instrumental in establishing the Tortugas Ecological Reserve, see Box 10.1). Further, documentation of long-term changes that emerge from resource users themselves, rather than government agencies or other institutions, can also be perceived by some stakeholders as more valid and trusted. For example, Kittinger documented stark declines in habitat quality and fisheries catches through participatory, community-led survey efforts with elders in Hawaii (Kittinger 2013). This information formed the basis for a community-based fisheries planning effort and was perceived by community members as more reliable and legitimate than other information. Similarly, holders of historical information can be asked to select areas they think are important for conservation. Such an approach was taken on the north coast of British Columbia, Canada (Ban et al. 2008), where indigenous people identified key areas they thought were important for conservation (Ban et al. 2009).

Describing the Context for Conservation Areas

Historical perspectives can play a critical role in characterizing the social, political, and economic setting for conservation planning and threats to natural features (e.g., species and ecosystems) of conservation interest (Pressey and Bottrill 2009, Ban et al. 2013). For example, as part of a stakeholder-led process to designate a network of marine conservation zones throughout England, the background report described hundreds of years of degradation and encouraged the incorporation of historical information to help set conservation and recovery objectives for marine ecosystems (Natural England and JNCC 2010, Thurstan et al. 2010). Similarly, providing historical context for conservation planning may also engender greater support for conservation among stakeholders if the need for action is articulated in a clear message or story of decline. Stories are important motivators for action (Leslie et al. 2013),

BOX 10.1 Viewpoint from a Practitioner: The Historical Perspective—A Key to Success in the Florida Keys

Billy Causey

NOAA's Florida Keys National Marine Sanctuary (FKNMS) learned an important lesson about incorporating the historical perspectives of its users into resource management. In 1990, NOAA was mandated by the U.S. Congress to "consider temporal and geographical zoning, to ensure protection of sanctuary resources." This phrase had different meanings to different stakeholders and created an air of suspicion, mistrust, and hostility that continued throughout the conservation planning process. As such, when the FKNMS Final Management Plan was implemented in July 1997, its marine zoning scheme did not truly represent the perspectives of the Keys' waterfront community. More than 6,000 written comments were received on the plan, the majority of which supported far more protection than was established with the new zoning design. In response, sanctuary managers made a commitment to undertake an inclusive public process to design the (Dry) Tortugas Ecological Reserve. The Tortugas 2000 Working Group, which included members of the FKNMS Sanctuary Advisory Council and representatives from a wide range of waterfront professions, was established to develop the reserve. The input of commercial fishermen, who were the primary users of the Tortugas area, was given equal importance to that of scientists and managers on the working group. Oceanographic, biological, and ecological information provided a compelling case for establishing a marine reserve in the Tortugas, but it was the socioeconomic and historical perspectives that allowed the working group to identify the most environmentally and economically significant areas for protection. For example, commercial fishermen shared their past experiences and observations

of incredible fish-spawning aggregation sites, such as Riley's Hump, where, historically, different species of fish would congregate on certain moon phases to spawn in mass. The comprehensive planning process was a success that culminated with the implementation of the 151 square-nautical-mile (518 km²), fully protected Tortugas Ecological Reserve in 2001. The ecological reserve and adjacent Dry Tortugas National Park Research Natural Area (also designed by the working group) have provided enormous conservation and fishery benefits. An analysis of socioeconomic and scientific information published in February 2013 found that after the ecological reserve was designated,

- Overfished species increased in abundance and size inside the reserve and throughout the region;

- Annual spawning aggregations, once thought to be wiped out from overfishing, began to recover inside the reserve;

- Commercial catches of reef fish in the region increased, and continue to do so; and

- No financial losses were experienced by local commercial or recreational fishermen.

These positive trends would not have been achieved without the historical perspectives provided by fishermen and other users of this unique area. The protection of the ocean wilderness known as "the Tortugas" for future generations is a testament to their dedication.

Billy Causey is Superintendent of the Florida Keys National Marine Sanctuary, National Oceanic and Atmospheric Administration.

and those that are embedded in historical context can be quite powerful (see chapter 12, this volume).

While historical information is commonly used or alluded to in describing declines of ecosystems or resources, it could also be used to provide context about the social, political, and economic setting (e.g., demographic changes to human coastal populations, and improvements to fisheries vessels that allow areas farther offshore to be harvested). Such information provides planners with the ability to project future trajectories of change or to identify emerging threats. Similarly, historical information can be used to describe past conservation efforts and ascertain whether conservation outcomes have been achieved. Important questions for conservation planners might include the following. Which past conservation efforts have been successful, and why? What is the historical interest of different stakeholder groups in the region, and how has it changed over time? What kinds of conservation measures are culturally appropriate, and how have these changed over time? If historical information is not readily available at this stage to answer these questions, the data collection phase provides the opportunity for collating it, and for feeding it back into this part of the process.

Developing a Vision for the Planning Region

A broad vision for the planning region needs to be agreed upon for refinement into qualitative goals about biodiversity (e.g., representing all habitat types and ensuring persistence of species), ecosystem services, livelihoods, and other concerns (Pressey and Bottrill 2009). Conservation planning is, by its nature, forward looking—concerned with measures that should be taken in the future (see Box 10.2). Conservation planning is also typically focused on mitigating anticipated adverse changes caused by current and future human activities. This emphasis points to an important implicit assumption in the visions that underpin planning efforts: given the present or recent conditions of ecosystems and populations, are these conditions environmentally or socially adequate? Our argument here is that this assumption will often cause us to aim too low, limited by shifting baselines that lack historical context (Pauly 1995, Knowlton and Jackson 2008; for examples, see Table 10.2). Historical perspectives can fundamentally change the conservation vision for a region, providing a window into possibilities for the future (Jackson 2001; Table 10.2).

Goals are important because they define the data that should be collated in subsequent stages and form the basis of the quantitative objectives for planning. Like visions of the region, goals are better informed and more ambitious when based on historical perspectives (see Table 10.2). The most obvious role of historical perspectives in shaping goals is to shed light on past losses of biodiversity and other features that can help identify previous attainable baselines for at least some species or ecosystems (Kittinger et al. 2011, Lybolt et al. 2011). Other contributions of historical data to defining goals are to demonstrate the potential benefits of effective management of marine ecosystems (Jackson 2001) and, by recording cycles of loss and recovery of species or assemblages, indicate at what level modern recovery is possible (Kittinger et al. 2011). In Table 10.2, we summarize examples of goals drawn from

BOX 10.2   Viewpoint from a Practitioner: Back to the Future—Integrating Past, Present, and Future

Charles (Bud) Ehler

If we do not learn from history, we shall be compelled to relive it. True. But if we do not change the future, we shall be compelled to endure it. And that could be worse.

ALVIN TOFFLER, *Future Shock,* 1970

Most strategic planning, including marine spatial planning, boils down to four simple questions: (1) Where are we today? (2) Where do we want to be? (3) How do we get there? and (4) What have we accomplished?

Time is as important as space in marine spatial planning. As the authors of this chapter point out, we need better knowledge of the past to know where we have come from and why we are where we are today—the first fundamental question of any planning process. However, more questions await.

We also need to understand the past and present to anticipate the future—where do we want to be?—the second fundamental planning question. Planning is about taking decisions today to get where we want to be tomorrow, however "tomorrow" is specified—5, 10, or 25 years. A number of different trajectories into the future are possible—and should be consistent with historical data. Natural scientists can help describe where we might be by extrapolating existing trends into the future. But we cannot "observe" the future, so scientists are usually reluctant to advocate for futures that are not data driven or value free and that are often highly uncertain. Identifying alternative futures and choosing which one (the vision) we want to move toward—a social and political choice—is most effectively done in cooperation with stakeholders and politicians, not only planners or natural scientists.

The next question is how do we get there. What management measures do we need to put in place to achieve the desired vision? What incentives do we need to change the behavior of individuals and institutions over time? What institutional arrangement has the authorities that are required to implement appropriate incentives to change behavior? These are questions that need be addressed in a management plan for the marine area. We want to achieve through planning a sustainable future, while acknowledging that finding and striking that balance will never be easy and will always involve both high uncertainty and social values.

The development and implementation of integrated marine spatial plans has been limited. Only a handful of plans have been approved and implemented so far—Belgium, the Netherlands, Germany, Norway, Australia, and a few coastal states in the United States. Of these, only the Netherlands and Norway have analyzed historical data and then attempted to look forward by constructing alternative scenarios while developing management plans for their marine regions. None of the plans have laid out a clear process for answering the last fundamental question—what have we accomplished?—but that question is too soon in almost every case. Clearly we have come a long way in learning how to manage marine areas, but we have a very long way to go toward implementing practical planning processes and techniques that integrate the past, present, and future.

Charles (Bud) Ehler is President of Ocean Visions Consulting, Paris, France.

historical data and give examples of how these goals would have been less effective had they not considered historical perspectives. Of course, conservation goals need not necessarily strive to recreate a particular past state, because such a state might no longer be attainable or even desirable (e.g., changes in climate might make it impossible to attain past abundances). Inevitably, shaping goals with historical information requires a choice by the planning team and stakeholders about which historical state is appropriate or achievable. Such decisions are inherently based on stakeholder values and may vary depending on which species and ecosystems are being considered. Importantly, historically informed goals should not be too easily dismissed as infeasible until all views and potential management approaches have been discussed.

Gathering Data for the Planning Process

Spatially explicit data on social and economic characteristics and biodiversity of regions, along with explicit objectives (below), are basic requirements of systematic conservation planning. Social data include variables such as demography, socioeconomic conditions, institutional arrangements, data on extractive uses and other activities, costs of conservation, and constraints on, and opportunities for, conservation actions. Biodiversity data include representation units, special elements, focal species, and ecological processes (Pressey and Bottrill 2009). Below, we discuss how these data-provisioning stages can be enhanced by including historical information.

Collecting Social Data

Social information includes current and past social attributes of human communities within and adjacent to a planning region. Commonly sought information in marine conservation planning includes human use patterns (e.g., where fishing occurs), socioeconomic data on livelihoods and ocean industries (e.g., employment rates and economic contribution of ocean industries), and demographic and cultural profiles of coastal communities. Social assessments provide valuable information to planners and managers seeking to evaluate potential impacts of proposed conservation actions. For example, in the public process to design the Tortugas Ecological Reserve in the Florida Keys, historical perspectives provided by fishermen and other users of this unique area were key in developing a management plan (Box 10.1). Such assessments can provide historical context on how human uses—and, by extension, ecosystem goods and services—have changed through time. These assessments often include historical context on changes in livelihoods, industries, and ocean use patterns (e.g., Levine and Allen 2009, Pomeroy et al. 2010).

Historical data on human activities can guide the planning process in at least three ways. First, historical reconstructions can help define the array of activities to which planning must respond, including the history of specific threats (and those that might be emerging), their current intensity in temporal as well as spatial contexts, and insights into direct versus indirect drivers of change. For example, Kittinger et al. (2011) characterized human threats and their underlying social drivers in coral reef ecosystems in the Hawaiian archipelago.

This reconstruction revealed the linkage between social drivers, direct human threats, and resultant ecological outcomes, thereby informing appropriate responses. Second, historical information can help determine the extent of threats, including whether specific threats are external or internal to the footprint of a planning region. For example, Roff et al. (2013) found that local stressors associated with changing water quality had large impacts on nearshore coral communities in the Great Barrier Reef region, long before the effects of climate change were documented, and similar results were found in Panama (Cramer et al. 2012). Third, historical reconstructions of human activities and threats can be coupled with assessments of ecological outcomes, providing information on current trajectories of specific features of interest, in turn refining conservation objectives and decisions about specific conservation actions.

Collecting Data on Biodiversity and Other Natural Features

Biodiversity data can be grouped into two categories: (1) fine-scale data on distributions of key focal species and subpopulations; and (2) coarse-scale information such as habitat types, bioregions, or ecozones that serve as surrogates for species or subpopulations when those data are not available. These data are typically synthesized from a broad array of sources, including available ecological survey data, models, or surrogate data that predict spatial distribution patterns. Historical perspectives can contribute to fine- and coarse-scale data in several ways. For fine-scale data, historical estimates of previous abundance of selected species can be used to reconstruct past biomass, diversity, or long-term changes (e.g., reef sediment cores; Rosenberg et al. 2005; Table 10.2). Historical data may also provide information on species that may otherwise be overlooked because present low abundances are erroneously perceived to be normal. For example, Roman and Palumbi (2003) used genetic analyses to show that pre-exploitation North Atlantic whale populations may have been underestimated nearly tenfold compared with previous estimates from historical logbook records (Roman and Palumbi 2003; Table 10.2). Similarly, genetic and historical analyses of dugong populations along the east coast of Australia suggest that past populations were significantly larger than current ones (Jackson et al. 2001, Marsh et al. 2005, McDonald 2005).

Historical data can also be used to determine the previous types, past distribution, and conditions and trajectories of change of communities and habitats in the planning region (Pandolfi and Jackson 2006), all of which can elucidate the recovery potential for existing habitats and ecosystems (Egan and Howell 2005, Beller et al. 2011, Whipple et al. 2011). For instance, data on historical extent, type, and linkages between wetland habitats around San Francisco Bay were gathered by the San Francisco Estuary Institute to enable planners to produce feasible goals and objectives for restoration and conservation (Grossinger et al. 2005).

Historical data can go beyond identifying baselines for species or habitats by also providing critical information on natural variability of ecosystems and the ecological processes relevant for maintaining biodiversity and resilience of ecosystems. This can help differentiate natural variability from changes that have occurred as a result of human impacts (e.g., Guzman et al.

2008, Lybolt et al. 2011). Taken together, historical data allow planners and managers to realize the past condition and the future recovery potential for marine ecosystems.

Translating Vision and Data into Actions

When the context is understood and the data have been collected, the marine conservation planning process needs to be realized. This can happen in several ways. First, the goals for the planning region ought to be converted into quantitative conservation objectives for each conservation feature (e.g., each species or habitat type). Such objectives might be, for example, to protect, in an MPA, 500 ha of eelgrass in three separate locations, or 23,000 individuals of a species. Next, a gap analysis is needed to gauge the extent to which existing MPAs or other conservation actions are already achieving the objectives. Then the location and configuration of additional MPAs or other conservation actions need to be vetted through a stakeholder engagement process. Finally, the MPAs need to be implemented, considering appropriate and feasible actions and institutional arrangements.

Developing Quantitative Conservation Objectives

Objectives describe goals for specific components of ecosystem patterns (e.g., species occurrences and marine habitat distributions) or processes (e.g., connectivity between reefs, and maximum extent of disturbance over some time frame; Pressey et al. 2007).

Objectives should ideally be set in relation to estimated historical extents or abundances of features of interest. For example, objectives can be pinned to historical estimates for the recovery of marine habitats or to the recovery of abundances for key species. Objectives set as percentages of historical distributions automatically compensate for losses by representing proportionally larger percentages of current distributions, depending on the extent of loss (Pressey et al. 2003). Considering a potentially perverse outcome of setting objectives as percentages of current distributions or population sizes is also important: objectives might become smaller and more easily achieved even as the features of interest experience further declines.

An example from the terrestrial realm illustrates the value of objectives being formulated in relation to historical distributions. In the Cape Floristic Region of South Africa (Figure 10.2), objectives for protection are larger, and more realistic in terms of conservation needs of the target species, for habitats that have undergone more loss of native vegetation. Thus, when historical distributions are taken into account, restoration objectives might be necessary, and these can be directed at species (Didier et al. 2009), habitats (Bryan et al. 2011), or, with minor adaptations to protection objectives, processes (Pressey et al. 2007; for more detailed discussion of restoration, see chapters 8 and 9, this volume).

Reviewing Achievement of Objectives

In its basic form, this stage represents an analysis of gaps (Caicco et al. 1995), or a tallying of the extent to which each objective has been achieved through existing management actions, thereby indicating the need for additional management measures. Even nonspatial

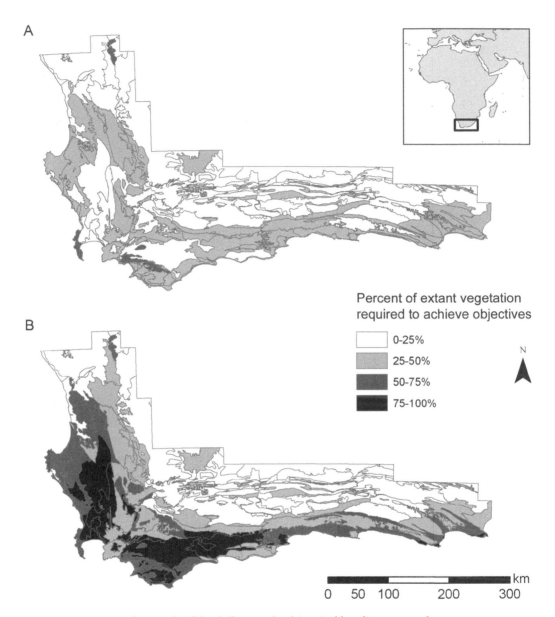

FIGURE 10.2 Terrestrial example of the difference that historical baselines can make to assessment of conservation requirements. The figure shows the Cape Floristic Region of South Africa, a global biodiversity hotspot, with outlines of 102 broad habitat units (BHUs) defined by vegetation and physical variables. Quantitative conservation objectives for BHUs were taken from Pressey et al. (2003). Conservation planning requirements are shown (A) without and (B) with historical baselines. (A) Percentage objectives applied only to extant vegetation cover in BHUs, reflecting the hypothetical lack of historical baselines of preclearing extent of BHUs. This plan is flawed because the original pre-European extent of native vegetation was not considered, especially for areas that were extensively cleared. (B) Percentage objectives applied to the original pre-European extent of native vegetation in BHUs, informed by historical baselines, then expressed as percentages of extant native vegetation required. Conservation requirements for many BHUs are larger in panel B, especially for BHUs that have been extensively cleared (shaded black).

information could be important in this stage. For example, historical information can be extremely valuable in assessing the effectiveness of past and current conservation measures, and information on the relative contributions of different management actions (either positive or negative in terms of their success) can be integrated into gap analysis (e.g., Mills et al. 2011). For instance, Lotze (see chapter 2, this volume) has shown that conservation efforts for coastal seas and estuaries over the past century have enabled partial recovery of some animals (i.e., pinnipeds, otters, and birds) but that these successes did not extend to recovery of ecosystem structure and function.

Selecting Additional Conservation Areas and Applying Conservation Actions to Selected Areas

Once it is clear what the gaps are, additional conservation areas need to be identified and implemented. Historical information can inform the selection of conservation actions in several ways. First, spatial information on the previous distributions of habitats and species will indicate candidate areas to achieve conservation (or restoration) objectives. Second, historical information can highlight management potential of areas that might not otherwise have been considered for conservation. For example, paleoecological data indicate that some subtropical waters once supported coral reefs (Greenstein and Pandolfi 2008). As the oceans warm under climate change, such areas might once again become important habitats for corals, including those for which conditions elsewhere might have become less favorable. Third, the historical significance of some areas can facilitate conservation management that might be difficult elsewhere. For example, Midway Atoll in the northwestern Hawaiian Islands was protected as a national wildlife refuge prior to the establishment of the Papahānaumokuākea Marine National Monument. The historical significance of Midway Atoll as the site of an important World War II battle resulted in additional protections and a special management plan for the atoll that was nested within the larger site-management plan (U.S. Fish and Wildlife Service et al. 2008).

Ensuring That the Expanded Conservation System Is Maintained, Monitored, and Reviewed

MPAs and other conservation actions instigated through the planning process have to be managed to achieve the objectives for which they were established (Pressey and Bottrill 2009). Through monitoring and evaluation, practitioners can ensure that the conservation actions they are implementing are effective and, if not, can adjust their strategies. This adaptive management framework depends on careful review of progress and a plan to adjust strategies as needed.

Maintaining and Monitoring MPAs

Probably the main role of historical information in supporting maintenance and monitoring of marine MPAs is assessing current progress in relation to historical context. When monitoring is undertaken in specific areas, whether they are preexisting management sites or new areas implemented as part of the planning process, new information will inevitably come to light that was not apparent from initial data. Historical information in the form of systematic field surveys, opportunistic field sampling, or archival photographs could indicate whether current characteristics such as coral cover are different from previous characteristics (Hughes et al. 2011). This information, in turn, could inform managers about the extent to which return to historical conditions is feasible by managing the area itself, or whether restoration depends on external factors beyond their control, such as supply of sediment and nutrients from catchments. Archival remote sensing and aerial photographs could also provide information on previous conditions within and around management areas to guide management responses.

Barriers to Incorporating Historical Data into Conservation Planning

Historical data are fundamental to our perceptions of what was natural in the sea. However, despite the stated aim of conservation planning to articulate goals and set priorities, relatively little attention has been given to use of historical data in planning. Several reasons help explain this disparity. First, conservation planners may not be familiar with the data or methods of historical ecology and, hence, may place low priority on the extra effort and cost needed to include historical information. Marine historical ecology is a relatively new research field, and scholars working in this area use a variety of multidisciplinary techniques that differ from traditional approaches and datasets more familiar to planners. Concerns might exist over the quality or type of the data, which are often anecdotal, qualitative, or nonspatial in nature.

Another impediment to using historical information is that societal values can shift over time, and conservation planners and some stakeholders may not be interested in the past or may be unwilling to fully consider the enormity of past decline. In some cases, specific stakeholders may not want to restore ecosystems or resources to previous states, even if it was possible (though usually it is not). For example, in the lobster fishery in Maine (USA), centuries of intense fishing have extirpated most apex predators, resulting in the development of a profitable lobster fishery. To the communities involved in this fishery, the past state may be less desirable than the current profitable one (Steneck et al. 2011). Similarly, given urbanization and migration to coastal areas, some coastal communities have changed so quickly that the collective social conceptualization of the past, and their relationship with the environment, has also changed. Such change can result in the perception of multiple baselines for the same resource among stakeholders, a problem that is compounded when recovery is longer than human generation times.

While we recognize these barriers, we believe that the advantages of incorporating historical perspectives outweigh the potential downsides. Integrating historical perspectives will allow planners to better understand the social and ecological dynamics that have led to the current state of a planning region and hence, we believe, will increase the likelihood of successful conservation efforts. Marine conservation planning is gaining momentum as countries and regions try to achieve global and regional conservation targets, providing ample opportunity to integrate historical information into these processes. Furthermore, the field of marine historical ecology, though rapidly expanding, is relatively new and is not yet part of conservation planning practice. As conservation planning and historical ecology become more connected, best practices and innovative methods will need to be developed to integrate these fields.

OPPORTUNITIES

As we have shown in this chapter, information on historical changes in marine ecosystems and human communities has the potential to be integrated into all of the stages of systematic conservation planning (Figure 10.1 and Table 10.1), with substantial opportunities to improve the effectiveness of conservation planning processes. Below, we outline significant opportunities for marine historical ecology to improve planning practice and conservation outcomes.

Historical data can inform our understanding of changes in the state or condition of ecosystems or a particular set of ecosystem constituents (e.g., species and habitats), but it can also provide substantive information on changes in ecological processes that are key to the structure and functioning of healthy ecosystems (Figure 10.3). Radical changes in ecosystem state are often referred to as "regime shifts," the transitions between which can be sudden and unexpected but are often the result of long-term processes best identified through historical analyses. For planners and managers, one of the major implications of regime shifts is defining the "safe operating space"—that is, the suite and intensity of human actions that can be accommodated without tipping an ecosystem into an undesirable state. Historical data on population declines and habitat losses can help characterize the current state, but thresholds are difficult to define (McClanahan et al. 2011), and hence it is prudent to avoid pushing this envelope and instead plan for maintaining ecological processes that are known to be critical for ecosystem resilience (Figure 10.3).

Retrospective analyses of historical changes in ocean systems can also provide some guidance on the state of key processes. For example, the regime shift from coral- to algal-dominated states in Caribbean coral reefs were the result of long-term declines in the abundance and diversity of herbivores (Hughes 1994, Jackson et al. 2001). Habitat loss, coupled with loss of functional diversity and intensity of herbivory—a key ecological process for reef resilience—resulted in collapse of coral reef ecosystems (Figure 10.3B). Other systems that have exhibited worrying trajectories but have avoided catastrophic regime shifts can also provide lessons on how to avoid thresholds. For example, the decline of Pacific pinniped communities resulted in major loss of kelp forest along the west coast of North America, but species protections enabled recovery (Estes et al. 1998). In this case, it is possible that a permanent regime shift was avoided by the availability of suitable pinniped habitat (coastal rocky reefs), which provided the ecological building blocks necessary for the recovery of pinnipeds and kelp forest ecosystems (Figure 10.3C; Steneck et al. 2003).

Marine historical ecology also allows planners and stakeholders to develop more realistic conservation goals and objectives that are grounded in reasonable reconstructions of the

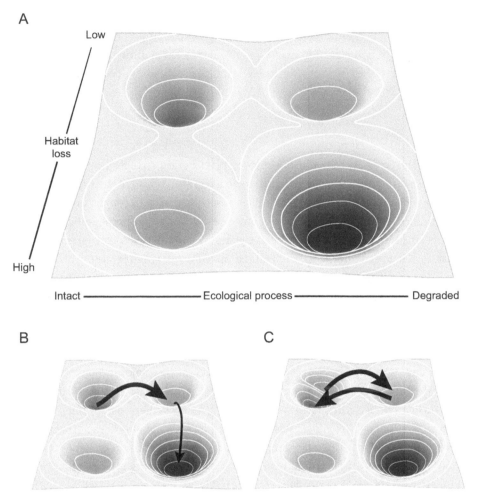

FIGURE 10.3 A heuristic model describing the relationship between habitat loss and ecological processes. (A) Different basins of attraction illustrate different stable states of ecosystem condition. In the upper left, habitat loss is low and ecosystem processes are intact. As a managed population or habitat degrades, management actions that stop or reverse this decline will enhance system resilience, but those that fail to stop loss of habitat or ecosystem processes will place the system in jeopardy of converting into a degraded state. (B) An example of a historical trajectory of degradation: herbivore populations in Caribbean coral reef ecosystems declined because of centuries of human impacts (arrow from left to right), resulting eventually in a rapid loss of herbivory (arrow from top right to bottom right) and an eventual change from a stable coral-dominated to a stable algal-dominated system. (C) An example of recovery: pinniped populations in nearshore kelp forest ecosystems along the west coast of North America were decimated during the sealing and whaling era of the nineteenth century, resulting in loss of predation and increases in herbivores that grazed kelp forest systems intensively (arrow from left to right). Protection of pinnipeds in the twentieth century resulted in restoration of predation, reduction in herbivory, and the subsequent return of kelp forests (arrow from right to left), although a novel pinniped community emerged as a result of this recovery, as indicated by a bifurcated basin in the upper left.

past. Historical data can be used not only to develop qualitative goals grounded in historical understandings but also, in some cases, to develop quantitative objectives based on historical trajectories and measures of success to track progress toward these objectives. Without a historical perspective, planning processes may proceed on a superficial understanding of current conditions or, worse yet, may establish arbitrary objectives that are either not grounded in reality or cannot feasibly result in the desired changes, given the trajectory of the target system. Objectives based on long-term perspectives can allow for more accurate scenarios of future changes and more prescriptive management approaches. In this way, planners and managers will be equipped to develop conservation strategies better attuned to ongoing and expected environmental and social changes.

Finally, incorporating historical data does not represent a fundamental departure from current conservation planning practice. As we have illustrated, historical data are already being used and have the potential for increased use to improve conservation planning practice and social and ecological conservation outcomes. The techniques and methodologies in marine historical ecology are growing in familiarity, and datasets are growing in number, type, and complexity (Lotze and Worm 2009). Further, regional coverage is expanding and more refined trajectories of change at spatial scales better suited for conservation planning are growing in number. Moreover, marine historical ecology is increasingly being oriented to facilitate uptake by conservation practitioners (Lybolt et al. 2011, Rick and Lockwood 2013).

CONCLUSIONS

Without a long-term historical perspective, planners, managers, and stakeholders will have only a rudimentary understanding of the systems they seek to protect and enhance. The danger of excluding long-term data is that planners and managers may misdiagnose the direct and underlying causes of decline in natural features and the real rates of ecological change. Conservation strategies based on the limited perspective of recent observations alone may limit understanding about the achievable goals for restoration and management of ocean environments (Jackson et al. 2001). Furthermore, historical perspectives—which invoke the living memory of stakeholders with intergenerational knowledge—can provide a powerful mechanism to engage all stakeholders in developing a shared vision in a conservation planning process. Thus, marine historical ecology has much to offer to marine conservation planning and should be considered in all conservation planning exercises.

ACKNOWLEDGMENTS

We thank Rodolphe Devillers, Rebecca Weeks, and Adrian Arias for helpful suggestions. N.C.B. and R.L.P. thank the Australian Research Council for support. N.C.B., J.M.P., R.L.P., and R.H.T. acknowledge support from the ARC Centre of Excellence for Coral Reef Studies. J.N.K. thanks the Center for Ocean Solutions at Stanford University for support.

REFERENCES

Airamé, S., Dugan, J., Lafferty, K. D., et al. (2003) Applying ecological criteria to marine reserve design: a case study from the California Channel Islands. *Ecological Applications* 13, S170-S184.

Almany, G., Connolly, S., Heath, D., et al. (2009) Connectivity, biodiversity conservation and the design of marine reserve networks for coral reefs. *Coral Reefs* 28, 339–351.

Ban, N. C., Mills, M., Tam, J., et al. (2013) Towards a social-ecological approach for conservation planning: embedding social considerations. *Frontiers in Ecology and the Environment* 11, 194–202.

Ban, N. C., Picard, C. R., and Vincent, A. C. J. (2008) Moving towards spatial solutions in marine conservation, with indigenous communities. *Ecology and Society* 13, 32.

Ban, N. C., Picard, C. R., Vincent, A. C. J. (2009) Comparing and integrating community-based and science-based approaches in prioritizing marine areas for protection. *Conservation Biology* 23, 899–910.

Ban, N. C., Pressey, R. L., and Weeks, S. (2012) Conservation objectives and sea surface temperature anomalies in the Great Barrier Reef. *Conservation Biology* 26, 799–809.

Beller, E., Grossinger, R., Salomon, M., et al. (2011) Historical ecology of the lower Santa Clara River, Ventura River, and Oxnard Plain: an analysis of terrestrial, riverine, and coastal habitats. SCCWRP Technical Report 0662. San Francisco Estuary Institute, Oakland, CA.

Bryan, B. A., Crossman, N. D., King, D., and Meyer, W. S. (2011) Landscape futures analysis: assessing the impacts of environmental targets under alternative spatial policy options and future scenarios. *Environmental Modelling & Software* 26, 83–91.

Caicco, S. L., Scott, J., Butterfield, B., and Csuti, B. (1995) A gap analysis of the management status of the vegetation of Idaho (USA). *Conservation Biology* 9, 498–511.

Cardinale, M., Bartolino, V., Llope, M., et al. (2011) Historical spatial baselines in conservation and management of marine resources. *Fish and Fisheries* 12, 289–298.

Convention on Biological Diversity (2010) Conference of the Parties 10 (COP 10) Decision X/2. Strategic Plan for Biodiversity 2011–2020. www.cbd.int/decision/cop/default.shtml?id= 12268.

Cramer, K. L., Jackson, J. B. C., Angioletti, C. V., et al. (2012) Anthropogenic mortality on coral reefs in Caribbean Panama predates coral disease and bleaching. *Ecology Letters* 15, 561–567.

Didier, K. A., Glennon, M. J., Novaro, A., et al. (2009) The landscape species approach: spatially-explicit conservation planning applied in the Adirondacks, USA, and San Guillermo-Laguna Brava, Argentina, landscapes. *Oryx* 43, 476–487.

Egan, D., and Howell, E. (Eds.) (2005) *The Historical Ecology Handbook: A Restorationist's Guide to Reference Ecosystems.* Island Press, Washington, DC.

Estes, J. A., Tinker, M. T., Williams, T. M., and Doak, D. F. (1998) Killer whale predation on sea otters linking oceanic and nearshore ecosystems. *Science* 282, 473–476.

Fernandes, L., Day, J., Kerrigan, B., et al. (2009) A process to design a network of marine no-take areas: lessons from the Great Barrier Reef. *Ocean & Coastal Management* 52, 439–447.

Fernandes, L., Day, J., Lewis, A., et al. (2005) Establishing representative no-take areas in the Great Barrier Reef: large-scale implementation of theory on marine protected areas. *Conservation Biology* 19, 1733–1744.

Green, A., Smith, S. E., Lipsett-Moore, G., et al. (2009) Designing a resilient network of marine protected areas for Kimbe Bay, Papua New Guinea. *Oryx* 43, 488–498.

Greenstein, B. J., and Pandolfi, J. M. (2008) Escaping the heat: range shifts of reef coral taxa in coastal Western Australia. *Global Change Biology* 14, 513–528.

Grossinger, R. M., Askevold, R. A., and Collins, J. N. (2005) T-sheet user guide: application of the historical U.S. Coast Survey maps to environmental management in the San Francisco Bay area. SFEI Report No. 427. San Francisco Estuary Institute, Oakland, CA.

Guzman, H. M., Cipriani, R., Jackson, and J. B. C. (2008) Historical decline in coral reef growth after the Panama Canal. *AMBIO: A Journal of the Human Environment* 37, 342–346.

Halpern, B. S., and Warner, R. R. (2002) Marine reserves have rapid and lasting effects. *Ecology Letters* 5, 361–366.

Harrison, H. B., Williamson, D. H., Evans, R. D., et al. (2012) Larval export from marine reserves and the recruitment benefit for fish and fisheries. *Current Biology* 22, 1023–1028.

Hughes, T. P. (1994) Catastrophes, phase shifts, and large-scale degradation of a Caribbean coral reef. *Science* 265, 1547–1551.

Hughes, T. P., Bellwood, D. R., Baird, A. H., et al. (2011) Shifting base-lines, declining coral cover, and the erosion of reef resilience: comment on Sweatman et al.(2011). *Coral Reefs* 30, 653–660.

Jackson, J. B. C. (2001) What was natural in the coastal oceans? *Proceedings of the National Academy of Sciences USA* 98, 5411–5418.

Jackson, J. B. C., Kirby, M. X., Berger, W. H., et al. (2001) Historical overfishing and the recent collapse of coastal ecosystems. *Science* 293, 629–637.

Kelleher, G., and Kenchington, R. A. (1992) Guidelines for establishing marine protected areas: a marine conservation and development report.

Kittinger, J. N. (2013) Participatory fishing community assessments to support coral reef fisheries comanagement. *Pacific Science* 67, 361–381.

Kittinger, J. N., Pandolfi, J. M., Blodgett, J. H., et al. (2011) Historical reconstruction reveals recovery in Hawaiian coral reefs. *PLoS ONE* 6, e25460.

Knowlton, N., and Jackson, J. B. C. (2008) Shifting baselines, local impacts, and global change on coral reefs. *PLoS Biology* 6, e54.

Leslie, H. M., Goldman, E., McLeod, K. L., et al. (2013) How good science and stories can go hand in-hand. *Conservation Biology* 27, 1126–1129.

Levine, A., and Allen, S. (2009) American Samoa as a fishing community. NOAA Technical Memorandum NMFS-PIFSC-19.

Levy, J., and Ban, N. C. (2013) A method for incorporating climate change modelling into marine conservation planning: an Indo-west Pacific example. *Marine Policy* 38, 16–24.

Lotze, H. K., and Worm, B. (2009) Historical baselines for large marine animals. *Trends in Ecology & Evolution* 24, 254–262.

Lybolt, M., Neil, D., Zhao, J., et al. (2011) Instability in a marginal coral reef: the shift from natural variability to a human-dominated seascape. *Frontiers in Ecology and the Environment* 9, 154–160.

Lynam, T., De Jong, W., Sheil, D., et al. (2007) A review of tools for incorporating community knowledge, preferences, and values into decision making in natural resources management. *Ecology and Society* 12, article 5.

Margules, C. R., and Pressey, R. L. (2000) Systematic conservation planning. *Nature* 405, 243–253.

Marsh, H., De'Ath, G., Gribble, N., and Lane, B. (2005) Historical marine population estimates: triggers or targets for conservation? The dugong case study. *Ecological Applications* 15, 481–492.

McClanahan, T. R., Graham, N. A. J., MacNeil, M. A., et al. (2011) Critical thresholds and tangible targets for ecosystem-based management of coral reef fisheries. *Proceedings of the National Academy of Sciences USA* 108, 17230–17233.

McClenachan, L. (2009) Documenting loss of large trophy fish from the Florida Keys with historical photographs. *Conservation Biology* 23, 636–643.

McDonald, B. (2005) Population genetics of dugongs around Australia: implications of gene flow and migration. PhD dissertation, James Cook University, Queensland, Australia.

Mills, M., Jupiter, S. D., Pressey, R. L., et al. (2011) Incorporating effectiveness of community-based management in a national marine gap analysis for Fiji. *Conservation Biology* 25, 1155–1164.

Mitchell, R. K., Agle, B. R., and Wood, D. J. (1997) Toward a theory of stakeholder identification and salience: defining the principle of who and what really counts. *Academy of Management Review* 22, 853–886.

Mora, C., and Sale, P. F. (2011) Ongoing global biodiversity loss and the need to move beyond protected areas: a review of the technical and practical shortcomings of protected areas on land and sea. *Marine Ecology Progress Series* 434, 251–266.

Natural England and JNCC (2010) The Marine Conservation Zone Project: Ecological Network Guidance. Natural England and JNCC, Sheffield and Peterborough, UK.

Pandolfi, J. M., Bradbury, R. H., Sala, E., et al. (2003) Global trajectories of the long-term decline of coral reef ecosystems. *Science* 301, 955–958.

Pandolfi, J. M., and Jackson, J. B. C. (2006) Ecological persistence interrupted in Caribbean coral reefs. *Ecology Letters* 9, 818–826.

Pandolfi, J. M., Jackson, J. B. C., Baron, N., et al. (2005) Are US coral reefs on the slippery slope to slime? *Science* 307, 1725–1726.

Pauly, D. (1995) Anecdotes and the shifting baseline syndrome of fisheries. *Trends in Ecology & Evolution* 10, 430.

Pauly, D., Christensen, V., Guenette, S., et al. (2002) Towards sustainability in world fisheries. *Nature* 418, 689–695.

Pomeroy, C., Thomson, C., and Stevens, M. (2010) California's North Coast fishing communities: historical perspective and recent trends. California Sea Grant Technical Report T-072.

Pomeroy, R., and Douvere, F. (2008) The engagement of stakeholders in the marine spatial planning process. *Marine Policy* 32, 816–822.

Pressey, R. L., and Bottrill, M. C. (2009) Approaches to landscape- and seascape-scale conservation planning: convergence, contrasts and challenges. *Oryx* 43, 464–475.

Pressey, R. L., Cabeza, M., Watts, M. E. J., et al. (2007) Conservation planning in a changing world. *Trends in Ecology & Evolution* 22, 583–592.

Pressey, R. L., Cowling, R. M., and Rouget, M. (2003) Formulating conservation targets for biodiversity pattern and process in the Cape Floristic Region, South Africa. *Biological Conservation* 112, 99–127.

Pyne, S. J. (1999) Attention! All keepers of the flame. *Whole Earth Magazine* (Winter), 1–2.

Ravnborg, H. M., and Westermann, O. (2002) Understanding interdependencies: stakeholder identification and negotiation for collective natural resource management. *Agricultural Systems* 73, 41–56.

Reed, M. S. (2008) Stakeholder participation for environmental management: a literature review. *Biological Conservation* 141, 2417–2431.

Reed, M. S., Graves, A., and Dandy, N., et al. (2009) Who's in and why? A typology of stakeholder analysis methods for natural resource management. *Journal of Environmental Management* 90, 1933–1949.

Rick, T. C., and Lockwood, R. (2013) Integrating paleobiology, archeology, and history to inform biological conservation. *Conservation Biology* 27, 45–54.

Roberts, C. M., Andelman, S., Branch, G., et al. (2003) Ecological criteria for evaluating candidate sites for marine reserves. *Ecological Applications* 13, S199-S214.

Roff, G., Clark, T. R., Reymond, C. E., et al. (2013) Palaeoecological evidence of a historical collapse of corals at Pelorus Island, inshore Great Barrier Reef, following European settlement. *Proceedings of the Royal Society of London Series B* 280, article 20122100.

Roman, J., and Palumbi, S. R. (2003) Whales before whaling in the North Atlantic. *Science* 301, 508–510.

Rosenberg, A. A., Bolster, W. J., Alexander, K. E., et al. (2005) The history of ocean resources: modeling cod biomass using historical records. *Frontiers in Ecology and the Environment* 3, 78–84.

Sala, E., and Knowlton, N. (2006) Global marine biodiversity trends. *Annual Review of Environment and Resources* 31, 93–122.

Steneck, R. S., Graham, M. H., Bourque, B. J., et al. (2003) Kelp forest ecosystems: biodiversity, stability, resilience and future. *Environmental Conservation* 29, 436–459.

Steneck, R. S., Hughes, T., Cinner, J., et al. (2011) Creation of a gilded trap by the high economic value of the Maine lobster fishery. *Conservation Biology* 25, 904–912.

Stewart, G. B., Kaiser, M. J., Côté, I. M., et al. (2009) Temperate marine reserves: global ecological effects and guidelines for future networks. *Conservation Letters* 2, 243–253.

Thurstan, R. H., Brockington, S., and Roberts, C. M. (2010) The effects of 118 years of industrial fishing on UK bottom trawl fisheries. *Nature Communications* 1, 15.

University of Queensland (2009) Scientific principles for design of marine protected areas in Australia: a guidance statement. Ecology Centre, University of Queensland, Brisbane, Australia.

U.S. Fish and Wildlife Service, NOAA, and State of Hawaii (2008) Papahānaumokuākea Marine National Monument. Management plan. http://sanctuaries.noaa.gov/management/mpr/papahanaumokuakeamp.pdf.

Whipple, A., Grossinger, R., and Davis, F. (2011) Shifting baselines in a California oak savanna: nineteenth century data to inform restoration scenarios. *Restoration Ecology* 19, 88–101.

Willis, K., Bailey, R., Bhagwat, S., and Birks, H. (2010) Biodiversity baselines, thresholds and resilience: testing predictions and assumptions using palaeoecological data. *Trends in Ecology & Evolution* 25, 583–591.

Wood, L. J., Fish, L., Laughren, J., and Pauly, D. (2008) Assessing progress towards global marine protection targets: shortfalls in information and action. *Oryx* 42, 340–351.

Worm, B., Barbier, E. B., Beaumont, N., et al. (2006) Impacts of biodiversity loss on ocean ecosystem services. *Science* 314, 787–790.

Zeller, D., Booth, S., Davis, G., and Pauly, D. (2007) Re-estimation of small-scale fishery catches for U.S. flag-associated island areas in the western Pacific: the last 50 years. *Fishery Bulletin* 105, 266–277.

PART FOUR

ENGAGING THE PUBLIC

Lead Section Editor: LOREN MCCLENACHAN

All of the issues addressed in this volume—restoring ecosystems, managing fisheries, and recovering endangered species—are informed by science but ultimately play out in the public sphere. Historical ecology can provide information on long-term environmental change and baselines, but how that information is received and implemented is a question of policy choice and public engagement.

In many ways, historical ecology has grown up in the public eye, with extensive media coverage of the often shocking results of historical ecological research. For example, a *Science* paper by Jeremy Jackson and colleagues, "Historical overfishing and the recent collapse of coastal ecosystems," was named the most important scientific discovery in 2001 by *Discover* magazine. Ram Myers and Boris Worm's 2003 *Nature* paper "Rapid worldwide depletion of predatory fish communities" was profiled on the cover of *Newsweek* magazine, which asked the question "Are the oceans dying?" The dramatic results from these and other historical ecology research papers grabbed the media's attention, which was quick to make links to the need for conservation. But would these headlines and popular news pieces be enough to engage the public for longer than an airplane ride or a stint in the dentist's waiting room? Would the public be shocked into caring enough to motivate conservation or would they quickly move on to other concerns or diversions?

Randy Olson, a marine ecologist turned filmmaker, took up the cause of creating a deeper public interest in historical ecology, founding the Shifting Baselines website (http://shifting-baselines.org). He and his team filled the site with humorous public service announcements (PSAs) that were populated by Hollywood actors. In one, the "Tiny Fish PSA," two men enthuse about a 2-inch fish, crowing that "I haven't seen a fish this big in years!" Shifting Baselines packaged a dark hilarity to draw in and educate a wider public about ocean decline. Their PSAs made explicit links to the need for conservation action, like support for marine protected areas and reduced fishing effort. Their work also inspired grassroots efforts to engage the public; shiftingbaselines.org held annual PSA contests to motivate individuals and university students to upload their own shifting-baseline messages to the public.

While Olson and the Shifting Baselines team endeavored to expand the reach of historical ecology's message, others worked to deepen the information communicated to the public, and to link it to the human experience in the past and the future. In his 2007 book *Unnatural History of the Sea,* Callum Roberts synthesized the work of marine historical ecologists for the subset of the public that was thirsty for details about long-term ocean change and its link to global expansion, exploration, and exploitation. James MacKinnon's 2013 book *The Once and Future World* used historical ecology to ask how knowledge of the past changes the way we view our role in nature and society's conservation goals. With these additions to the public discourse, public engagement with historical ecology moved from shocking accounts of decline to dark humor and finally to the essential question: How does knowledge about the past inform what we want from nature, as individual humans and as a society?

In answering this question, historical ecology provides much in the way of hope and wonder, evoking a time in which ecosystems were intact, fish populations were robust, and now depleted species were abundant. In doing so, it offers a more expansive view of ecosystem potential and what conservation might achieve if successful. Intact coastal habitats like wetlands and oyster reefs provide essential ecosystem services; robust fish stocks support vital fishing communities; abundant marine animals in their natural environment provide intangible benefits to the humans who encounter and marvel at them. Historical ecology highlights these benefits in a place-specific way and reminds people how healthy coasts and oceans benefit people and what specific changes have occurred in their own communities.

Historical ecology has emerged at a time in which society is increasingly connecting people in new and exciting ways, and the pace and scale of communications are increasing exponentially. This context provides a rich forum for action and public engagement. While some mechanisms for effective communication and engagement are timeless, historical ecology has been part of conservation initiatives that have arisen as a result of a globally connected society.

Chapters in this section address the ways in which historical ecology provides a portal to the past and the ways in which imagining the bounty of oceans past reconnects people to nature, to place, and to resources and environments. Marzin, Evans, and Alexander explore the potential of "new media" to engage a broader range of the public who may not traditionally be interested in the relationship between people and the sea. In many ways, engaging the public has never been easier, as the rise of new media provides many points of access and new ways to frame conservation messages. These same properties make it even more challenging to reach an engaged audience, as the public is distracted with an overwhelming amount of information and choices about how to spend their time. In the final chapter in this volume, James MacKinnon describes the ways in which narratives of marine historical ecology inspire hope, and their potential to enliven and reconnect people to the natural environment and the social history of place. Viewpoint boxes further investigate topics including the role of art, museum spaces, and government initiatives in engaging the public with these important issues.

The questions raised in this section remain open and, in many ways, depend on what you and others do with the knowledge gained from this volume. Will you be shocked into caring enough to engage in conservation or will you quickly move on to other concerns or diversions? How will the knowledge about the past inform what you want from nature, for yourself, and for the generations to come?

REFERENCES

Jackson, J. B. C., Kirby, M. X., Berger, W. H., et al. (2001) Historical overfishing and the recent collapse of coastal ecosystems. *Science* 293, 629–637.

MacKinnon, J. B. (2013) *The Once and Future World*. Random House, New York, NY.

Myers, R., and Worm, B. (2003) Rapid worldwide depletion of predatory fish communities. *Nature* 423, 280–283.

Roberts, C. (2007) *Unnatural History of the Sea*. Island Press, Washington, DC.

ELEVEN

Engaging Public Interest in the Ocean of the Past

The Promise of New Media

CATHERINE MARZIN, SIAN EVANS, and KAREN ALEXANDER

Marine historical ecology powerfully frames ocean issues. It reveals rich new storylines and opportunities for new constituencies in the public to identify with a specific place or time in history. Historical anecdote, personal experience, and imagery, which are less polarizing than some conservation messages, can create interest among people who may not naturally care about fish, the sea, or the health of the marine environment. Using the concept of framing, this chapter describes mechanisms to engage the public via old and new media. Exciting the imagination on an individual level, marine historical ecology can create a sense of ownership and recognition of the need for ocean stewardship. New outreach strategies can be developed, utilizing all the tools of new media, including discoverability, social participation, and mobility. Historians and scientists can harness the power of history and scientific observation in new arcs of storytelling across new media platforms—social, textual, and visual.

INTRODUCTION

[C]oncentrating on documentable and remembered phenomena, the living ocean, in all its vastness and vulnerability, becomes connected to human societies in intimate and time-specific ways. That seems like a story worth telling, a sea story with the ocean included.

W. JEFFREY BOLSTER, *The Mortal Sea: Fishing the Atlantic in the Age of Sail,* 2012

The destiny of mankind has always been bound to the sea. Marine historical ecology has generated a growing body of evidence that the sea and its resources have shaped and been shaped by human history. In his 1995 paper, "Anecdotes and the shifting baseline syndrome of fisheries," Daniel Pauly argued for the perspective of time and the informed long views that integrating historical perspectives into fisheries science can bring (Pauly 1995). Ecologists, fisheries scientists, and historians have since provided evidence that there is ecological information buried in historical sources that can describe this change and inform conservation and management more broadly (see, e.g., chapter 4, this volume). For the nonacademic public, general-interest books, articles, and movies have also begun to tell this story. In the past 30 years, projects including Farley Mowat's *Sea of Slaughter* (1984), Sebastian Junger's *The Perfect Storm* (1997), Mark Kurlansky's *Cod* (1997) and *The Big Oyster* (2006), Randy Olson's Shifting Baselines website (http://shiftingbaselines.org), Charles Clover's *The End of the Line* (2004), and Callum Roberts's *The Unnatural History of the Sea* (2007) revolve around the subject of the oceans, fishing, and long-term human impact on marine life. These projects often describe monumental story arcs of lineages of fishermen and working waterfronts, changing communities, misbegotten perceptions, and lost natural resources. Other points of access to marine historical ecology provide touchstones for identification and empathy, including cultural landscapes, historical narratives and trends, and aesthetic and environmental stewardship concerns. As well, primary source materials of marine historical ecology—original documents that are contemporary with the things and events they describe—can provide unique and diverse connections to the sea. These framings of history and science within a dramatic narrative present many opportunities to educate and engage the public.

Although marine historical ecology has demonstrated the need for conservation action, the task of engaging the public in substantive, ongoing dialogue connected to advocacy and policy is incomplete. The first step in creating active constituencies is engagement and building the knowledge that fuels concern. For marine historical ecologists who wish to see their work inspire the public, one of the most effective things they can do is to become involved in new forms of communication. They can partner strategically with communities and advocacy organizations, many of which are well acquainted with translating interest into active participation, through targeted calls to action.

Today, effective outreach considers multiple points of access from multiple individual perspectives, using all media tools: old and new, analog and digital. Some in the nonacademic public may not be interested in a deep understanding of ocean science or history, but they can nonetheless develop awareness and a sense of pride, wonder, and ownership of marine resources. This chapter focuses on the emergence of new media, including digital, interactive, social, and mobile media, and the opportunities they offer to reach the public.

New media provide novel opportunities to reach beyond a centralized, same-minded audience. In marketing, the "long tail" approach is defined as selling a large number of items in small quantities to a large number of people. Anderson (2004) extended the long tail strategy to a new media setting, observing that new media increase the reach of a message by microtargeting different audiences. There are useful parallels for public education in marine historical ecology. Studies of science learning have shown that for the general public, the process of learning is lifelong, comes from multiple experiences, and is rooted in personal interests and motivations (Maibach et al. 2008, National Research Council 2009, Groffman et al. 2010). Individual concerns thus have a great bearing on the ability to listen to and absorb science. Similarly, in the context of marine historical ecology, scientists and historians work with a broad range of material, which can be tailored to speak to many individual perspectives. Current studies of science communications support the long tail approach, which ties effective learning to individual interests, often introduced from multiple perspectives.

TO FRAME IS TO EXPLAIN

Scientists can make greater efforts to reach nonscientific audiences and to think more deeply about the social networks that influence audiences (Groffman et al. 2010). Scientists who choose to take part in public engagement can do so in many ways, such as providing lectures, giving interviews to intermediary parties such as journalists, and writing popular science books (Bauer and Jensen 2011). The Internet has expanded these options, but to be successful, public engagement has to go beyond translating facts. It has to communicate science and its policy implications in compelling ways that speak to an audience's values, interests, and worldviews (Nisbet 2010).

Framing is a cognitive approach that packages complex information into a specific context geared for a receptive group (Bovens and Hart 1996). Employed as a communication strategy, framing selectively presents information for better reception, for example through a specific perspective, arrangement, context, or style. As with the frame around a painting, framing focuses a presentation, removes distractions, lays out boundaries, and creates new alignments for attention. A frame can communicate the essence of a message so that the audience immediately understands the reason why they care about an issue, who are the main players, and what is at stake (McCombs et al. 1997, cited in Dirikx and Gelders 2010).

Frames can have different functions: defining the problem, diagnosing causes, making moral judgments, and suggesting remedies (Entman 1993). Social scientists have assessed the efficacy of this communication technique for conservation issues. For example, Maibach et al. (2010) showed the effectiveness of engaging the public in the topic of climate change through health issues. They found that audiences responded better to a positive message through the health benefit associated with mitigation-related policy actions. In practice, this means that scientists can remain true to their underlying science while designing a message that will be personally relevant and meaningful to a diverse public (Nisbet 2009).

In communicating historical ecology to the general public, the "shifting baselines" concept is essentially one frame among many that can be used to present information about our changing oceans. Pauly (1995) first described this phenomenon in relation to fisheries scientists who had not taken into account changing generational perspectives, thereby grossly underestimating the nature and scale of change. This revelation sparked a new interdisciplinary endeavor, encouraging historians and scientists to take a longer view and to consider anecdotes about the past as valid types of ecological information. As a frame, the shifting baselines concept engages the public by shining a light on the role of generational amnesia and inviting members of the public to consider what may have changed in their own communities over recent and long-term time frames.

Put into practice, different framings of knowledge can drive public engagement in policy debate (Nisbet 2010). Framing has been employed to influence policy and regulatory changes for a long time. For example, in the nineteenth and early twentieth centuries, federal fishing regulations were routinely influenced by idealized public representations of fishing and fishermen (McKenzie 2012). At the time, artists and writers portrayed fishermen as embodying an American identity, and in the 1890s, a romanticized framing of traditional fishermen was used to successfully ban industrialized fishing from Maine state waters. However, 30 years later, the romanticized image was reframed as outmoded and fanciful in order to defeat Progressive Era conservation efforts (McKenzie 2012). In this case, the fishing community became the victims of framing employed by an opposing special interest group.

Marine historical ecological frames should not be confused with content or message. Frames are perspectives employed in settings where people of similar values and interests congregate. Working with frames requires awareness of what is inherently of interest to a niche audience outside one's own frame of reference. The marine historical ecologist interested in public engagement may actively seek out overlaps of interest and knowledge within specific constituencies. Used in this way, a framing strategy can appeal to individual interests with the goal of developing dialogue and connection. Framing is not the presentation of information; it is a strategy intended to find common language, in advance of any teaching, learning, or sharing of information. It is well suited to science learning, which, in academic situations, has been shown to be built upon preexisting affinities and affiliations within one's community, peers, and experience (National Research Council 2009, Groffman et al. 2010, Falk et al. 2012). Such learning is, in a word, personal.

Using Primary Sources to Connect, Communicate, and Illustrate the Changing Ocean

The historical ocean makes for great storytelling. Drawing out the human story is one way to capture and hold public attention. For example, archives overflow with images and texts depicting the historical fishing life of nineteenth-century New England (Figures 11.1 and 11.2). These primary sources can draw modern readers into a world that otherwise might not grab their attention. For example, curiosity, attention, and even emotion are evoked by the image of a young woman cleaning herring (Figure 11.3). Wearing a flowered hat, with layers of petticoats peeking out beneath her aproned skirt, she depicts a vivid moment from life at the same time that she illustrates historical fish processing in the 1800s. Similar primary sources include a free-form poem in a ship's log that describes cod fishing on the Maine coast in the month of June (Ash 1861) and a letter that describes the aftermath of a storm at sea (Altham 1624). In each case, nature becomes intertwined with the historical story. Human stories draw the public in and create an emotional and evocative link to people and places in the past.

FIGURE 11.1 Appeal to a sense of history: putting humanity into a landscape. A quiet scene on T-Wharf, Boston, one of the largest commercial wharves of its time, with fishing schooners tied up to the dock, sails furled, the warehouse surrounded by a forest of masts. On the pier, horses and wagons, long-shoremen and fishermen fill the road with activity, moving ice, fish, ship's stores, and supplies, preparing for the next avalanche of fish. (Source: National Archives and Records Administration, LICON RG 22-C. Records of the U.S. Fish and Wildlife Service. U.S. Commissioner of Fish and Fisheries. Cyanotypes: Commercial Fishing Activity in the United States, 1882–1891. 22-CF-22.)

Sometimes a primary source is graphic. Photographs alone can make a huge impact. In 2009, Loren McClenachan published an ecological analysis of 50 years of photographs taken on the same dock in Key West, documenting the decline in numbers of fish species and in the size of trophy fish (Figure 11.4, McClenachan 2009). The striking juxtaposition of photos over 50 years and the cautionary tales they told were picked up in popular publications and newspaper articles (Helmuth 2008, LaFee 2009). The series was chosen by *Wired* magazine as one of their favorite long-term datasets of 2011 (Keim 2011). The Smithsonian Institution has chosen to include these images in an update to the Sant Ocean Hall in the National Museum of Natural History (Box 11.1 and Figure 11.5).

While issues of environmental change are complex, the message itself can be simple, and humor can be a powerful communication tool to foster connections. Who can deny the effectiveness of Dan Piraro's cartoon "Find the Hidden History Lesson" (see Figure 12.3 in chapter 12, this volume)? His depiction of shrinking fish size vividly communicates the concept of shifted baselines. The depressing state of our ocean is clear, and it is made

FIGURE 11.2 Appeal based on aesthetics and the individual story: horses and carts moved products from place to place. Here, in an image of men transporting fish along the streets of Gloucester, onlookers are caught in time. Although industrial ice was transforming fish markets by 1900, nothing beat the ease of dried cod. (Source: National Archives and Records Administration, LICON RG 22-C. Records of the U.S. Fish and Wildlife Service. U.S. Commissioner of Fish and Fisheries. Cyanotypes: Commercial Fishing Activity in the United States, 1882–1891. 22-CF-275.)

compelling to a wide audience by the means used to communicate the message. This darkly humorous cartoon creates instant identification with fishermen, and secondarily with sympathy toward the fish, depicting an inverse relationship between the size of fish and the size of waistlines.

To engage nonscientific audiences, researchers can make use of stories and anecdotes that humanize the numbers and put them in context (Leslie et al. 2013). Stories of communities, cultures, and places are examples of powerful framing devices. Science asks us to put aside the emotive and subjective, but framing this material for the public allows us to circle back and use strong individual anecdotes to establish empathy and perspective on ocean changes and to create compelling cases for action (Olson 2009). As much as a sense of urgency, emotional identification fosters a personal connection to the issue.

Historical documents and maps are also tools that can be used to graphically connect a viewer to marine historical ecology, revealing early conceptions of the ocean, becoming sources of new data, and even changing management perspectives. Working from ships' logs from Beverly, Massachusetts, researchers from the University of New Hampshire's Gulf

FIGURE 11.3 Illuminating the personal creates a connection with history. Women played important roles in New England fisheries. In Maine, where herring was the principal catch around 1900, canneries often employed entire families. Men performed heavier labor. Women cleaned fish. Children ran errands and did small jobs to keep busy. Here, women in fashionable hats process small herring to be canned and sold as sardines. (Source: National Archives and Records Administration, LICON RG 22-C. Records of the U.S. Fish and Wildlife Service. U.S. Commissioner of Fish and Fisheries. Cyanotypes; Commercial Fishing Activity in the United States, 1882–1891. 22-CF-379.)

FIGURE 11.4 Fifty years of decline and change in sportfishing. Three photos convey the story of change. Trophy fish caught on Key West charter boats in (A) 1957, (B) 1980s, and (C) 2007. (Source: Monroe County Public Library and Loren McClenachan.)

BOX 11.1 Viewpoint from a Practitioner: Seeing Is Believing, Sharing Is a Must!

Jill Johnson

To a young child, standing next to a 7-foot fish is a pretty cool experience. To the parents, it is a great photo opportunity. To the grandparents, a memory emerges of a fishing trip to the tropics many decades ago. To the teenager, look out social media—this shot is going to be Tweeted and Instagrammed and Cinemagrammed and maybe even posted on Facebook (although that is sooooo passé and my mom is using that for crying out loud!). To the teacher, the barcode is going to take them to a wealth of information on the exhibition's website focusing on historical marine ecology or "shifting baselines." They will find curriculum already packaged, video, photographs, big content pages full of resources—all by scanning this grocery-looking code with their smart phone for further exploration at a later date that they can share with their tech-savvy middle school kids.

The new "shifting baselines" exhibit at the Smithsonian Institution's Sant Ocean Hall has engaged a variety of visitors of all ages, from all walks of life, from Nebraska to Hong Kong, with a piece of history and with concepts that are difficult to convey verbally: changes to size, abundance, and diversity of marine species and ecosystems. But visually, these concepts just click; visitors can compare the past to the present conditions and do so as part of a fun, social experience. In the design of the exhibit, we included a photo of a really big fish found today in a Marine Protected Area and a video featuring a family that really cared about what was happening to the big fish in their local waters. So the good news being taken away from the experience as a whole is that ocean stewardship matters and personal engagement makes a difference in preserving and restoring marine ecosystems of the past.

Jill Johnson is Exhibit Developer, Sant Ocean Hall, National Museum of Natural History, Smithsonian Institution.

FIGURE 11.4 *(continued)*

of Maine Cod Project were able to describe dramatic ocean changes, finding that the cod biomass on the Scotian Shelf has been reduced by two orders of magnitude since 1852 (Rosenberg et al. 2005). Inspired by such work, the National Oceanic and Atmospheric Administration's (NOAA) Office of the National Marine Sanctuaries recognized the benefit of letting primary source documents tell the history of the ocean "through the eyes of fishermen." This frame, built around the personal stories and documents of early fishermen, helped stimulate support from the Sanctuaries office for new research (see Box 11.2). Up until that time, NOAA's Office of National Marine Sanctuaries had separated the human history of the oceans from the scientific research and management of natural resources.

FIGURE 11.5 Historical ecology exhibit in the Smithsonian Institution's Sant Ocean Hall (photo courtesy of Jim DiLoreto and Smithsonian Institution).

Historical documents in the form of seventeenth-century maps also illustrate changing marine perspectives (Figure 11.6). On his original 1616 map of New England, explorer John Smith placed a marker, a large ship, at the entrance of Massachusetts Bay near Cape Cod (Smith 1616). A later revision of this map shows a large school of fish below the prow of the ship (Barbour 1986). Over the next 100 years, cartographers embellished the original Smith map, essentially advertising to Europeans the marine bounty of Gulf of Maine waters. Using these maps and their compelling research findings, the Gulf of Maine Cod Project researchers were able to make a compelling case for the Office of the National Marine Sanctuaries to invest in historical ecology. Because of this and other research, a historical perspective has now been incorporated in the management of the Stellwagen Bank National Marine Sanctuary

Using Primary Sources to Foster Cultural Identification and Connection to Place

Data derived from primary source documents testify to the thoughts, feelings, and day-to-day concerns of the past. Diaries, records of sales, species weights, charts recording expeditions, and ships' logs recording weather, catches, and daily life at sea all connect us to people and communities of the past. Images and narratives make the oceans' past personal

BOX 11.2   Viewpoint from a Practitioner: Contribution of Marine Historical Ecology
to Marine Protected Areas

Daniel J. Basta

The concept of "shifting baselines" is one of the most important recent ideas to be introduced into the management of natural resources—and, in fact, the consciousness of society. It explains the tendency in each of us to not only have different impressions of what "the good old days" were actually like, but to ignore (or not to understand) their relationship to today—and the relationship of today to tomorrow. Few of us know enough about the natural history of the places we care for and our relationship to that history. Marine historical ecology uses the geography of time and place to help us understand how the history of people, natural resources, and the ocean are intertwined.

The U.S. National Marine Sanctuaries were conceived to protect special biological, ecological, and cultural or historic places in our nation's ocean. But protecting them requires that we use them to help understand the larger relationships and trends that ultimately affect all places. These special places, National Marine Sanctuaries, provide a lens through which to understand how the oceans, atmosphere, living creatures, and human societies shaped each other through time. Historical ecology provides a structure that enables us to put a social context to changes occurring in the ocean and, through a focus on discrete places, allows us to realize and understand our own individual actions that affect the ocean and, ultimately, our lives.

Since 1972, when the National Marine Sanctuaries Act was enacted, 13 National Marine Sanctuaries and one National Marine Monument have been designated and managed by NOAA's Office of National Marine Sanctuaries. The system encompasses 170,000 square miles of ocean, from the Atlantic to the Pacific, and also includes some of the most remote island groups in the world. Marine historical ecology allows us to understand the full range of factors, acting over time, that have conditioned these 14 special places and made them as we know them today, and that are likely to determine their fate in the future.

Through the lens of marine historical ecology, we have a new way to engage the public, reframe the conversation, and connect with a broad range of audiences across the country. In many ways, this approach makes resource management, conservation, and protection a personal matter. It creates conversations with people who may never visit a marine sanctuary yet may connect with them through common cultures, ethnicities, or other aspects of shared history.

Consequently, our experience is that the application of the historical ecology perspective may be an important breakthrough, if it can find widespread acceptance within the larger natural-resource-management community.

Daniel J. Basta is Director of the Office of National Marine Sanctuaries, National Ocean Service, National Oceanic and Atmospheric Administration.

(Figure 11.7). For example, although the image portrayed in Figure 11.8 of Portuguese fishermen outside an outfitter's store in Gloucester contains no information about ocean ecology, it creates an affinity with Portuguese-Americans curious about their forebears. Portuguese immigrants in the 1800s were mackerel fishermen and whalemen. Historical connections to the environment are more likely to resonate with a community when they appeal through cultural identification.

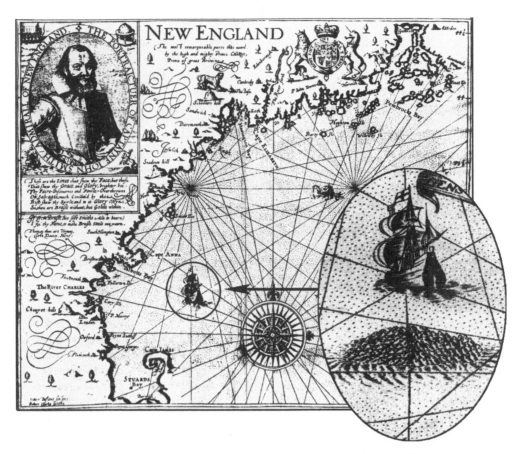

FIGURE 11.6 Early advertising for America's marine bounty: John Smith's map of New England, 1616. This map reflects Europe's increasing interest in New World fishery resources. In the early 1600s, explorer John Smith speculated that the cod in the Gulf of Maine waters would eventually bring England more wealth than all the gold in New Spain. Here, a primary source image becomes a framing device connecting marine protected areas with explorers in early American history (Smith 1616, Barbour 1986). (Source: Osher Map Library, University of Southern Maine, barcode no. 12548.0001. Image assembled by Karen Alexander.)

Historical sources and stories can create an affinity with place. On the coast of California in the 1800s, one of the few activities in which Chinese immigrants could make a good living was fishing. Chinese fishermen settled in Monterey Bay in the mid-1800s (Figure 11.9), developing a major commercial fishery in the waters of what is today's Monterey Bay National Marine Sanctuary. Dried fish and abalone from China Point were sold as far away as Asia. Although the community prospered, it was slowly pushed out by new European immigrants who took over most fisheries by 1900 (McEvoy 1986, Palumbi and Sotka 2010). China Point is no longer a Chinese fishing community, but marine historical ecology can tell the story of generational and cultural change in this place.

FIGURE 11.7 The human side of fishing: a snapshot of a past way of life. Men unload dried cod on a Gloucester wharf. While Boston specialized early in fresh and frozen fish, Gloucester stuck longer with salt cod. Dried fish lost most of their volume. If fresh, the stacks on the wharf and in the wheelbarrow would have towered above the fishermen's heads. (Source: National Archives and Records Administration, LICON. RG 22-C. Records of the U.S. Fish and Wildlife Service. U.S. Commissioner of Fish and Fisheries. Cyanotypes: Commercial Fishing Activity in the United States, 1882–1891. 22CF-342.)

To understand coastal, inland, native, and immigrant communities, and the many evolutions of their connection to the sea, is to understand a sense of place. One does not need to be a diver or boater to participate on a community level. Dialogue with local communities and stakeholders can take place around issues of culture as well as environment and resources. Stories of communities, cultures, and places will offer great opportunities as framing devices to communicate historical ecology to the general public.

HOW TO CONVEY THE MESSAGE: CREATING LOYALTIES, FOSTERING ACTION

Today's media landscape straddles both old and new technologies. Media change as technology and practices change. Marine historical ecologists may enjoy partnering outside their

FIGURE 11.8 Appeal through cultural identification: Portuguese fishermen in front of an outfitter's store in Gloucester, Massachusetts, 1882. Portuguese immigrants were active in New England fin-fishing as early as the 1830s and began whaling in the 1880s. Portuguese fishing families brought strong community loyalties and distinctive fishing traditions, reflected in their cultural heritage today. (Source: National Archives and Records Administration, LICON. RG 22-C. Records of the U.S. Fish and Wildlife Service. U.S. Commissioner of Fish and Fisheries. Cyanotypes: Commercial Fishing Activity in the United States, 1882–1891. 22CF-259.)

own communities and learning new methods from commercial and marketing analysis, artists, bloggers, and impassioned advocates. Images and videos are fast rivaling text as the primary content in digital stories set in specific places and times. The nature of browsing, linking, and content curation encourages digital users to choose their own paths of knowledge and information. Through these mixed modes of information sharing, the public will experience marine historical ecology through both old and new media.

Defining Old and New Media

"Old media" are often defined as linear forms of communication that existed before the advent of the Internet. Linear media involve a style of consumption routinely controlled by the maker. A viewer reads or watches from start to finish and is not encouraged to deviate, stop, edit, or rearrange the sequence. Broadcast and cable television, radio, movies, newspapers, magazines, and books are generally long form, concentrated on a single thesis or through line. The tone is often authoritative, conveying generalized experiences or representing group perspectives as opposed to personal opinion. Old media might be a book, a

FIGURE 11.9 Appeal through a connection to place: Chinese fishermen settled in Monterey Bay in the mid-1800s and developed the first commercial fishery. Dried fish and abalone were sold as far away as China. Here, warehouses and fishermen's dwellings alike perched on stilts along the shore. (Source: National Archives and Records Administration, LICON. 22-FF. Series FFA. Records of the Fish and Wildlife Service. Cyanotypes: Cruises of the Albatross, Fish Hatcheries, and Marine Specimens, 1879–1922. 22-FFA.)

feature-length film seen in a public theater, or a museum exhibit. Most often, the viewer is intended to be a passive consumer.

"New media" is a broad term applied to both content and delivery mechanisms, including a range of ever evolving digital devices and mobile platforms. According to Blodget and Cocotas (2012), about 33% of the world's population, or 2 billion people, used the Internet in 2012. As of August 2012, 85% of all American adults, both male and female, used the Internet, and 84% of those routinely looked for information on a hobby or interest (Pew Research Center 2012). Nearly four in five active Internet users visit social media networks and blogs (Nielsen/McKinsey Company–NM Incite 2011). Media platforms, including hundreds of paid, free, subscription-based, or advertising-based portals, rival print as publishers of text. Image and video are joining the public conversation. Video consumption on mobile devices almost doubled in 2012 (Blodget and Cocotas 2012). More and more users post images with a minimum of attached text. By the end of 2012, about 40 million users per week engaged in picture-based social media—a strong trend toward personal engagement and participation via images (Anonymous 2012, Rodriguez 2012).

Much of the new media world is social: exchanging comments, linking to people and places, and discussing content in various forms. In this setting, the originator loses absolute control of how materials are consumed, because content can be taken out of context, manipulated, and transformed. Images, for example, may be improperly cited or used incorrectly. On the other hand, new media's great strength is in the connection. In return for potential pitfalls, new audiences are exposed to new topics. New media publishers can be strategic about how materials are framed and where, when, and how materials appear. Engaging the public in historical marine ecology requires engaging the public through these expanding new media formats. Content will be consumed on a constant, continuous basis in many disparate but loyal community dialogues. To some extent, publishers in this realm relinquish control of narrative and metadata. In exchange, they gain the broad reach of a long tail strategy.

Old and new media communicate to different audiences, in different ways. Marine historical ecology can cross all these spectrums, but no one medium is superior to another. Each has its strengths and can lead to the other. The emotion and perspectives of new media uphold the scholarship and authority of old media, working hand-in-hand. Old media connect to the already committed, appealing to large groups, such as the membership of a conservation organization or academic discipline. Through social dialogue and shared content, new media reach out to new audiences, including casual readers and friends of friends. New media's emphasis on relationship, emotion, image, and story offers many platforms for information about the changing state of the ocean.

Three Key Tools in New Media

Three major aspects of new media are emerging as tools for public engagement: search, social participation, and design for mobile consumption. These tools can be capitalized upon by marine historical ecologists seeking public engagement.

The Search Power of New Media

The search power of new media is amplified by linking, discoverability, and navigation. Links point to other sources of related information. This approach to publishing leads interested viewers to build their own linear experiences, which, in turn, may include more concentrated engagement in books, papers, and studies, which are all examples of old media. A tourist at Acadia National Park may start by searching the Internet for "Acadia," then click on a description of changing ecosystems for ocean life in the area, and move seamlessly to a scientific description or video about the devastation of eelgrass in Frenchman's Bay, ultimately ending up engaged in the Mount Desert Island Biological Laboratory's community eelgrass restoration efforts.

Linking connects fragmented pieces of information often found in disparate locations. Site optimization by the publisher increases the likelihood of content being found by users through searching—a functionality known as "discoverability." Navigation—the means by which users choose their path to media—affects both the design of the structural interface and the structure of the text.

A positive aspect of linking fragmented information is the ability for users to discover new information through keywords. Keywords make media discoverable, through tagging with category names or keywords. For example, an image that includes two African-American men captioned "Whalemen, New Bedford, 1882" has resided for the past hundred years in a U.S. Fish Commission photo album, seen by a scant hundred individuals over that time (Figure 11.10). A viewer would have had to have permission to enter the archive, and the persistence to plow through the archive to discover this cryptic image and caption. Once it has been digitized, recaptioned, and placed on the Internet as "Group including two African-American whalemen, New England, late 1800s," a far larger community of interested and diverse individuals can access this image. With keyword tagging for "African-American," "Seamen," "Whalemen," "New England," "whaling," "portraiture," "late 1800s," "New Bedford," people interested in the history of fishing, the history of African Americans, the history of Massachusetts, or the history of New England will discover this image through a browser search. The image can be linked back to its original collection context for more images and information, and links can lead to additional resources such as an intensive, specialized blog post by a passionate or professional curator, books by authors specializing on this topic, or websites featuring informative personal family narratives by descendants or peers.

Social Participation and Engagement in New Media

The social aspects of new media range from dialogue to crowdsourcing, crowdfunding, and live event integration, among many others. Digital media are increasingly used simultaneously, instantaneously, and interactively. Participation is the essence of the experience, through decisions to "like," "comment," or "share" content. Users can "click-through" or "link." They decide to watch all or part of videos and bookmark or forward them. Participants take possession of material as they repost with comments, reshaping content in novel ways. They appropriate, remix, or mash up images, videos, and comments by others. Audiences decide to become subscribers or return visitors and contributors. In all these processes, viewers form and demonstrate differing levels of loyalty, and they learn.

Users may also increase their commitment as they engage. Each time a viewer clicks a link, he or she is "participating." "Liking" the Papahānaumokuākea Marine National Monument Facebook page expands the number of people who will see comments and notices. Each "liker" knows that he or she is one of a group that is aware of the 140,000 square miles of marine monument in the Northwestern Hawaiian Islands. The simple act of "liking" inaugurates membership in both virtual and geographic communities.

Each time a viewer clicks, he or she is expecting a return, which the creator must plan for and deliver. The expectation may be for further information, sources, opinions, GPS-based content, archival footage, a larger collection of photos (held on personal sites, published by

FIGURE 11.10 Finding African-American fishermen. In the era of Jim Crow, this 1882 professional portrait includes two African-American whalemen in New Bedford, Massachusetts, working in a trade where color mattered little if one pulled one's weight. Free African Americans, escaped slaves, Africans, and Cape Verdeans worked in the maritime industries before the Civil War, particularly on whaling vessels. (Source: National Archives and Records Administration, LICON. RG 22-FF. Series FFC. Records of the Fish and Wildlife Service. Cyanotype: Cruises of the Albatross, Fish Hatcheries, and Marine Specimens, 1879–1922. 22-FFC.)

image services such as Flickr or in the archives of institutions) or linking to a book, blogs, historical societies, a historical museum, an archive, or an academic department. Through these mechanisms, the public becomes involved in active listening in a new online dimension.

Taken another step, user participation involves the solicitation of ideas, content, support, and action from the online community, known as "crowdsourcing." In this model, as applied to marine historical ecology, the viewer is not only an active listener, but he or she may effectively collaborate with scientists and historians by sharing his or her own heritage, family lore, and "history in the attic." In contributing directly to research and discovery, personal history becomes a public activity reaffirming connections to the sea. Or it can take the form of "citizen science" as in the case of the Zooniverse project, in which users are invited to contribute their time to a range of scientific and humanities projects, including the identification of galaxies, the search for planets near stars, and transcribing historical weather data from historical ships' logs (www.zooniverse.org). iNaturalist provides a similar mechanism for citizen science and public engagement (www.inaturalist.org).

Forms of user participation in new media are still being invented as the lines between scholarship and commentary, entertainment, and advertising blur. Utilizing the multiple approaches of framing, for which new media are so well suited, historians and scientists can see storylines from their scholarship moving out into the general public. Marine historical ecology on the net will be transformed into nonlinear, highly visual, fragmented, and reframed forms of content but will nevertheless find new life and reach other audiences.

Design for Mobile Media

Communicating digitally with the public requires shorter and more visual statements; thus, content design features prominently in new media. Structure, arrangement, and navigation become important aspects of the learning experience, providing tools by which scientists can reach the public. Consumers of new media are more dependent on subheads, often skimming long sections of text. One common format is the "listicle," a list of bulleted point items. Listicles are easily scanned and therefore are more likely to be read. Impressions—and decisions to leave a site—are often made within as little as 10 seconds (Liu et al. 2010). Clear, visual choices for navigation are critical. Effective sites plan for fragmented use, employing communication points that are short, clear, and upfront. Brevity helps a user commit to dig deeper or more widely. Simple menus and graphic-laden design that can be easily scanned increase a viewer's sustained attention. Design of image, video, and content support rich engagement, dialogue, and, ultimately, the viewer's receptivity to calls to action.

Online and interactive experiences on location—on a phone or tablet—are referred to as "mobile." By physical definition these differ from desktop engagement, with often small images or blocks of text, constrained vertically, horizontally, or in depth. Consumers look up a fact on Google on their phone, or consult a Google or Apple map for the "bigger picture" of a geographic area, when standing on a beach or in the middle of a meeting. As the new media environment becomes predominantly mobile, the design and content requirements will continue to create new forms of engagement. Mobile media platforms have contributed to, and perhaps accelerated, the fragmentation of information. Geospecific materials from historical ecology are particularly suited to these moments of live consultation and can be discovered through linking via the integration of GPS coordinates. All these functionalities increase options for personalization and connection.

FIGURE 11.11 A mobile app goes viral: Whale Alert links near real-time data from acoustic buoys listening for right whale calls to a tablet or smartphone. Maps charting whale activity outside Boston Harbor are now available to anyone in the world on a mobile phone or tablet. (Source: NOAA.)

ENGAGING MARINE HISTORICAL ECOLOGY IN MOBILE MEDIA

Marine historical ecology does not currently have a large presence in new media. However, the following are two successful examples of mobile applications for marine conservation issues that are successful models for engaging the public. Both deliver information and functionality, are highly graphic, and have implicit or explicit calls to action.

Whale Alert

Conservationists are just beginning to explore the use of mobile devices in the protection of marine species and areas. NOAA worked with digital charting company Earth NC to develop

FIGURE 11.11 *(continued)*

Whale Alert, a mobile application based on NOAA's data on right whale traffic in the Boston Harbor. The application was intended to reduce right whale strikes by providing a display on a ship's bridge showing the whale's presence to captains transiting the shipping lanes in and around Stellwagen Bank National Marine Sanctuary. While the original intent was to reduce potential animal–ship collisions, the free Whale Alert mobile app was downloaded by 15,000 users within 1 week of launch, demonstrating reach to audiences not even conceived by developers (Figure 11.11). The underlying science supporting the application is complex, integrating constantly updated data (acoustics, charts, whale-movement tracking data), but the interface is simple to use and highly visual.

Seafood Watch

As described above, new media's success is often built on exchange, a return to the user for their loyalty. The Monterey Bay Aquarium's Seafood Watch mobile application is a successful example of loyalty converted to action. The user can access information on seafood meal alternatives immediately—standing in line at a supermarket or while ordering from a restaurant menu. At the same time that the user is given information on food choices, he or she is presented with information about other fish, such as the sustainability of other fisheries, illustrations of the species, common names, locations, means of cultivation, and whether the species is wild caught. Neither history nor ecology per se, the application is a participatory tool for consumers that simultaneously involves users in the diversity of marine life and in issues of sustainability (Figure 11.12).

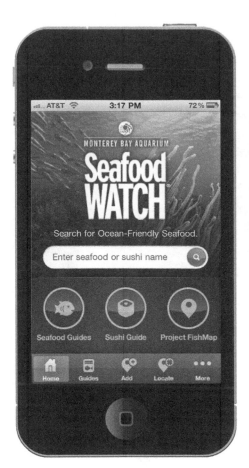

FIGURE 11.12 New media engagement leading to user action: the mobile application "Seafood Watch: Choices for Healthy Oceans" is a tool for finding alternative food choices in a simple but graphic context. Originally a printed wallet card, the mobile application exemplifies user participation and conversion of information transmission to new media. (Source: Monterey Bay Aquarium/Seafood Watch Program.)

In new media's social model, the viewers have varying opportunities for contribution. They can leave their name, quote or recommend a resource, and comment in their own space or in the blogs, pages, or streams of others. Actions lead to a sense of belonging with people, known only digitally, with similar interests or styles. Our two examples above show that large numbers of people who would not otherwise know or communicate with each other can be united in support of marine conservation. New media develop new audiences through the accretion of fledgling loyalties. Those loyalties will have to be converted to sponsorship, volunteer action, event attendance, lifestyle changes, and advocacy, but connection is the first step.

CONCLUSIONS: ENGAGING THE INTEREST OF THE GENERAL PUBLIC

As an interdisciplinary research field, marine historical ecology offers a complex and diverse array of perspectives on the ocean. Pauly (1995) opened our eyes to the generational amnesia of shifting baselines. Scientists have since raised the alarm on the current state of the oceans, in part using analyses of historical records. Historians have reconceptualized the ocean as a historical place populated by people as well as creatures. For example, Matthew McKenzie (2012) warned of how the disconnect between image and reality has influenced the management of marine resources in the past (see Box 11.3). All this work has taken place in old rather than new media.

Communicating with the general public requires new strategies. Communication scholarship shows that for the general public, science learning takes place most successfully in highly individualized or personal environments, building on preexisting sympathies and communities. The concept of framing suggests that marine historical ecology is best communicated one-on-one, over time. The stylistic, emotional, and narrative appeals of the first-person story, the localized place, the familiar, and the connected are opportunities to reach many disparate groups of relatively small numbers (the long tail strategy). However, the strategic placement of pieces of the story in discrete segments, images, and multiple settings necessitates relinquishing control of the story as a whole. The integrity of scientific and historical results can be maintained in a new media environment even as users transform content to choose what best speaks to them.

Marine historical ecology presents many reasons for taking advantage of new media's discoverability, social participation, and design for a mobile public. In sum, these experiences create context and learning among new audiences. The promise of new media for historical ecology is this long tail accumulation and the opportunity to build new constituencies.

Capturing public attention is the first step in public engagement. By strategically emphasizing human stories, managers, scientists, and historians have many opportunities to enhance awareness of the ocean world and create emotional connections with the public. These new loyalties may ultimately lead to a better informed and participating public, which is, after all, the bedrock for any management action.

ACKNOWLEDGMENTS

We could not have done this work without the kind, competent, and generous assistance of Anne D. Marzin, Karan Sheldon, and William B. Leavenworth.

REFERENCES

Altham, E. (1624) [Letter to James Sherley.] In *Three Visitors to Early Plymouth* (S. V. James, Ed.). Plymouth Plantation, Plymouth, MA, 1963. pp. 42–47.

Anderson, C. (2004) The long tail. *Wired.* www.wired.com/wired/archive/12.10/tail.html.

Anonymous (2012) Instagram and Pinterest the new global stars of social. http://bundlr.com /clips/50ac194c598b4a0002000281.

Ash, B. (1861) Log of the Frenchman's Bay schooner GLIDE. National Archives and Records Administration, Waltham, ME, RG 36: Box 5, E-104.

Barbour, P. L. (Ed.) (1986) *The Complete Works of Captain John Smith (1580–1631).* Published for the Institute of Early American History and Culture, Williamsburg, VA, by the University of North Carolina Press, Chapel Hill, NC.

Bauer, M. W., and Jensen, P. (2011) *The mobilization of scientists for public engagement. Public Understanding of Science* 20, 3–11.

Blodget, H., and Cocotas, A. (2012) The state of the Internet [slide deck]. Business Insider Australia. www.businessinsider.com.au/state-of-internet-slides-2012–10#-1.

Bolster, W. J. (2012) *The Mortal Sea: Fishing the Atlantic in the Age of Sail.* Harvard University Press, Cambridge, MA.

Bovens, M. A. P., and Hart, P. (1996) *Understanding Policy Fiascoes.* Transaction, New Brunswick, NJ.

Clover, C. (2004) *The End of the Line: How Overfishing Is Changing the World and What We Eat.* Ebury, London, UK.

Dirikx, A., and Gelders, D. (2010) To frame is to explain: a deductive frame-analysis of Dutch and French climate change coverage during the annual UN Conferences of the Parties. *Public Understanding of Science* 19, 732–742.

Entman, R. M. (1993) Framing: toward clarification of a fractured paradigm. *Journal of Communication* 43, 51–58.

Falk, J. H., Randol, S., and Dierking, L. D. (2012) Mapping the informal science education landscape: an exploratory study. *Public Understanding of Science* 21, 865–874.

Groffman, P. M., Stylinski, C., Nisbet, M. C., et al. (2010) Restarting the conversation: challenges at the interface between ecology and society. *Frontiers in Ecology and the Environment* 8, 284–291.

Helmuth, L. (2008) Seeing is believing: photographs and other historical records testify to the former abundance of the sea. *Smithsonian* (September).

Junger, S. (1997) *The Perfect Storm: A True Story of Men against the Sea.* Thorndike Press, Thorndike, ME.

Keim, B. (2011) Transcending time: great long-term datasets. *Wired,* October 17.

Kurlansky, M. (1997) *Cod: A Biography of the Fish that Changed the World.* Walker, New York, NY.

Kurlansky, M. (2006) *The Big Oyster: History on the Half Shell.* Ballantine Books, New York, NY.

LaFee, S. (2009) Fish story. *San Diego Union Tribune,* May 16.

Leslie, H. M., Goldman, E., McLeod, K. L., et al. (2013) How good science and stories can go hand-in-hand. *Conservation Biology* 27, 1126–1129.

Liu, C., White, R. W., and Dumais, S. (2010) Understanding web browsing behaviors through Weibull analysis of dwell time. In *SIGIR '10: Proceedings of the 33rd International ACM SIGIR conference on Research and Development in Information Retrieval.* Association for Computing Machinery, Geneva, Switzerland. pp. 379–386.

Maibach, E. W., Nisbet, M., Baldwin, P., et al. (2010) Reframing climate change as a public health issue: an exploratory study of public reactions. *BMC Public Health* 10, 299.

Maibach, E. W., Roser-Renouf, C., and Leiserowitz, A. (2008) Communication and marketing as climate change-intervention assets a public health perspective. *American Journal of Preventive Medicine* 35, 488–500.

McClenachan, L. (2009) Documenting loss of large trophy fish from the Florida Keys with historical photographs. *Conservation Biology* 23, 636–643.

McCombs, M. E., Shaw, D. L., and Weaver, D. (1997) *Communication and Democracy: Exploring the Intellectual Frontiers in Agenda-Setting Theory.* Lawrence Erlbaum, Mahwah, NJ.

McEvoy, A. F. (1986) *The Fisherman's Problem: Ecology and Law in the California Fisheries, 1850–1980.* Cambridge University Press, Cambridge, UK.

McKenzie, M. (2012) Iconic fishermen and the fates of New England fisheries regulations, 1883–1912. *Environmental History* 17, 3–28.

Mowat, F. (1984) *Sea of Slaughter.* Atlantic Press, Boston, MA.

National Research Council (2009) *Learning Science in Informal Environments: People, Places, and Pursuits.* National Academics Press, Washington, DC.

Nielsen/McKinsey Company–NM Incite (2011) State of the media: the social media report—Q3 2011. www.nielsen.com/us/en/reports/2011/social-media-report-q3.html.

Nisbet, M. C. (2009) Communicating climate change why frames matter for public engagement. *Environment: Science and Policy for Sustainable Development* 51, 12–23.

Nisbet, M. C. (2010) Framing science: a new paradigm in public engagement. In *Communicating Science: New Agendas in Communication* (L. Kahlor and P. Stout, Eds.). Routledge, New York, NY. pp. 40–67.

Olson, R. (2009) *Don't Be Such a Scientist: Talking Substance in an Age of Style.* Island Press, Washington, DC.

Palumbi, S. R., and Sotka, C. (2010) *The Death and Life of Monterey Bay: A Story of Revival.* Island Press, Washington, DC.

Pauly, D. (1995) Anecdotes and the shifting baseline syndrome of fisheries. *Trends in Ecology & Evolution* 10, 430–430.

Pew Research Center (2012) Demographics of Internet users, August 2012. Pew Research Center, Internet and American Life Project.

Roberts, C. (2007) *The Unnatural History of the Sea*. Island Press, Washington, DC.

Rodriguez, S. (2012) Pinterest, Instagram continue meteoric growth. Los Angeles Times Online. http://articles.latimes.com/2012/aug/29/business/la-fi-tn-pinterest-instagram-growth-20120829.

Rosenberg, A. A., Bolster, W. J., Alexander, K. E., et al. (2005) The history of ocean resources: modeling cod biomass using historical records. *Frontiers in Ecology and the Environment* 3, 84–90.

Smith, J. (1616) A description of New England. Printed by H. Lownes for R. Clerke, London, UK.

TWELVE

Choice without Memory

Uncovering the Narrative Potential of Historical Ecology

J. B. MACKINNON

Marine historical ecologists frequently cite the need for better storytelling to make their research meaningful to society at large. A review of literature from this discipline points to a promising area of narrative focus. The view of nature as a relatively fixed commodity predominates in public discourse. A historical perspective, however, suggests that the natural world that surrounds us can be seen, in large part, as a product of choice, with the tragic decline of the marine environment the result of choices made without the benefit of social memory. This understanding places a high value on historical context and offers a forward-looking alternative to familiar narratives of past and present guilt and greed. Most importantly, it affirms that the debate over marine conservation is fundamentally about individual and collective values, in which everyone has a stake.

In July 2010, a gray whale (*Eschrichtius robustus*, Eschrichtiidae) entered False Creek, a narrow ocean inlet that reaches into the urban heart of Vancouver, British Columbia. The whale's appearance among the glassy high-rises attracted and thrilled people from every walk of life—in one online video, for example, construction workers still in their helmets, boots, and tool belts can be seen crossing a bridge at a full run to get a better look at the passing cetacean.

Most Vancouver residents and media outlets interpreted the whale's appearance as a "once-in-a-lifetime" experience (Figure 12.1). Only a small minority of people are aware that gray whales, humpback whales (*Megaptera novaeangliae*, Balaenopteridae), and minke whales (*Balaenoptera acutorostrata*, Balaenopteridae) were once an ordinary presence in the

FIGURE 12.1 May 2010: A "once-in-a-lifetime" gray whale visit to Vancouver, Canada (courtesy of Gail Edwin-Fielding, http://gailatlarge.com).

area (Winship 1998). Even a century ago, whale populations in the waters around Vancouver had been reduced by whaling (Nichol et al. 2002; Figure 12.2). Still, at least 200 and as many as 600 humpback whales were then present as a resident population (Merilees 1985). Although the number of minke and gray whales that passed through each year is unknown, genetic research now indicates that ≤96,000 gray whales—three times as many as today— may once have migrated up the Pacific coast of North America (Alter et al. 2007). To those who are aware of this history, the whale that visited Vancouver in 2010 represented a rare and perhaps bittersweet opportunity to witness living memory.

What is the potential contribution of historical ecology to the public conversation in such a situation? This question goes to the heart of a critical challenge within the discipline today. Marine historical ecologists frequently cite the need for better storytelling to make their research meaningful to society at large; it has become commonplace for such scientists to call for "a very different perception of nature" (Lotze 2009), for example, or "new mental pictures" (Lang 2009) and "new sea stories" (Bolster 2009). A "pending revolution" in conservation is seen to be at stake (Duarte et al. 2009).

Historical ecology indeed has a novel story to tell. Perhaps surprisingly, however, this new narrative does not center on past states of nature. Instead, it derives its importance from the timeless and universal concept of choice. Through the lens of historical ecology, the natural world that surrounds us can be seen largely as the product of choices, with the tragic decline of our environment the result of choices made without the benefit of memory.

FIGURE 12.2 Circa 1895: Whaling severely reduced Vancouver's resident humpbacks (image E-06690 courtesy of Royal BC Museum, BC Archive).

Marine historical ecology has proved to be a paradoxical science. On the one hand, it has laid the groundwork for a renewed conservation vision rooted in the extraordinary ecological wealth of the past (Jackson and Hobbs 2009); on the other, it has identified the greatest barrier to that vision, which is collective amnesia and its manifold effects (Saenz-Arroyo et al. 2005). In region after region, historical evidence paints a picture of past ecological abundance and diversity that astonishes the modern mind, from Caribbean seas teeming with ≤90 million green turtles (*Chelonia mydas*, Cheloniidae), to South Pacific reefs with at least 5× the fish biomass of typical reefs today, to North Sea waters where bluefin tuna (*Thunnus thynnus*, Scombridae), at best a rare visitor in the present day, turn up regularly in fishers' nets (McClenachan et al. 2006, MacKenzie and Myers 2007, DeMartini et al. 2008). Significant research effort, however, has also focused on fisheries scientist Daniel Pauly's idea of "shifting baseline syndrome," which posits that each generation perceives the natural world of its time as normal, and measures changes in the environment against that standard (Pauly 1995, Jackson et al. 2011). With each generation, the baseline for the normal state of nature is reset, with the result that accumulating injuries to the integrity of ecosystems go largely unnoticed over time.

Collective ecological amnesia has proved so pervasive as to apply even to itself. While Pauly's insight was undoubtedly his own, it was not sui generis. Almost exactly the same idea was expressed less than 10 years earlier by the ecologist Raymond Dasmann in the seminal book *Ends of the Earth: Perspectives on Modern Environmental History:*

But one adjusts to slow, deleterious changes in the environment and begins to accept them as normal. Young people, growing up in the smog, have no basis for believing that things were better in the past, and could be better in the future if certain actions were taken. The abnormal is accepted as normal and becomes the standard by which future change is measured. Even those who have lived under better conditions seem to forget what it was like to have clean air. . . . Environmental events that approach the level of catastrophes, at least on a local scale, tend to be discarded from active memory unless they occur with a high frequency. (Dasmann 1988)

We seem capable, in other words, even of forgetting what we have remembered we forgot. By the time of Pauly's insight, however, shifting baseline syndrome was clearly an idea whose time had come. By all accounts unaware of each other's efforts, between 1994 and 1995 the environmental theorist Raymond Rogers, the child psychologist Peter Kahn, Jr., and the fisheries scientist Pauly each published works recognizing collective ecological amnesia as a significant factor shaping the human relationship to the natural world (Rogers 1994, Kahn and Friedman 1995, Pauly 1995).

The fact that any specific cohort of people can be expected to lose sight of past natural conditions over a few generations, or even within the life spans of individuals, is now widely supported (Papworth et al. 2009). Research has identified shifting baseline syndrome in groups as varied as fishermen in the Gulf of Mexico, birdwatchers in Yorkshire, and bush-meat hunters in Equatorial Guinea (Sáenz-Arroyo et al. 2005, Papworth et al. 2009). The syndrome has been found to apply to children and to resource professionals, to those with little daily interaction with the natural environment and those whose daily livelihoods depend on it (Kahn and Friedman 1995, Pauly 1995, Sáenz-Arroyo et al. 2005). Indeed, there is some evidence of generational ecological amnesia among other species. For example, although gray whales in Laguna San Ignacio, Baja California Sur, Mexico, were considered wary and aggressive "devil fish" during the whaling era, their protected descendants appear to be increasingly curious and "friendly" in interactions with people (PBS 2012); on the other hand, wolves that no longer face the dire persecution of the past may be becoming bolder and more prone to consider humans as potential prey (McNay 2002).

One effect of shifting baseline syndrome has been to encourage the perspective that nature is a relatively fixed commodity. When people measure changes to their surroundings primarily against their own life experience, they necessarily perceive only a narrow band-width of the transformations that may have occurred over time. This is a reasonable explanation for the fact that many people express disinterest in the natural world; failing to recognize degradation that occurred in the distant past, they may see "nature" as generally unimpressive, uninteresting, not useful, and not visually arresting. Alternatively, such shortsightedness can discourage conservation efforts, especially in marine environments, which at the surface level may appear similar regardless of their ecological state. A poll in Seattle, Washington, for example, found that >70% of local residents considered Puget Sound—a water body that has been affected by human activities for millennia, and has suffered dramatic declines in marine mammals, diadromous fish, and sea birds—to be in good environmental health and not in need of restoration (Moore and Dehlin 2006).

Historical ecology, on the other hand, has shown that the spectrum of ecosystem states can be very broad indeed. In one compelling example, an attempt to determine gradients of difference between coral reefs at various degrees of remove from human influence found that the most remote reefs were so much more ecologically rich than their heavily affected counterparts that they could be said to operate according to different ecological principles (DeMartini et al. 2008, Wang et al. 2009). By moving public understanding of nature away from the narrow perspective of individual generations, the natural world of today can be seen as the end result of a series of actions that, because of a long-term pattern of social forgetting, were informed very little, if at all, by historical context. These actions were effectively choices, unconsciously made.

Nature as choice is a fundamentally new narrative, and one with ringing repercussions. Perhaps the most obvious way to dismiss historical ecology is as a nostalgic science, either stuck in the past or longing to make the impossible journey back though time. When the natural world that surrounds us is framed as the product of choice, however, historical ecology is revealed to be unexpectedly forward looking.

Consider again the story of the gray whale that visited the city of Vancouver. For the majority who lack historical ecological context, the event was an anomaly with no conservation implications whatsoever. Yet even historical awareness of the past presence of whales in the area is not enough; such knowledge may serve only to emphasize the ecological impoverishment of the present day. It is only when historical context is pressed into the service of decision making about the future that it fully acquires its narrative relevance and force. Citizens equipped with the necessary context, for example, are able to consider whether the regular companionship of whales might once again be encouraged through appropriate ecological conservation and restoration measures. Historical perspective broadens the boundaries of what is understood to be the normal, possible, or desirable state of nature; without it, the choice to live with or without whales in Vancouver effectively does not exist.

In engaging the public, it is crucial that this narrative of choice be consistent with findings both of past ecological wealth and of collective amnesia. Current conservation campaigns frequently promote an "us-and-them" division between "good" conservationists and "bad" resource users whose greed and lack of restraint has led and continues to lead to the degradation of the natural world. There can be little doubt that economic self-interest is a major driver of environmental destruction, frequently encouraging individuals and whole communities to ignore or downplay evidence that ecological harms will follow on particular human activities. Even the selective representation of nature in wildlife documentaries and other media can amount to deliberate ignorance, as discussed by Jim Toomey in his "Viewpoint from a Practitioner" (Box 12.1). Yet evidence in support of shifting baseline syndrome suggests that a widespread, long-term pattern of social forgetting may have played as great or greater a role in environmental decline than these more deliberate acts of denial. This alternative perspective is broadly inclusive in both a negative and a positive sense with

Jim Toomey

Apparently, there is no scientific basis to the parable of the boiling frog. As the story goes, a frog in a pot of water that is slowly heating up on a stove will not notice the gradual temperature change, and it will eventually allow itself to be boiled alive. Lab experiments prove the contrary: if the water gets hot enough, the frog will jump out. Evidently, frogs are more intelligent than humans. Or at least, they are more willing to acknowledge change when it's in their better interest. It's true—denial of change is human nature. One could even say that it's a form of optimism. Things are not getting worse; they are more-or-less the same as they were yesterday. I look in the mirror every morning and say something like that to mirror-me. Sure, I'm a day older, but I look the same. And if I extrapolate this trend, say, another 50 years, as I do every morning, I conclude that I'm one of the very few humans on the planet who has achieved immortality. Then I look at my wedding photos and I confront mortality again. As storytellers, one of our most important roles is to be a cultural memory. To be the wedding album. But it can be one of the most self-destructive things we do. My readers expect mild amusement when they read my comic strip, and on the occasion that I attempt anything that resembles a "message," I do so at my own peril. Filmmakers, particularly those who make nature documentaries, particularly those who make underwater nature documentaries, want to bring a world of wonder and beauty to their audiences. I'm an amateur underwater photographer myself, and when I visit a dive site, I look high and low for that perfect picture. Nobody wants to see a photo of a plastic bottle. So, as artists, filmmakers, and storytellers, many of us are guilty of propagating the myth that nature isn't changing. How to tell the story of environmental degradation, and still keep the attention of your audience, particularly in this era of limitless entertainment options, is one of the greatest challenges we face as storytellers.

Jim Toomey is a syndicated cartoonist, creator of *Sherman's Lagoon*.

regard to conservation. We have all been affected by environmental amnesia and can all be a part of what one commentator has called "a reminding" (D. Elverum, personal communication).

Historical ecology is not a panacea, and environmental choices that are informed by a deep-time perspective are by no means guaranteed to result in greater ecological integrity. In fact, one of the most challenging lessons of historical ecology from a conservation perspective is the fact that societies can survive and even thrive in a degraded state of nature in relation to the past, and in many places have done so for centuries if not millennia. In one of the most striking case studies to date, anthropologists Terry Hunt and Carl Lipo revisited the narrative of intertwined social and ecological collapse associated with Easter Island (Hunt and Lipo 2011), which has been repeatedly presented as a warning to the world at large about the potential consequences of overtaxing the available resources of a finite world (Bahn and Flenley 1992, Wright 2004, Diamond 2011). Presenting new evidence, the authors posit that social collapse did not occur on Easter Island, and instead the island's people ingeniously sustained a relatively stable population for ≥500 years. They did so, however, in an environment that rapidly degraded as a result of direct and indirect human impacts, ultimately losing nearly all of its forest cover and >20 species of tree and woody shrub, including giant palms (*Paschalococos disperta,* Arecaceae) that once numbered in the millions. At least six land birds, several seabirds, and an unknown number of other native species were lost to extinction, yet the human carrying capacity of the island may actually have increased through sophisticated advances in agriculture. If true, this new Easter Island narrative is an inspiring story of human resilience but offers little succor to the conservationist.

It is possible, then, that given a wider choice of states of nature, communities of people who live today in a depauperate natural state will choose to continue doing so. It is probable, too, that certain kinds of choice are more likely to be made than others. The absence of one or another species may be readily understandable to the general public, but many ecological changes are much harder to understand, as with the cascading transformations that can follow on the removal of predators and other strongly interactive species:

> There is little public awareness of impending biotic impoverishment because the drivers of collapse are the *absence* of essentially invisible processes . . . and because the ensuing transformations are slow and often subtle, involving gradual compositional changes that are beyond the powers of observation of most lay observers. (Terborgh and Estes 2010)

The tendency of the public to prefer certain species—often described as "charismatic megafauna"—over others is also likely to persist in a conservation narrative more influenced by historical ecology (Walpole and Leader-Williams 2002). The idea of returning whales to Vancouver's inshore waters, to continue with that example, may prove much more appealing than the restoration of the Olympia oyster (*Ostreola conchaphilia,* Ostreidae), another species that was extirpated from the area in historical times (Jacobsen 2009). On the other hand, any effort to bring gray whales back to Vancouver would likely involve ecological restoration of the benthic environment, which could result in improved habitat for Olympia oysters. It is likely, too, that a more historically informed debate will contain surprises. For example, research has shown that the snapper *Pagrus auratus* (Sparidae), the most common fish in northern New Zealand waters, has changed significantly not only in abundance, but also in behavior—another possible example of shifting baseline syndrome among nonhuman species. Now considered a "spooky" fish that avoids even divers, *P. auratus* was once frequently seen at or near the surface by observers in boats and even from shore, a phenomenon similar to that described by Fiorenza Micheli and Paolo Guidetti in their "Viewpoint from a Practitioner" (Box 12.2). It is speculated that the possibility of once again being able to see large fish such as snapper may prove to be "highly valued" by the public (Parsons et al. 2009); it will not be possible to say either way until the discussion is undertaken.

These and other potential obstacles should not be considered discouraging. Regardless of the complexities, historical ecology serves the overarching purpose of greatly broadening the spectrum of possibility when it comes to ecological conservation and restoration. Most

BOX 12.2   Viewpoint from Practitioners: What Does Ancient Art Tell Us about Past Marine Ecosystems?

Fiorenza Micheli and Paolo Guidetti

In several regions around the world, human interactions with marine ecosystems precede by hundreds or thousands of years any available quantitative information documenting the effects of such uses. Yet people witnessed these past ecosystem states and described them in graffiti, mural paintings, writings, and other works of art. Thus, ancient art can provide a link between prehistoric and modern evidence of marine ecosystem change.

Fishing has occurred for millennia along the coasts of the Mediterranean Sea. In this region, we used ancient works of art (Roman mosaics), combined with ancient writings, archaeozoological records, and monitoring of modern marine reserves, to examine shifts in sizes and habitat use of the dusky grouper (*Epinephelus marginatus*), a large, long-lived, slow-growing, protogynous (i.e., sex reversal from female to male) predatory fish that has been decimated in recent decades by commercial and recreational fishing and is now categorized as "Endangered" on the IUCN Red List.

In 10 of 23 Roman mosaics that represent groupers, dating from the first to fifth centuries CE, this fish is portrayed as being very large, approximately the maximum size reported from archaeozoological data and from marine reserves established for several decades. These mosaics also represent fishing scenes where groupers are caught using poles or harpoons from boats at the water's surface, a technique that would surely yield no grouper catch today, given that field surveys of grouper abundance and distribution within marine reserves show that these fish, particularly the larger individuals, are mostly seen below 10–15 m depth. These representations suggest that groupers were, in ancient times, so large as to be portrayed as "sea monsters" and that their habitat use and depth distribution have shifted in historical times; apparently this species lived in shallow waters, where it is now rare if not completely absent. This notion is also supported by written evidence from ancient Roman sources:

Ovid (in *Halieuticon Liber*) and Pliny the Elder (in *Historia Naturalis*) wrote that groupers were fished by anglers in shallow waters and that fish were so strong (i.e., big?) as to break fishing lines.

Ancient works of art, together with other qualitative and quantitative sources of information, thus suggest that shallow nearshore Mediterranean ecosystems have lost large, high-level predators along with their ecological roles. This loss is unaccounted for in current models of food-web dynamics and community responses to disturbance in rocky reefs. Reconstructing "pristine" patterns of habitat occupation of marine predators is crucial for understanding species' roles and food-web structure as they existed in the past, and for setting appropriate conservation and fisheries management goals. For example, reconstruction of species' maximum sizes, habitat use, or geographic distribution may help set targets for assessing recovery. In the case of the dusky grouper in the Mediterranean, the presence of large groupers at shallow depths could be used as an indicator of population recovery following regulation of fishing (e.g., within marine reserves). Currently, the performance of marine reserves in achieving their biological goals of recovering depleted species is generally assessed through comparison with reference, exploited ecosystems, or by examining population trajectories through time, but true targets are not available. Ancient art and writings provide such baselines, albeit only qualitatively.

Historical ecology approaches that utilize ancient art can also be effective for communicating conservation issues because of the appeal and curiosity that the distant past and ancient cultures evoke. Documentation of the extent of depletion of exploited populations based on works of art can thus complement other, more quantitative sources of information and perhaps play a unique role in reaching a broader audience to create awareness about the status of key marine species.

Fiorenza Micheli is Professor at the Hopkins Marine Station, Stanford University. Paolo Guidetti is Professor of Ecology at the University of Nice.

FIGURE 12.3 Find the hidden history lesson (*Bizarro* by Dan Piraro, copyright 2009; courtesy of the artist).

importantly, it affirms that the debate over marine conservation is fundamentally about individual and collective values, in which everyone has a stake.

It is often the case that public dialogues about resource management policy and conservation goals are seen by the public as essentially technical, the legitimate domain only of those who have scientific expertise or economic interests at issue. Historical ecology, however, permits a broader overview within which regulatory, scientific, and economic debates can take place. The fundamental question it raises is neither technical nor dependent for expertise on lived experience, but rather value based and universal: "What kind of nature do you want to live with?" (Figure 12.3).

As one reply to this question, consider the following quotation from an 18-year-old interviewed during research into the attitudes to nature of young people living along the Rio Tejo in Lisbon, Portugal:

> I heard that some time ago, when there was none of that pollution, the river was, according to what I heard, was pretty, there were dolphins and all swimming in it. I think it should have been pretty to see. Anyone would like to see it. (Kahn and Kellert 2002)

Anyone would like to see it: there can be no more hopeful expression of the potential power of historical awareness. Ultimately, life under shifting baselines, with nature understood as a choice, will remain complex. Much of that complexity, however, will result from opening the possibilities of conservation far wider than is conventional today. When it comes to our perception of the natural world, baselines have been shifting for millennia; today, paradigms are shifting as well.

ACKNOWLEDGMENTS

I thank Jack Kittinger and Loren McClenachan for inviting me into this project, and Anne Collins of Random House Canada, Courtney Young of Houghton Mifflin Harcourt, and Andrew Blechman of *Orion* magazine for their support in the development of this material.

REFERENCES

Alter, S. E., Rynes, E., and Palumbi, S. P. (2007) DNA evidence for historic population size and past ecosystem impacts of gray whales. *Proceedings of the National Academy of Sciences USA* 104, 15162–15167.

Bahn, P. G., and Flenley, J. (1992) *Easter Island, Earth Island.* Thames & Hudson, London, UK.

Bolster, J. (2009) Oral comments in plenary. Proceedings of the Oceans Past II: Multidisciplinary Perspectives on the History and Future of Marine Animal Populations Conference, Vancouver, BC, 26–28 May, 2009.

Dasmann, R. F. (1988) Toward a biosphere consciousness. In *Ends of the Earth: Perspectives on Modern Environmental History* (D. Worster, Ed.). Cambridge University Press, Cambridge, UK. pp. 277–288.

DeMartini, E. E., Friedlander, A. M., Sandin, S. A., and Sala, E. (2008) Differences in fish-assemblage structure between fished and unfished atolls in the northern Line Islands, central Pacific. *Marine Ecology Progress Series* 365, 199–215.

Diamond, J. (2011) *Collapse: How Societies Choose to Fail or Succeed.* Penguin, New York, NY.

Duarte, C. M., Conley, D. J., Carstensen, J., and Sánchez-Camacho, M. (2009) Return to Neverland: shifting baselines affect eutrophication restoration targets. *Estuaries and Coasts* 32, 29–36.

Hunt, T., and Lipo, C. (2011) *The Statues that Walked.* Free Press, New York, NY.

Jackson, J. B. C., Alexander, K. E., and Sala, E. (Eds.) (2011) *Shifting Baselines: The Past and the Future of Ocean Fisheries.* Island Press, Washington, DC.

Jackson, S. T., and Hobbs, R. J. (2009) Ecological restoration in the light of ecological history. *Science* 325, 567–568.

Jacobsen, R. (2009) *The Living Shore: Rediscovering a Lost World.* Bloomsbury, New York, NY.

Kahn, P. H., Jr., and Friedman, B. (1995) Environmental views and values of children in an inner-city Black community. *Child Development* 66, 1403–1417.

Kahn, P. H., Jr., and Kellert, S. R. (2002) *Children and Nature: Psychological, Sociocultural, and Evolutionary Investigations.* MIT Press, Cambridge, MA.

Lang, M. (2009) Histories of Cornwall's large marine predators 1602–1878: perspectives for restoration (oral presentation). Proceedings of the Oceans Past II: Multidisciplinary Perspectives on the History and Future of Marine Animal Populations Conference, Vancouver, BC, 26–28 May.

Lotze, H. (2009) Lessons from the past: emerging patterns of historical declines in large marine animals (plenary talk). Proceedings of the Oceans Past II: Multidisciplinary Perspectives on the History and Future of Marine Animal Populations Conference, Vancouver, BC, 26–28 May, 2009.

MacKenzie, B. R., and Myers, R. A. (2007) The development of the northern European fishery for north Atlantic bluefin tuna *Thunnus thynnus* during 1900–1950. *Fisheries Research* 87, 229–239.

McClenachan, L., Jackson, J. B. C., and Newman, M. J. H. (2006) Conservation implications of historic sea turtle nesting beach loss. *Frontiers in Ecology and the Environment* 4, 290–296

McNay, M. E. (2002) Wolf–human interactions in Alaska and Canada: a review of the case history. *Wildlife Society Bulletin* 30, 831–843.

Merilees, B. (1985) The humpback whales of Strait of Georgia. *Waters: Journal of the Vancouver Aquarium* 8.

Moore, B., and Dehlin, J. (2006) Puget sound residents survey. Moore Information, Portland, OR.

Nichol, L. M., Gregr, E. J., Flinn, R., et al. (2002) British Columbia commercial whaling catch data 1908 to 1967: a detailed description of the B.C. historical whaling database. Canadian Technical Report of Fisheries and Aquatic Sciences 2396.

Papworth, S. K., Rist, J., Coad, L., and Milner-Gulland, E. J. (2009) Evidence for shifting baseline syndrome in conservation. *Conservation Letters* 2, 93–100.

Parsons, D. M., Morrison, M. A., MacDiarmid, A. B., et al. (2009) Risks of shifting baselines highlighted by anecdotal accounts of New Zealand's snapper (*Pagrus auratus*) fishery. *New Zealand Journal of Marine and Freshwater Research* 43, 965–983.

Pauly, D. (1995) Anecdotes and the shifting baseline syndrome of fisheries. *Trends in Ecology & Evolution* 10, 430.

PBS (2012) Destination Baja. *Saving the Ocean.* http://video.pbs.org/video/2286020756/.

Rogers, R. A. (1994) *Nature and the Crisis of Modernity: A Critique of Contemporary Discourse on Managing the Earth.* Black Rose, Montreal, QC.

Sáenz-Arroyo, A., Roberts, C. M., Torre, J., et al. (2005) Rapidly shifting environmental baselines among fishers of the Gulf of California. *Proceedings of the Royal Society of London Series B* 272, 1957–1962.

Terborgh, J., and Estes, J. A. (2010) Conclusion: our trophically degraded planet. In *Trophic Cascades: Predators, Prey, and the Changing Dynamics of Nature* (J. Terborgh and J. A. Estes, Eds.). Island Press, Washington, DC. pp. 353–367.

Walpole, M. J., and Leader-Williams, N. (2002) Tourism and flagship species in conservation. *Biodiversity and Conservation* 11, 543–547.

Wang, H., Morrison, W., Singh, A., and Weiss, H. (2009) Modelling inverted biomass pyramids and refuges in ecosystems. *Ecological Modelling* 220, 1376–1382.

Winship, A. (1998) Pinnipeds and cetaceans in the Strait of Georgia. In *Back to the Future: Reconstructing the Strait of Georgia Ecosystem* (D. Pauly, T. J. Pitcher, and D. Preikshot, Eds.). *Fisheries Centre Research Reports* 6(5), 51–55.

Wright, R. (2004) *A Short History of Progress.* House of Anansi Press, Toronto, ON.

INDEX

Aarhus Convention, 68

abalone fishermen and sea otter recovery, 2

abiotic factors and ecosystem service provision, 191–192

abundance: catch reconstruction and fisheries conservation, 124–125; data-limited management systems and fisheries stock assessment, 109–111, 111*fig.*; and endangered species recovery, 16; fish abundance and catch per unit effort (CPUE), 93–94; habitat degradation and restoration motivations, 166, 166*fig.*; statistical fisheries stock assessment models, 92. *See also* population abundance

access, data access and sharing, 67–68, 72

Acipenser brevirostrum, 172

Acipenser oxyrhynchus, 172

Acropora cervicornis, 112

Acropora palmata, 112

Adélie penguin, 47

Albula spp., 98*fig.*, 99*table*

Alligator mississippiensis, 23–24

alligator populations, 23–24

Alligator sinensis, 24

Alopias vulpinus, 74–75

Alosa spp., 26

American alligator, 23–24

anchovy-scale deposits and population reconstructions, Santa Barbara Channel, 105, 105*fig.*

applied historical ecology and ecosystem restoration, 165*table*, 176–177

aquaculture, customary versus conventional fisheries management, 137, 139*table*

archaeology and human impacts on pinnipeds, 46–52, 49*fig.*, 50*table*, 55–57

archival data research: fisheries catch data reconstruction, 122–123, 243, 246–247*fig.*; framing of knowledge, 241–251, 243–247*fig.*, 250–253*fig.*, 261; systematic conservation planning, 213*table*

Arctocephalus forsteri, 19

Arctocephalus townsendi, 41, 41*fig.*, 42, 43*table*, 46, 49, 50, 50*table*

Ardea herodias, 21

Aristotle, 70

Asia-Pacific region, marine protected areas (MPAs) and comanagement outcomes, 143–145, 144*fig.*

Atlantic haddock, 25, 25*fig.*

Atlantic salmon, 26

Atlantic sturgeon, 172

Atlantic swordfish, 24–25

Atractoscion nobilis, 25, 25*fig.*

"Back to the Future" ecosystem reconstruction modeling, 95, 176

Balaena mysticetus, 17*fig.*, 19

Balaenoptera bonaerensis, 20, 213*table*, 265

Balaenoptera borealis, 20

Balaenoptera musculus, 19, 20

Balaenoptera physalus, 20, 213*table*

Baltic ringed seal, 19

barangays and hybrid marine resource management in the Philippines, 149

basking shark, 13

conservation and management: birds, 21–23, 22*fig.*; challenges overview, 7–9; data reconstruction and synthesis, 72; endangered species recovery and, 13–14; fisheries catch data reconstruction, 124–125, 128, 129–131; fishes, 24–27, 25*fig.*; historical human–environmental relationships and, 56; historical knowledge and, 10, 27–31, 29*fig.*, 56, 161–162; invertebrates, 27, 28*fig.*; long-term population trend estimates, 28–30, 29*fig.*; management strategy planning, 31–32; marine mammals, 16–21, 17*fig.*; meaningful targets for, 30; recovery measurement and interpretation, 15–16, 31–32; recovery timelines, 26–27, 31; reptiles, 23–24, 24*fig.*; single-species approach to conservation, 13. *See also* ecosystem restoration; endangered species recovery; fisheries conservation; pinniped recovery; systematic conservation planning

Convention on International Trade in Endangered Species (CITES), 24

coral reefs: Caribbean coral reef ecosystems, 161; Caribbean parrotfish and coral reef ecosystem stock assessments, 112, 227, 228*fig.*; global reef assessment and systematic conservation planning, 215*table*; historical degradation and recovery of, 168; historical ecology and, 4, 5; Marquesan coral reef deposits and small-scale fisheries management, 143; precontact Hawaiian customary reef management systems, 141–143, 142*fig.*; reef paleontology and systematic conservation planning, 212*table*

CPUE. *See* catch per unit effort (CPUE)

Crassostrea virginica. See eastern oyster restoration case study

crocodilian populations, 23–24

crowdsourcing, 256–257

customary fisheries management: customary management systems versus conventional fisheries management, 136–137, 138–139*table*, 140–141; precontact Hawaiian customary reef management systems, 141–143, 142*fig.*

Dasmann, Raymond, 267–268

data-limited management systems, fisheries stock assessments, 92–94, 109–111, 111*fig.*

dataset analysis. *See* historical baseline reconstruction

denitrification: ecosystem service provision and, 191–192; estuary filtration goals, 195–196

Dermochelys coriacea, 23, 24*fig.*

Diadema antillarum, 112

digitization of historical literature, 69

DiPasquale, Nick, 170*fig.*

discoverability and search power of new media, 254–255, 256*fig.*

"distant mirror" perspective on ecological reconstruction, 176

DNA testing: genetic diversity in whales and systematic conservation planning, 213*table*; historical fisheries stock assessments, 105–106; museum collection discoveries, 71

dusky grouper, 272

Easter Island, 270–271

eastern oyster restoration case study, 169, 192–201, 193*fig.*, 194*fig.*, 197*fig.*, 198*fig.*; estuary filtration goals, 195–198, 197*fig.*, 198*fig.*; habitat quality and restoration, 193, 195; multiple ecosystem services and conservation values, 198–199; natural history and abundance, 192–193, 194*fig.*; restoration goals and ecosystem service production estimates, 199, 200–201

ecoinformatics, 68–69

ecological amnesia, 267–270

economic importance of small-scale fisheries, 135–136

ECOPATH, ECOSIM, and ECOSPACE ecosystem models, 95–96

ecosystem resilience: fisheries stock assessments and, 111–113; systematic conservation planning and, 227–229, 228*fig.*

ecosystem restoration: applied historical ecology and, 176–177; baselines and goals for, 10–11, 163–165, 164*fig.*; benefits of using historical data for, 164–165, 165*table*, 178, 179–180; definitions of, 163–164, 178; historical data limitations, 169–170, 171*table*; human impacts on marine ecosystems, 166–168, 166*fig.*, 167*table*; novel ecosystems, 179, 180; restoration targets and strategies, 168–173, 170*fig.*; sustainability of, 175–178; understanding landscape connections and, 173–175, 174*fig. See also* historical ecosystem service provision; systematic conservation planning

gaspereau, 26

gear restrictions: comanagement resilience and durability, 151–153; customary versus conventional fisheries management, 137, 138*table*

genetics and systematic conservation planning, 213*table*

Georges Bank, 25, 27

giant leatherback sea turtle, 23, 24*fig.*

giant palm, 271

gill net bans, 25

glaucous-winged gull, 71

global fisheries catch reconstructions and stability estimates, 129–130

goliath grouper, 26, 30

Google Book Library Project, 69

gray seal, 17*fig.*, 18–19

gray whale, 17*fig.*, 19, 29, 30, 265–266, 265*fig.*, 269

great auk, 8

Great Barrier Reef nearshore coral communities, 222

great blue heron, 21

great white egret, 22

green turtle, 23, 24*fig.*, 214*table*, 267

grouper: dusky grouper, 272; Florida and Cuba grouper fisheries historical catch records and stock assessment, 97, 99–100, 100*fig.*; goliath grouper, 26, 30; Gulf grouper, 69; Nassau grouper, 99–100, 100*fig.*; Roman mosaics of, 272

GRUND (National Group for Demersal Resource Evaluation), 68

Guadalupe fur seal, 41, 41*fig.*, 42, 43*table*, 46, 49, 50, 50*table*

Guadalupe Island, 42, 51

Gulf Coast restoration, 200

Gulf grouper, 69

habitat loss: ecosystem resilience and systematic conservation planning, 227–229, 228*fig.*; restoration motivations and, 166, 166*fig.*

Haematopus ostralegus, 21, 22*fig.*

Halichoerus grypus, 17*fig.*, 18–19

Haliotis rufescens, 168

hammerhead sharks, 74–75

harbor seal: biogeography and natural history of, 41, 41*fig.*, 42, 43*table*, 46, 49, 50*table*; Wadden Sea recovery, 17*fig.*, 18–19, 30

Hawaii: Hawaiian commercial fisheries historical catch records and stock assessment, 97, 98*fig.*, 99*table*; marine protected areas (MPAs) and comanagement outcomes, 143–145, 144*fig.*; Papahānaumokuākea Marine National Monument, 107, 149–150, 225, 255; precontact Hawaiian customary reef management systems, 141–143, 142*fig.*; reef assessment and systematic conservation planning, 215*table*, 221–222; unfished reference areas and fisheries stock assessments, 107, 108–109*fig.*, 109

hawksbill turtle, 23, 122

Heath, Harold, 1

herring fishery, 242, 245*fig.*

herring gull, 21, 22*fig.*

hierarchical modeling, 76–77

historical baseline reconstruction: about, 14, 63–64, 78–79; Bayesian analysis, 70, 75–77; combining datasets, 70, 72–77; comparisons across datasets, 64–65; conservation and management implications, 78; data access and sharing, 67–68, 72; discussion and further work stimulation, 77–78; ecoinformatics and text data mining, 68–69; heterogeneous dataset integration, 65; hierarchical modeling, 76–77; historical data integration and interpretation, 66–67, 67*fig.*, 70, 71; interview surveys, 69–70; long-term population trend estimates, 28–30, 29*fig.*; meta-analysis, 73–75, 73*fig.*, 77; multifaceted data interpretation, 65–66; population and endangered species recovery, 16. *See also* stock assessment

historical cultural periods and human impacts on marine ecosystems, 166–168, 166*fig.*, 167*table*

historical data limitations for ecosystem restoration, 169–170, 171*table*

historical ecology. *See* marine historical ecology

historical ecosystem service provision: application of habitat data to estimation of, 189–191, 190*fig.*; data challenges, 191–192; data required for estimation of, 189; eastern oyster restoration case study, 192–201, 193*fig.*, 194*fig.*, 197*fig.*, 198*fig.*; quantification of ecosystem services, 187–188; restoration goals and, 188–189; RESTORE Act and Gulf Coast restoration, 200

historical human–environmental relationships and resource management, 56

small-scale fisheries management: comanagement design and implementation, 145–146, 147–148*table*, 149–150, 156; comanagement hybridization success, 151–154, 154*fig.*, 155*table*; comanagement outcomes, 143–145, 144*fig.*; customary management systems versus conventional fisheries management, 136–137, 138–139*table*, 140–141; economic importance of small-scale fisheries, 135–136; historical ecology and, 88, 154–155; Marquesan coral reef deposits, 143; precontact Hawaiian customary reef management systems, 141–143, 142*fig.*

smalltooth sawfish, 26

smelt (Osmeridae), 52

Smith, John, 248, 250*fig.*

Smithsonian Institution, 6, 243, 246, 248*fig.*

snapper, 271

Sneaker Index, Chesapeake Bay water quality, 170*fig.*

social data and systematic conservation planning, 217, 218, 221–222

social participation and engagement in new media, 255–257

soldierfishes (Holocentridae), 97, 99*table*

Somateria mollissima, 21, 22*fig.*

soupfin shark, 25, 25*fig.*

southern elephant seal, 46

"space-for-time" substitution, unfished reference areas and fisheries stock assessments, 106–107

spatial and species restrictions, customary versus conventional fisheries management, 137, 138*table*

sperm whale, 19

Sphyrna lewini, 26, 74–75

stability of global fisheries, 129

stable isotope analysis, 47, 51, 71

staghorn coral, 112

stakeholder identification and engagement: ecosystem service provision and, 199; systematic conservation planning, 217; use of historical data for restoration and, 165*table*

statistical fisheries stock assessment models, 92

statistical power of meta-analysis, 73

status assessments, species, 74

STECF (Scientific, Technical and Economic Committee for Fisheries), European Commission, 75

Steinbeck, John, 140

Steller sea cow, 8, 46

Steller sea lion, 41, 41*fig.*, 43*table*, 44, 49, 50, 50*table*, 53

Sterna albifrons, 21, 22*fig.*

stock assessment: Caribbean parrotfish and coral reef ecosystems, 112; catch reconstructions, 100–102, 214*table*; cod fishery historical catch records, 96–97; current state of knowledge and practice, 94; data-limited management systems, 109–111, 111*fig.*; data sources and limitations, 92–94; ecosystem resilience and, 111–113; fisheries conservation and, 8; Florida and Cuba grouper fisheries historical catch records, 97, 99–100, 100*fig.*; Hawaiian commercial fisheries historical catch records, 97, 98*fig.*, 99*table*; historical data interpretation and, 77, 91; paleoecological and archaeological evidence, 104–106, 105*fig.*; simulation modeling, 95–96; statistical stock assessment models, 92; traditional and local ecological knowledge (TEK and LEK), 95, 101, 102–104, 110, 113; unfished reference areas, 106–109, 108–109*fig.*

striped bass, 2, 3*fig.*

surveys, household income and expenditure surveys, 126

sustainability: data-limited management systems and fisheries stock assessment, 109–111, 111*fig.*; ecosystem restoration and, 175–178; fisheries conservation and, 8; precontact Hawaiian customary reef management systems, 141–143, 142*fig.*; statistical fisheries stock assessment models and maximum sustainable yields, 92; sustainable fisheries, 88

systematic conservation planning: assessment and goals expansion, 210*table*, 223, 225; data collection, 221–223; data incorporation barriers, 226–227; data sources and goals identification, 210*table*, 212–215*table*, 219–221; Florida Keys National Marine Sanctuary, 218; historical context for restoration, 165*table*, 211*fig.*, 217, 219; implementation of networked MPAs and, 208–209; importance of historical perspectives, 207, 209, 220; maintenance and monitoring of MPAs, 225–226; opportunities for improvement of, 227–229, 228*fig.*; planning framework summary, 209–211, 210*table*, 211*fig.*; quantitative conservation objectives, 223, 224*fig.*; scoping and costing process, 210*table*, 211*fig.*, 216; social data, 217, 218, 221–222; stakeholder identification and engagement, 210*table*, 217

Tadorna tadorna, 21, 22*fig.*
Tanzanian fisheries catch reconstructions, 129
TEK. *See* traditional and local ecological knowledge (TEK and LEK)
temporal restrictions, customary versus conventional fisheries management, 137, 138*table*
text data mining, 68–69
threadfin, 98*fig.*, 99*table*
Thunnus thynnus, 267
tidal marshlands restoration, San Francisco Bay, 173, 174*fig.*, 175, 178
tiger shark, 26
Tortugas Ecological Reserve, 217, 218, 221
traditional and local ecological knowledge (TEK and LEK): customary fisheries management and, 136–137, 138–139*table*; stock assessment, 95, 101, 102–104, 110, 113
translation software and data access, 69
trawl-survey data, 68, 77
trend data and single-species approach to conservation, 13
Triakis semifasciata, 25, 25*fig.*
Tuchman, Barbara, 176
turtles. *See* sea turtles
T-Wharf, Boston, 243*fig.*

underestimated and underreported catches, 92–93, 123–127, 128–130
unfished reference areas, stock assessment, 106–109, 108–109*fig.*
United Nations Food and Agriculture Organization (FAO) data, 119–121, 124, 128–129
The Unnatural History of the Sea (Roberts), 236, 240
urchin fishermen, 2

Uria aalge, 22–23
U.S. Endangered Species Act, 17, 23
U.S. Fish and Wildlife Service, 18
U.S. Lacey Act, 21
U.S. Marine Mammal Protection Act, 17
U.S. National Marine Sanctuaries Program, 6, 218, 249

Vanuatu, marine protected areas (MPAs) and comanagement outcomes, 143–145, 144*fig.*

walrus, 44, 45*fig.*, 46
water quality: Chesapeake Bay restoration efforts, 169–173, 170*fig.*; fish recovery and, 25*fig.*, 26, 172
Whale Alert app, 258–259, 258*fig.*, 259*fig.*
whales: blue whale, 19, 20; bowhead whale, 17*fig.*, 19; conservation efforts, 19–20; fin whale, 20, 213*table*; genetics and systematic conservation planning, 213*table*; gray whale, 17*fig.*, 19, 29, 30, 265–266, 265*fig.*, 269; humpback whale, 20, 213*table*, 265; minke whale, 20, 213*table*, 265; right whale, 19; sei whale, 20; sperm whale, 19
whaling: endangered species recovery and, 13–14, 19, 20; International Whaling Commission, 19, 129; New Bedford whalemen, 256*fig.*; Vancouver, BC area, 265–266, 267*fig.*
whitesaddle goatfish, 98*fig.*, 99*table*
white sea bass, 25, 25*fig.*
winter steelhead, 76
Wood, Gordon, 7

Xiphias gladius, 24–25

yellowhammer, 77

"zero" catch interpretations of missing data components, 130